Behavior Analysis and Substance Dependence

37

Topics in Organometallic Chemistry

Topics in Organometallic Chemistry
Recently Published Volumes

Asymmetric Catalysis from a Chinese Perspective
Volume Editor: Shengming Ma
Vol. 36, 2011

Higher Oxidation State Organopalladium and Platinum Chemistry
Volume Editor: A. J. Canty
Vol. 35, 2011

Iridium Catalysis
Volume Editor: P. G. Andersson
Vol. 34, 2011

Iron Catalysis – Fundamentals and Applications
Volume Editor: B. Plietker
Vol. 33, 2011

Medicinal Organometallic Chemistry
Volume Editors: G. Jaouen, N. Metzler-Nolte
Vol. 32, 2010

C-X Bond Formation
Volume Editor: A. Vigalok
Vol. 31, 2010

Transition Metal Complexes of Neutral η^1-Carbon Ligands
Volume Editors: R. Chauvin, Y. Canac
Vol. 30, 2010

Photophysics of Organometallics
Volume Editor: A. J. Lees
Vol. 29, 2010

Molecular Organometallic Materials for Optics
Volume Editors: H. Le Bozec, V. Guerchais
Vol. 28, 2010

Conducting and Magnetic Organometallic Molecular Materials
Volume Editors: M. Fourmigué, L. Ouahab
Vol. 27, 2009

Metal Catalysts in Olefin Polymerization
Volume Editor: Z. Guan
Vol. 26, 2009

Bio-inspired Catalysts
Volume Editor: T. R. Ward
Vol. 25, 2009

Directed Metallation
Volume Editor: N. Chatani
Vol. 24, 2007

Regulated Systems for Multiphase Catalysis
Volume Editors: W. Leitner, M. Hölscher
Vol. 23, 2008

Organometallic Oxidation Catalysis
Volume Editors: F. Meyer, C. Limberg
Vol. 22, 2007

N-Heterocyclic Carbenes in Transition Metal Catalysis
Volume Editor: F. Glorius
Vol. 21, 2006

Dendrimer Catalysis
Volume Editor: L. H. Gade
Vol. 20, 2006

Metal Catalyzed Cascade Reactions
Volume Editor: T. J. J. Müller
Vol. 19, 2006

Catalytic Carbonylation Reactions
Volume Editor: M. Beller
Vol. 18, 2006

Bioorganometallic Chemistry
Volume Editor: G. Simonneaux
Vol. 17, 2006

Surface and Interfacial Organometallic Chemistry and Catalysis
Volume Editors: C. Copéret, B. Chaudret
Vol. 16, 2005

Chiral Diazaligands for Asymmetric Synthesis
Volume Editors: M. Lemaire, P. Mangeney
Vol. 15, 2005

Bifunctional Molecular Catalysis

Volume Editors: Takao Ikariya · Masakatsu Shibasaki

With Contributions by

J.-E. Bäckvall · C.P. Casey · C. Gunanathan · T. Ikariya · J.-i. Ito · K. Kamata · M. Kanai · N. Kumagai · P. Li · S. Matsunaga · D. Milstein · N. Mizuno · H. Nishiyama · M. Shibasaki · M.C. Warner · K. Yamaguchi · H. Yamamoto

Springer

Editors
Prof. Dr. Takao Ikariya
Department of Applied Chemistry
Graduate School of Science
and Engineering
Tokyo Institute of Technology
2-12-1 Ookayama, Meguro-ku
Tokyo 152-8552
Japan
tikariya@apc.titech.ac.jp

Prof. Dr. Masakatsu Shibasaki
Institute of Microbial Chemistry
3-14-23 Kamiosaki
Shinagawa-ku
Tokyo 141-0021
Japan
mshibasa@bikaken.or.jp

ISBN 978-3-642-20730-3 e-ISBN 978-3-642-20731-0
DOI 10.1007/978-3-642-20731-0
Springer Heidelberg Dordrecht London New York

Library of Congress Control Number: 2011929692

Cover design: eStudio Calamar, Spain

Printed on acid-free paper

Springer is part of Springer Science+Business Media (www.springer.com)

Volume Editors

Prof. Dr. Takao Ikariya

Department of Applied Chemistry
Graduate School of Science
and Engineering
Tokyo Institute of Technology
2-12-1 Ookayama, Meguro-ku
Tokyo 152-8552
Japan
tikariya@apc.titech.ac.jp

Prof. Dr. Masakatsu Shibasaki

Institute of Microbial Chemistry
3-14-23 Kamiosaki
Shinagawa-ku
Tokyo 141-0021
Japan
mshibasa@bikaken.or.jp

Editorial Board

Topics in Organometallic Chemistry Also Available Electronically

Topics in Organometallic Chemistry is included in Springer's eBook package *Chemistry and Materials Science*. If a library does not opt for the whole package the book series may be bought on a subscription basis. Also, all back volumes are available electronically.

For all customers who have a standing order to the print version of *Topics in Organometallic Chemistry*, we offer free access to the electronic volumes of the Series published in the current year via SpringerLink.

If you do not have access, you can still view the table of contents of each volume and the abstract of each article by going to the SpringerLink homepage, clicking on "Chemistry and Materials Science," under Subject Collection, then "Book Series," under Content Type and finally by selecting *Topics in Organometallic Chemistry*.

You will find information about the

- Editorial Board
- Aims and Scope
- Instructions for Authors
- Sample Contribution

at springer.com using the search function by typing in *Topics in Organometallic Chemistry*.
Color figures are published in full color in the electronic version on SpringerLink.

Aims and Scope

The series *Topics in Organometallic Chemistry* presents critical overviews of research results in organometallic chemistry. As our understanding of organometallic structures, properties and mechanisms grows, new paths are opened for the design of organometallic compounds and reactions tailored to the needs of such diverse areas as organic synthesis, medical research, biology and materials science. Thus the scope of coverage includes a broad range of topics of pure and applied organometallic chemistry, where new breakthroughs are being made that are of significance to a larger scientific audience.

The individual volumes of *Topics in Organometallic Chemistry* are thematic. Review articles are generally invited by the volume editors.

In references *Topics in Organometallic Chemistry* is abbreviated Top Organomet Chem and is cited as a journal. From volume 29 onwards this series is listed with ISI/Web of Knowledge and in coming years it will acquire an impact factor.

Preface

Catalysts are crucial, not only for the production of many thousands of materials and products, including fuels required by modern society, but also for the reduction of water and air pollution, as well as to minimize the waste of natural resources and energy. Recent advances in green and sustainable science and technology demand more powerful and sophisticated catalysts, which have a tunable function. Much efforts have been paid to the development of well-defined bifunctional molecular catalysts based on the combination of Lewis acid and basic sites working in concert to attain highly efficient molecular transformation for organic synthesis. Bifunctional catalysts contain two or more sites for the activation of electrophiles and nucleophiles, and consequently these catalysts can efficiently promote a wide range of molecular transformation by effective accumulation and cooperative activation of reacting substrates on the neighboring active centers in the same molecules. Careful and precise tuning of the structures of the molecular catalysts, as well as the spatial organization of the functionality, is required to achieve the best catalysis performance.

Metal–ligand cooperating bifunctional molecular catalysts are also recognized as an alternative and indispensable strategy to realize highly effective molecular transformation. The non-innocent ligand therein directly participates in the substrate activation and subsequent reactions through various secondary interactions including hydrogen bonding. Therefore, the catalyst deactivation due to the acid–base neutralization or destructive aggregation can be minimized. Indeed, a number of biological catalyses in nature have been performed in a sophisticated manner on such bifunctional and even multifunctional systems with ligated peptide residues and prosthetic groups.

This unique concept of the recently developed bifunctional metal-based molecular catalysts leads to high reaction rates and excellent stereochemical outcome because the reactions proceed through a closed assembly of reactants and catalysts, providing a wide substrate scope and applicability in organic synthetic chemistry. This volume is intended to highlight the recent exciting advances in bifunctional catalysis with well-designed multimetallic systems, dinuclear, mononuclear transition metal-based molecular catalysts, and Lewis acid catalysts. All chapters have been written with a brief introduction on the definition and concept of bifunctionality by leading chemists in the corresponding research area. We wish to express our sincere thanks to them for their contributions and fruitful and exciting collaboration in publishing a book dealing with the most recent achievement in this field: Motomu Kanai, Shigeki Matsunaga, Naoya Kumagai, Madeleine C. Warner, Charles P. Casey, Jan-E. Bäckvall, Chidambaram Gunanathan, David Milstein, Noritaka Mizuno, Keigo Kamata, Kazuya Yamaguchi, Pingfan Li, Hisashi Yamamoto, Jun-ichi Ito, Hisao Nishiyama. We are also grateful for all help and supports from Springer, in particular, from Drs. Elizabeth Hawkins and Jutta Lindenborn.

We hope not only that this book will be useful to researchers and students already involved in organic synthesis and molecular catalysis, but also that it will attract scientists working in various field of chemistry.

Tokyo, Japan

Takao Ikariya
Masakatsu Shibasaki

Contents

Top Organomet Chem (2011) 37: 1–30
DOI: 10.1007/3418_2011_1

Published online: 12 March 2011

Multimetallic Multifunctional Catalysts for Asymmetric Reactions

Masakatsu Shibasaki, Motomu Kanai, Shigeki Matsunaga, and Naoya Kumagai

Abstract The utility of several types of multimetallic asymmetric catalysts is reviewed. The appropriate design of multidentate chiral ligands provides homobimetallic, heterobimetallic, and polymetallic catalysts upon complexation with metal cations. Cooperative work of multiple catalytically active sites in the asymmetric multimetallic catalysts allows for enhanced catalytic activity and stereoselectivity over monometallic catalysts. Facile systematic tuning of the asymmetric multimetallic catalysts by manipulation of multidentate ligand structure and combination of metal cations enables the generation of diverse set of catalysts having distinct three-dimensional and catalytic properties, boosting the optimization process to identify the best catalyst for a specific reaction of interest.

Keywords Dinucleating Schiff base · Heterobimetallic · Homobimetallic · Homopolymetallic · Rare earth metal

Contents

M. Shibasaki (✉) and N. Kumagai
Institute of Microbial Chemistry, Tokyo, 3-14-23 Kamiosaki, Shinagawa-ku, Tokyo 141-0021, Japan
e-mail: mshibasa@bikaken.or.jp

M. Kanai and S. Matsunaga
Graduate School of Pharmaceutical Sciences, The University of Tokyo, 7-3-1 Hongo, Bunkyo-ku, Tokyo 113-0033, Japan

1 Introduction

The development of multimetallic catalytic systems and their application to asymmetric catalysis is an emerging area in modern organic synthesis. The use of a multimetallic entity in catalyst design is a viable approach to the construction of a multifunctional catalyst, which can activate multiple substrates simultaneously [1]. Implementation of a rationally designed multidentate chiral ligand as a platform for metal cations produced a multimetallic asymmetric catalyst, in which multiple catalytically active metal centers are incorporated in an appropriate three-dimensional architecture (Fig. 1). Thus, developed multimetallic asymmetric catalysts generally exert multifunctionality to exhibit enhanced catalytic activity and stereoselectivity through cooperative activation of a substrate pair in a suitable spatial arrangement [2–8]. In addition, using the same or related chiral ligand platform, diverse set of multimetallic catalyst can be produced by combinatorial selection of incorporating metal cations. In this chapter, leading examples of homo- and heterobimetallic asymmetric catalysts using Schiff base ligand, amide-based ligand, and sugar-derived ligand platforms are reviewed.

2 Bimetallic Bifunctional Schiff Base Catalysis

2.1 Lewis Acid/Brønsted Base Heterobimetallic Schiff Base Catalyst

To develop heterobimetallic complexes, the design of a suitable multidentate ligand is important to control the position of the two different metals in the complex. The positions of the two metals have crucial effects on the reactivity as well as the

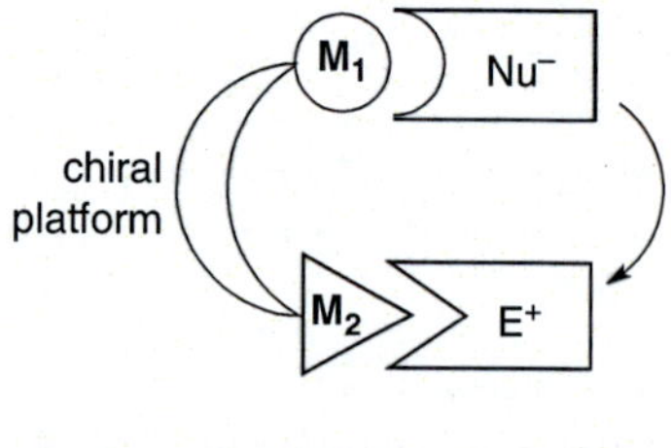

Fig. 1 Asymmetric multimetallic catalyst

stereoselectivity of the heterobimetallic complex. To realize bimetallic asymmetric catalysis using transition metal and rare earth metal combinations, we used new dinucleating Schiff bases **1** with additional phenolic hydroxyl groups in comparison with standard salens [9–11] (Fig. 2). We hypothesized that the Schiff bases **1** would incorporate a transition metal into the N_2O_2 inner cavity, and an oxophilic rare earth metal with a large ionic radius into the O_2O_2 outer cavity.

As expected, suitable selection of a metal combination for each targeted reaction was important to achieve high stereoselectivity. For a *syn*-selective nitro-Mannich-type reaction, a heterobimetallic complex prepared from $Cu(OAc)_2$, $Sm(O\text{-}i\text{Pr})_3$, and dinucleating Schiff base **1a** was the best [12]. Other metal combinations resulted in much lower selectivity. Either Cu or Sm alone also resulted in poor reactivity and selectivity. Thus, both Cu and Sm are essential for high catalytic activity and selectivity. After optimization studies, the addition of achiral phenol source and the use of oxo-samarium alkoxide, $Sm_5O(O\text{-}i\text{Pr})_{13}$, with a well-ordered structure, gave the superior reactivity and stereoselectivity. Under the optimized reaction conditions, 1–10 mol% of the Cu/Sm catalyst promoted asymmetric

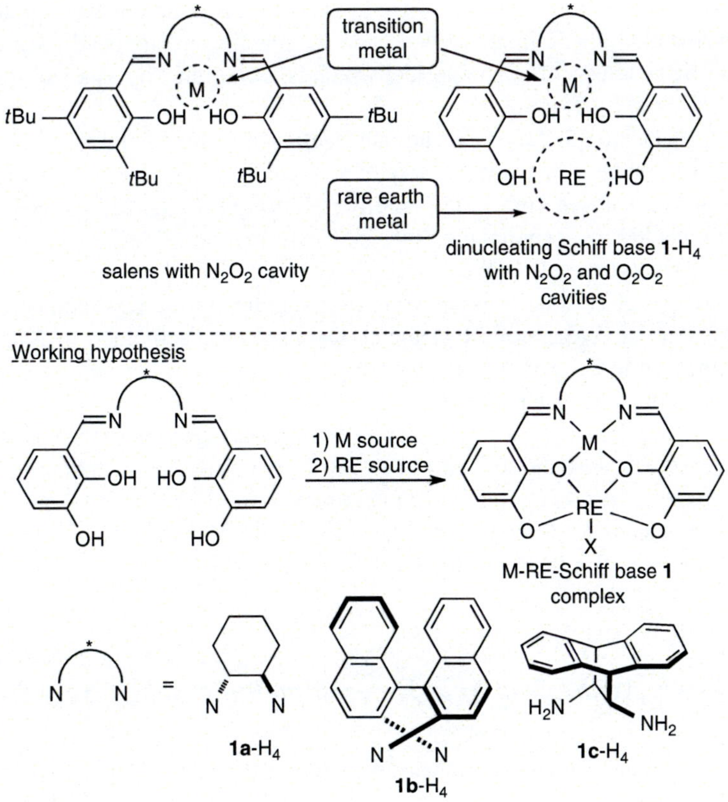

Fig. 2 Ligand design and working hypothesis for the formation of transition metal(M)/rare earth metal (RE) heterobimetallic complex

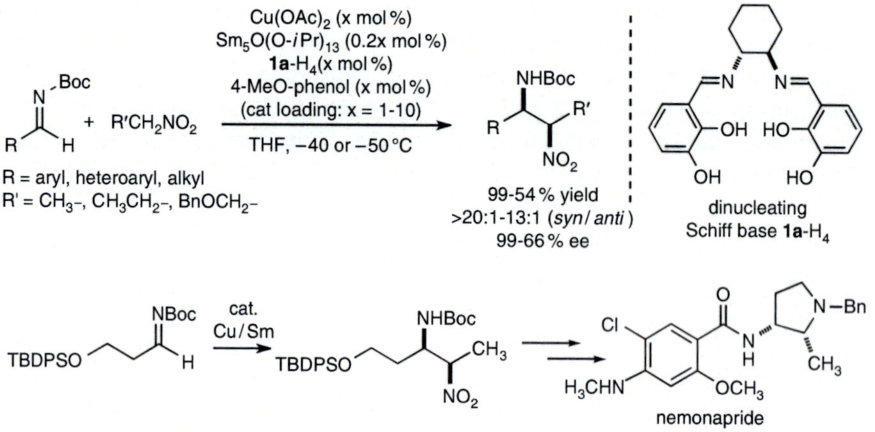

Scheme 1 Catalytic asymmetric nitro-Mannich-type reaction and its application to catalytic asymmetric synthesis of nemonapride

nitro-Mannich-type reaction, giving products with good yield, high *syn*-selectivity, and enantioselectivity (Scheme 1) [13]. The present reaction was applicable to functionalized aliphatic imine as well, and the reaction was applied for a catalytic asymmetric synthesis of nemonapride, which is used clinically as an antipsychotic agent.

On the basis of ESI-MS observation as well as positive nonlinear effects of this system, we assumed that μ-oxo-μ-aryloxy-trimer complex is the most enantioselective active species (Fig. 3). Therefore, $Sm_5O(O\text{-}i\text{Pr})_{13}$ with a well-ordered **structure would have beneficial effects for the formation of desired trimer species.** Postulated catalytic cycle of the reaction based on the initial rate kinetic studies and kinetic isotope effect studies is shown in Fig. 4. In this catalyst system, both Cu and Sm are essential. We assume that the cooperative dual activation of nitroalkanes and imines with Cu and Sm is important to realize the *syn*-selective catalytic asymmetric nitro-Mannich-type reaction. The Sm-aryloxide moiety in the catalyst would act as a Brønsted base to generate Sm-nitronate. On the other hand, Cu(II) would act as a Lewis acid to control the position of *N*-Boc-imine. Among possible transition states, the sterically less hindered TS-1 would be more favorable. Thus, the stereoselective C–C bond formation via TS-1 followed by protonation with phenolic proton affords *syn* product and regenerates the catalyst.

2.2 *Lewis Acid/Lewis Acid Heterobimetallic Schiff Base Catalyst*

In Sect. 2.1, heterobimetallic transition metal (Cu)/rare earth metal (Sm) system with Lewis acid/Brønsted basic properties was introduced. By suitably selecting metal combinations depending on the targeted reactions, variety of chiral bimetallic Lewis acid/Brønsted base bifunctional catalysts could be created, such as a Pd/La/**1a**

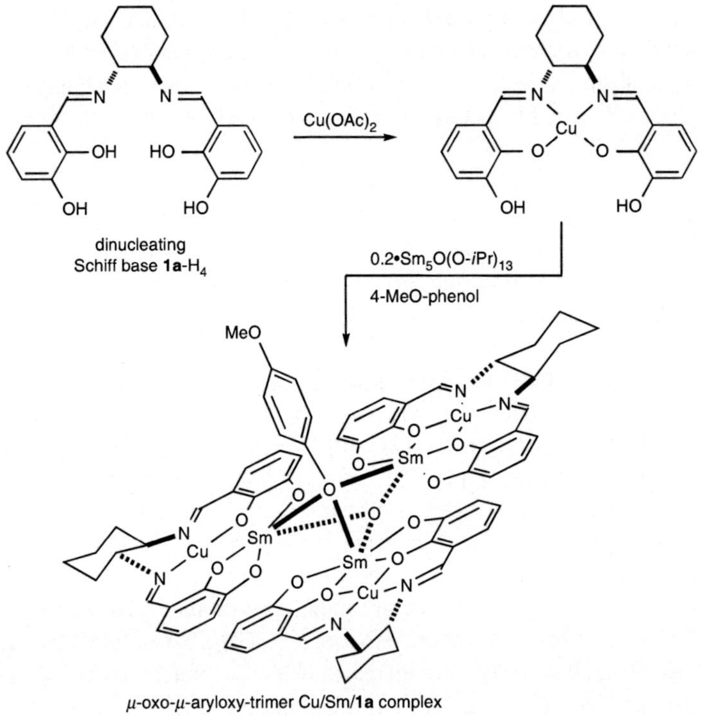

Fig. 3 Postulated structure of μ-oxo-μ-aryloxy-trimer Cu/Sm/**1a** active species

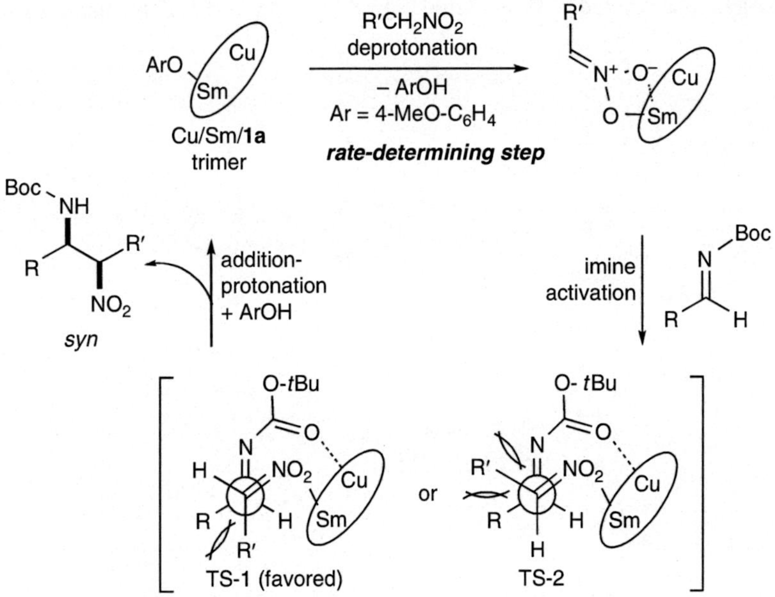

Fig. 4 Plausible catalytic cycle of Cu/Sm-catalyzed nitro-Mannich-type reaction

catalyst for *anti*-selective nitroaldol reaction [14]. In the Cu/Sm catalyst, Sm-aryloxide moiety functioned as a Brønsted base. In rare earth metal catalysis, the use of cationic rare earth metal, like rare earth metal triflate, as a Lewis acid is widely investigated [15]. If cationic rare earth metals can be incorporated into the outer O_2O_2 cavity of dinucleating Schiff bases, the bimetallic complex would work as a Lewis acid/Lewis acid cooperative catalyst [3].

To incorporate a cationic rare earth metal into the outer cavity of a dinucleating Schiff base, we used Schiff base **2** derived from *o*-vanillin (Fig. 5). We selected α-additions of α-isocyanoacetamides to aldehydes as a model reaction because a salen-Al catalyst gave only modest enantioselectivity and reactivity [16]. The working hypothesis for bimetallic Schiff base catalysis is shown in Fig. 6. We assumed that the bimetallic Schiff base complex would not only activate the aldehyde with one metal, but might also interact with the α-isocyanoacetamide through the other metal to effectively control the orientation of the two substrates in the enantio-discriminating step.

After screening suitable metal combinations for Schiff base **2**, Ga(O-*i*Pr)$_3$/Yb(OTf)$_3$ afforded promising results. The chiral diamine backbone affected both the reactivity and enantioselectivity, and the best reactivity and selectivity were achieved using Schiff base **2c** with an anthracene-derived diamine unit (Table 1, entry 3: 94% ee). After optimization of the solvent (CH_2Cl_2) and Ga/Yb ratio (1:0.95), product was obtained in >95% conversion and 96% ee after 24 h (entry 4). To confirm the utility of the Ga(O-*i*Pr)$_3$/Yb(OTf)$_3$ combination, several control experiments with the best ligand **2c** were performed (entries 5–13). Neither Ga-**2c** alone nor Yb-**2c** alone efficiently promoted the reaction (entries 5–6). With Ga(O-*i*Pr)$_3$ and other rare earth metal triflates, the reactivity decreased in correlation with the Lewis acidity of the rare earth metals (Yb > Gd > Nd > La) [17], while good to excellent enantioselectivity

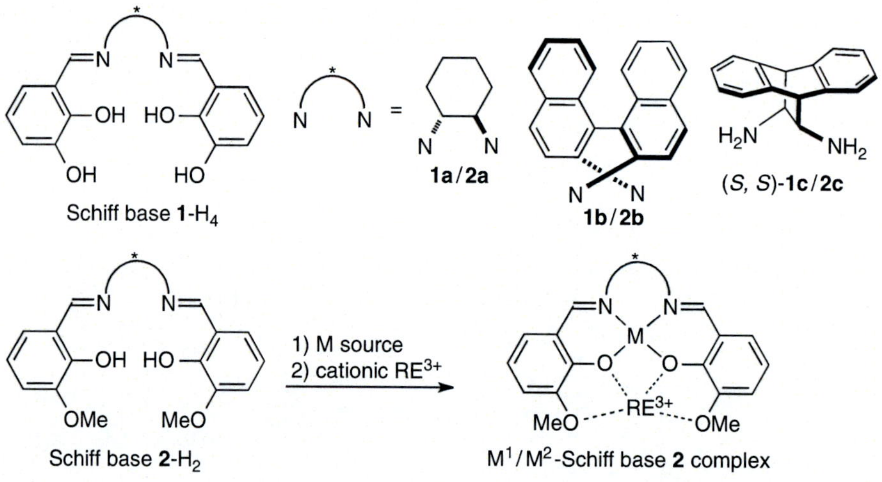

Fig. 5 Design of Schiff bases **2**-H$_2$ for the preparation of heterobimetallic complexes including cationic rare earth metals (RE^{3+})

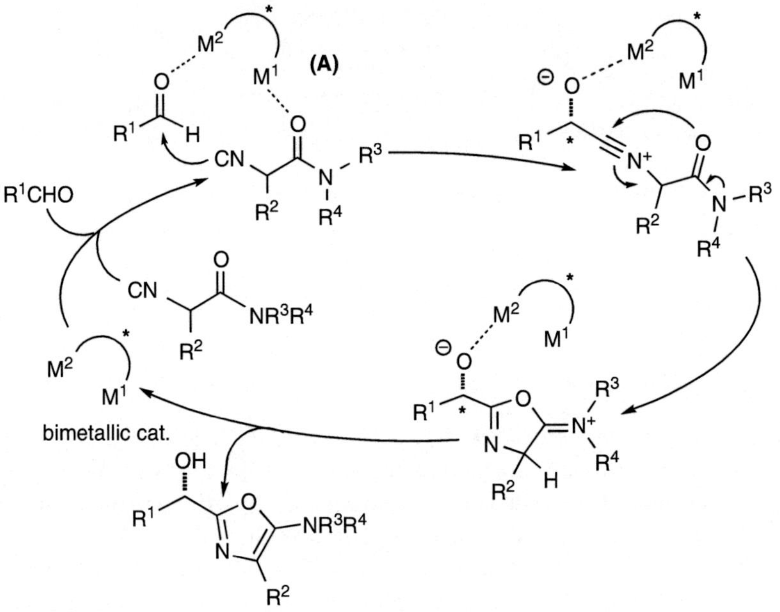

Fig. 6 Working hypothesis of bimetallic Schiff base catalysis on α-additions of α-isocyanoacetamides to aldehydes

Table 1 Screening of heterobimetallic M/cationic RE/Schiff base complexes for α-addition of isocyanide to benzaldehyde

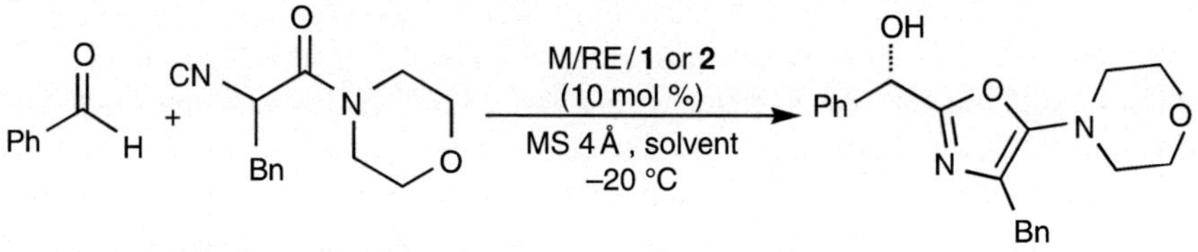

Entry	M source	RE source (× mol%)	Ligand	Time (h)	% Yield	% ee
1	Ga(O-*i*Pr)$_3$	Yb(OTf)$_3$ (10)	**2a**	74	55	29
2	Ga(O-*i*Pr)$_3$	Yb(OTf)$_3$ (10)	**2b**	71	27	78
3	Ga(O-*i*Pr)$_3$	Yb(OTf)$_3$ (10)	**2c**	74	67	94
4	Ga(O-*i*Pr)$_3$	Yb(OTf)$_3$ (9.5)	**2c**	24	>95	96
5	Ga(O-*i*Pr)$_3$	None	**2c**	24	Trace	–
6	None	Yb(OTf)$_3$ (10)	**2c**	24	Trace	–
7	Ga(O-*i*Pr)$_3$	Gd(OTf)$_3$ (9.5)	**2c**	24	62	96
8	Ga(O-*i*Pr)$_3$	Nd(OTf)$_3$ (9.5)	**2c**	24	33	95
9	Ga(O-*i*Pr)$_3$	La(OTf)$_3$ (9.5)	**2c**	24	11	89
10	Et$_2$AlCl	Yb(OTf)$_3$ (9.5)	**2c**	48	68	37
11	Al(O-*i*Pr)$_3$	Yb(OTf)$_3$ (9.5)	**2c**	24	>95	79
12	In(O-*i*Pr)$_3$	Yb(OTf)$_3$ (9.5)	**2c**	24	55	80
13	Ga(O-*i*Pr)$_3$	Yb(OTf)$_3$ (9.5)	**1c**	24	Trace	–

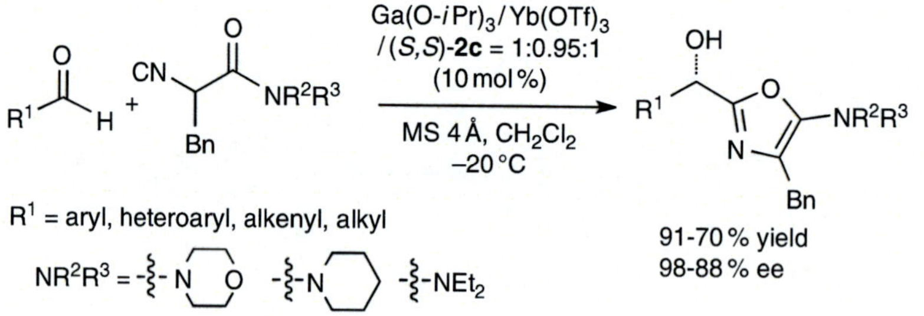

Scheme 2 Catalytic asymmetric α-addition of isocyanides to aldehydes using a heterobimetallic $Ga(O\text{-}i Pr)_3/Yb(OTf)_3$/Schiff base (*S*,*S*)-**2c** complex

was maintained in entries 7–9 (89–96% ee). We investigated Et_2AlCl, $Al(O\text{-}iPr)_3$, and $In(O\text{-}iPr)_3$ as other group 13 metals for the inner cavity (entries 10–12), but the results were less satisfactory than that in entry 3. In entry 13, Schiff base **1c** was used instead of Schiff base **2c**, but no reaction occurred. These results in entries 4–13 indicated that the $Ga(O\text{-}iPr)_3/Yb(OTf)_3$ combination and Schiff base **2c** were essential to obtain high reactivity and enantioselectivity in the present reaction. The optimized Lewis acid/Lewis acid bifunctional system was applicable to broad range of aryl, heteroaryl, alkenyl, and alkyl aldehydes, giving products in 98–88% ee (Scheme 2) [18].

2.3 *Homobimetallic Transition Metal Schiff Base Catalysts*

In Sects. 2.1 and 2.2, the dinucleating Schiff bases **1a** and **2c** selectively incorporated a transition metal or a group 13 metal into the N_2O_2 inner cavity and an oxophilic rare earth metal with a large ionic radius into the O_2O_2 outer cavity. To further expand the utility and diversity of the dinuclear Schiff base catalysis, we then decided to develop a new system that incorporates metals with smaller ionic radius than rare earth metals into the O_2O_2 outer cavity. On the basis of molecular modeling, we speculated that a Schiff base **1b** derived from 1,1′-binaphthyl-2,2′-diamine would be suitable because of the conformational difference between *trans*-1,2-diaminocyclohexane and 1,1′-binaphthyl-2,2′-diamine. In fact, treatment of the Schiff base **1b** with 2 equiv of $Ni(OAc)_2{\bullet}4H_2O$ afforded a homodinuclear Ni_2-Schiff base **1b** complex in 93% yield [19]. In a similar manner, homodinuclear Co_2-**1b** [19, 20] and Mn_2-**1b** [21] complexes were also synthesized in good yield (Scheme 3). The homodinuclear Ni_2-Schiff base **1b** complex was stable and storable under air at room temperature for more than 3 months without any loss of activity.

Mannich-type reactions of aryl, heteroaryl, and isomerizable alkyl *N*-Boc imines and α-substituted nitroacetates gave products in 91% to >99% ee and high *anti*-selectivity (Scheme 4) [22], which were successfully converted into α,β-diamino acids with α-tetrasubstituted stereocenter. Catalyst loading was successfully

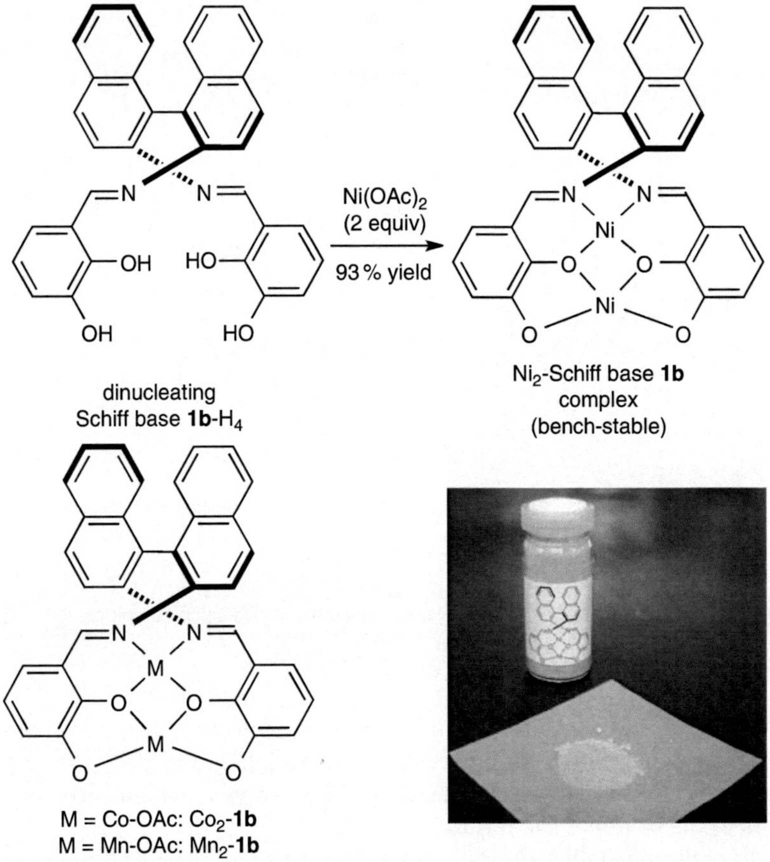

Scheme 3 Preparation of bench-stable Ni_2-Schiff base **1b**, Co_2-**1b**, and Mn_2-**1b** catalysts from (*R*)-binaphthyldiamine-based dinucleating Schiff base **1b**-H_4

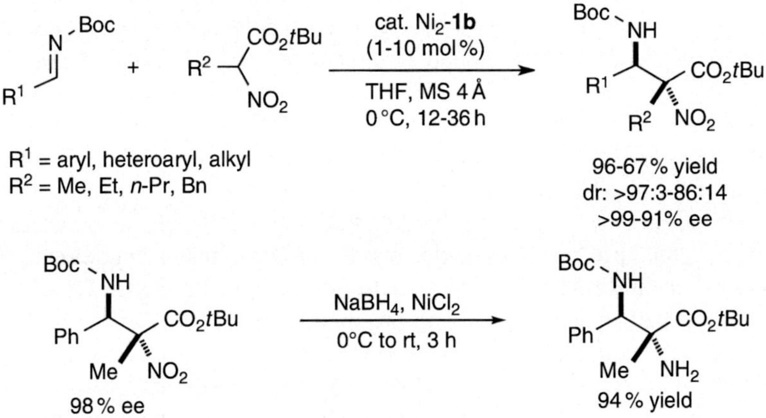

Scheme 4 Ni_2-**1b**-catalyzed direct asymmetric Mannich-type reaction of α-nitroacetates and transformation to α,β-diamino acid with α-tetrasubstituted stereocenter

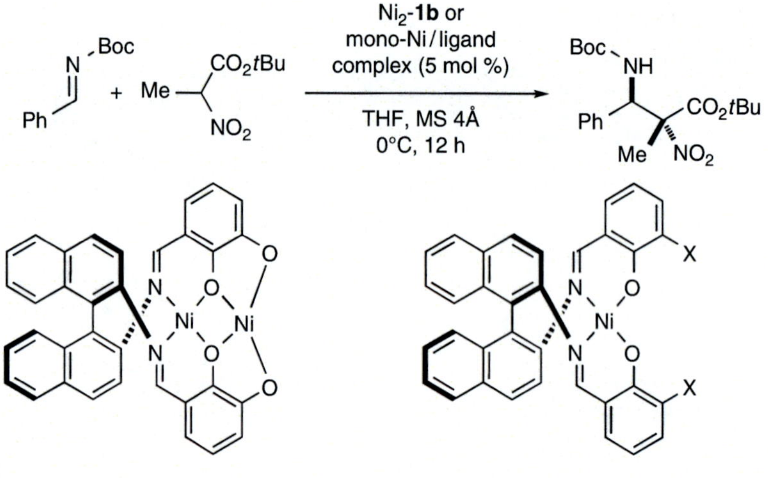

95 % yield, 91:9 dr, 98 % ee

X = OMe: 22 % yield, 57:43 dr, 42 % ee
X = H: 17 % yield, 68:32 dr, 3 % ee
X = *t*Bu: 58 % yield, 77:23 dr, 2 % ee

Scheme 5 Negative control experiments using mononuclear Ni-salen complexes

reduced to 1 mol%, while maintaining high enantioselectivity. As shown in control experiments in Scheme 5, monometallic Ni-salen complexes gave poor reactivity, diastereoselectivity, and enantioselectivity. Thus, cooperative functions of two Ni centers would be important. Postulated catalytic cycle is shown in Fig. 7. ^{1}H NMR analysis of the bimetallic Ni_2-**1b** complex does not show any peaks, suggesting that at least one of the Ni metal centers has nonplanar coordination mode. Based on the molecular model, we assume that the outer Ni center has *cis*-β configuration due to strain of the bimetallic complexes. In other words, one of the Ni–O bonds of the Ni in the outer O_2O_2 cavity is speculated to be in apical position. Thus, the weak Ni–O bond in the apical position would work as a Brønsted base to deprotonate α-nitroacetates to give the Ni-enolate intermediate in situ. The other Ni in the inner N_2O_2 cavity functions as a Lewis acid to control the position of imines, similar to conventional metal-salen Lewis acid catalysis. The C–C bond formation via the transition state (TS in Fig. 7), followed by protonation, affords the *syn*-adduct and regenerates the Ni_2-**1b** catalyst.

As shown in Fig. 8, the Ni_2-**1b** complex was applicable to broad range of catalytic asymmetric reactions under simple proton transfer conditions [19–30]. As donors in direct Mannich-type reactions, not only α-nitroacetates, but also malonates, β-keto esters [22], β-keto phosphonates [23], and α-ketoanilides [24] gave excellent enantioselectivity and diastereoselectivity. It is noteworthy that the Ni_2-**1b** complex also promoted vinylogous direct catalytic asymmetric Mannich-type reaction of an α,β-unsaturated γ-butyrolactam, giving synthetically versatile functionalized α,β-unsaturated γ-butyrolactam in 99% ee [25]. Because the Ni_2-**1b**

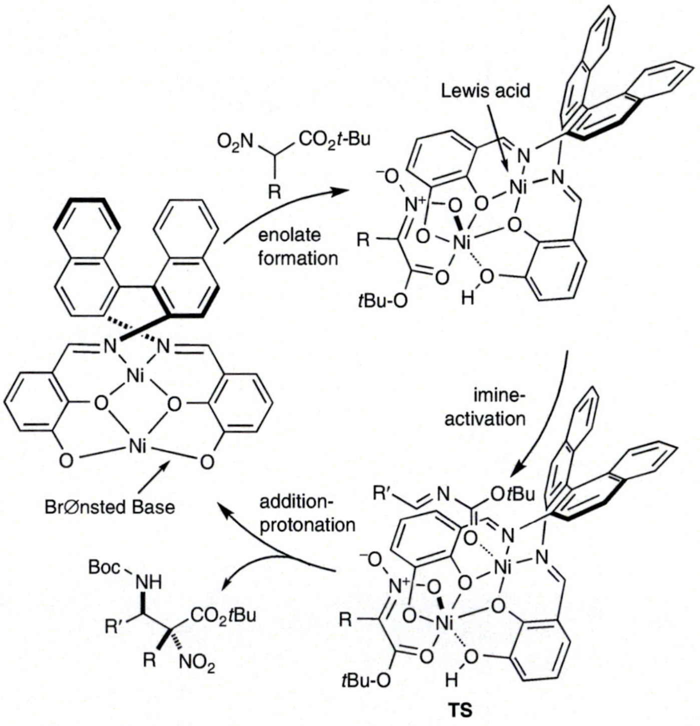

Fig. 7 Postulated catalytic cycle of Ni_2-**1b**-catalyzed asymmetric Mannich-type reaction

complex was bench-stable and can be used without regard to their exposure to air and moisture, hydroxymethylation of β-keto esters using formalin as an electrophile was also successfully achieved. The reaction proceeded with 0.1–1 mol% catalyst, and hydroxymethylated products were obtained in up to 94% ee and 94% yield (TON = up to 940) [26]. For asymmetric Michael reactions, suitable selection of metal was important depending on the combinations of nucleophiles and electrophiles. For Michael reactions of α-nitroacetates to vinylbisphosphonates [27], α-ketoanilides to nitroalkenes [28], and α,β-unsaturated γ-butyrolactam to nitroalkenes [25], the Ni_2 catalyst was the best. For asymmetric 1,4-addition of β-keto esters to alkynones [20], Co_2-**1b** catalyst was essential to achieve high enantioselectivity. The reaction proceeded in up to 99% ee under solvent-free conditions at room temperature. The Co_2-**1b** catalyst was also effective for Michael reaction of β-keto esters to nitroalkenes, and the reaction proceeded with as little as 0.1 mol% catalyst loading (TON = up to 980) [23]. The Mn_2-**1b** catalyst was the best in asymmetric 1,4-addition of *N*-Boc oxindoles to nitroalkenes for β-amino-oxindole synthesis, while the Ni_2-**1b** gave superior results for asymmetric 1,4-addition of *N*-Boc oxindoles for 3-amino-oxindole synthesis [29]. Applications of bifunctional Schiff base catalysis are further expanding such as for group 2 metals [30].

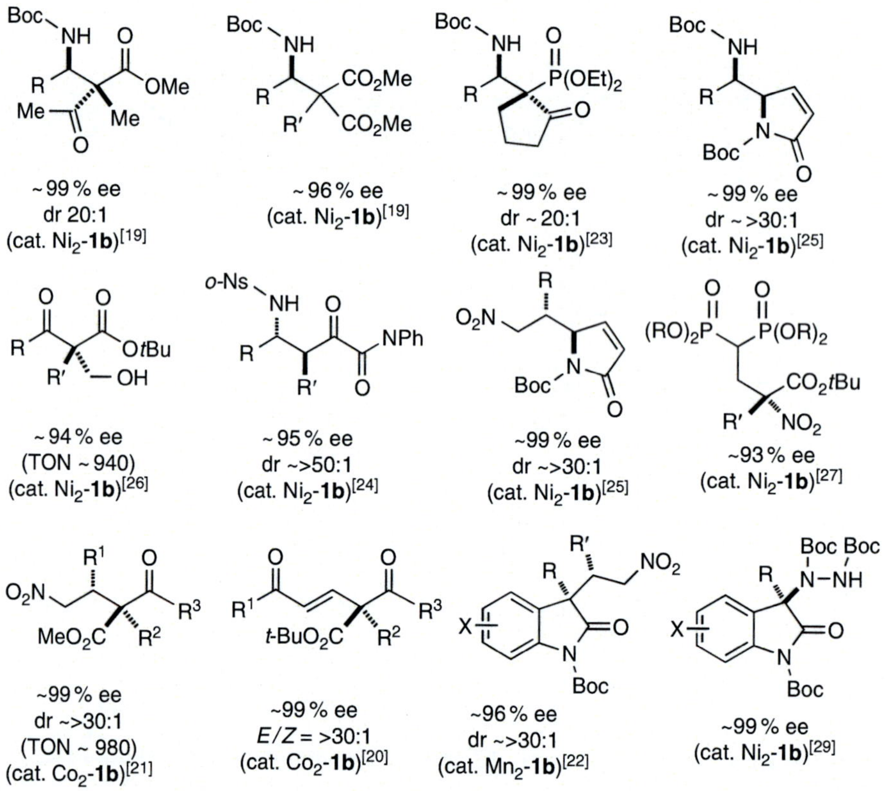

Fig. 8 Representative examples of catalytic asymmetric reactions promoted by homodinuclear bifunctional Schiff base catalysts

3 Heterobimetallic Catalysis with Amide-Based Ligand

3.1 Anti-*Selective Catalytic Asymmetric Nitroaldol Reaction*

The nitroaldol (Henry) reaction provides 1,2-nitro alkanols under atom-economical proton transfer conditions, which allows for easy access to highly versatile 1,2-amino alcohols (Scheme 6) [31]. A number of catalytic systems have been devised to render this useful C–C bond-forming reaction asymmetric; however, diastereoselectivity remained a longstanding problem, in particular for *anti*-selective reactions [14, 32, 33]. *syn*-Selective reaction can be achieved by a monometallic catalytic system as shown in Fig. 9a, where both an aldehyde and a nitronate coordinate to the metal center to give *syn* product due to steric repulsion [34–36]. To make the reaction proceed in *anti*-selective manner, different strategy in catalyst design is required [37, 38]. Simultaneous activation of both the aldehyde and the nitronate in an *anti*-parallel fashion can afford the *anti*-1,2-nitro alkanols preferentially (Fig. 9b). To attain the *anti*-parallel transition state, a heterobimetallic catalyst offers a suitable

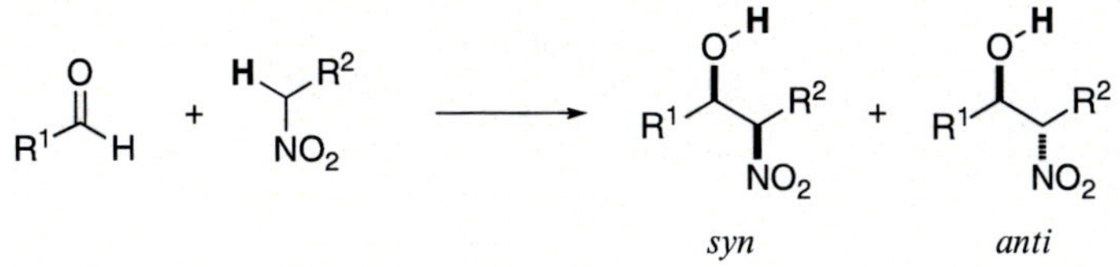

Scheme 6 Nitroaldol (Henry) reaction under proton transfer conditions

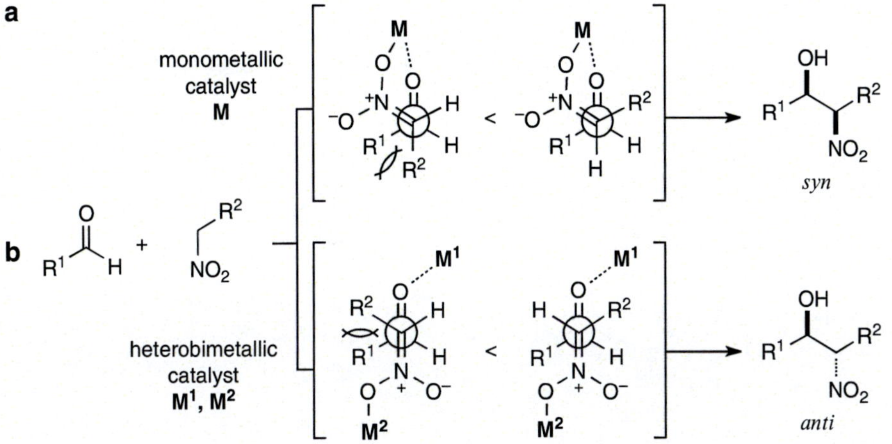

Fig. 9 Diastereoselectivity with (**a**) a monometallic and (**b**) a heterobimetallic catalyst

Fig. 10 Implementation of amide-based ligand as a platform for a heterobimetallic complex

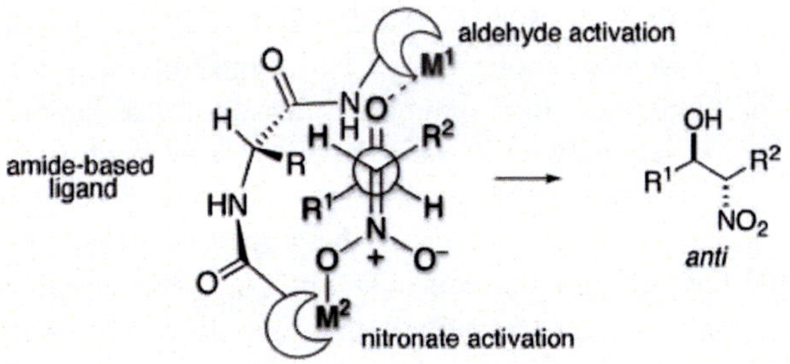

entity; two distinct metals activate the aldehyde and the nitronate independently. Exquisite spatial arrangement of two metals by a suitable design of the ligand is a key to the *anti*-selectivity. We have focused on the implementation of amide-based ligand to assemble two metals because the reasonable rigidity of amide functionality is expected to provide a platform favorable to an extended transition state architecture for *anti*-selective reaction (Fig. 10).

We designed the amide-based ligand **3a** bearing two aryloxide moieties to capture two distinct metal cations for the selective activation of the aldehyde and the nitroalkane (Fig. 11) [39]. A heterobimetallic catalyst prepared from ligand **3a**,

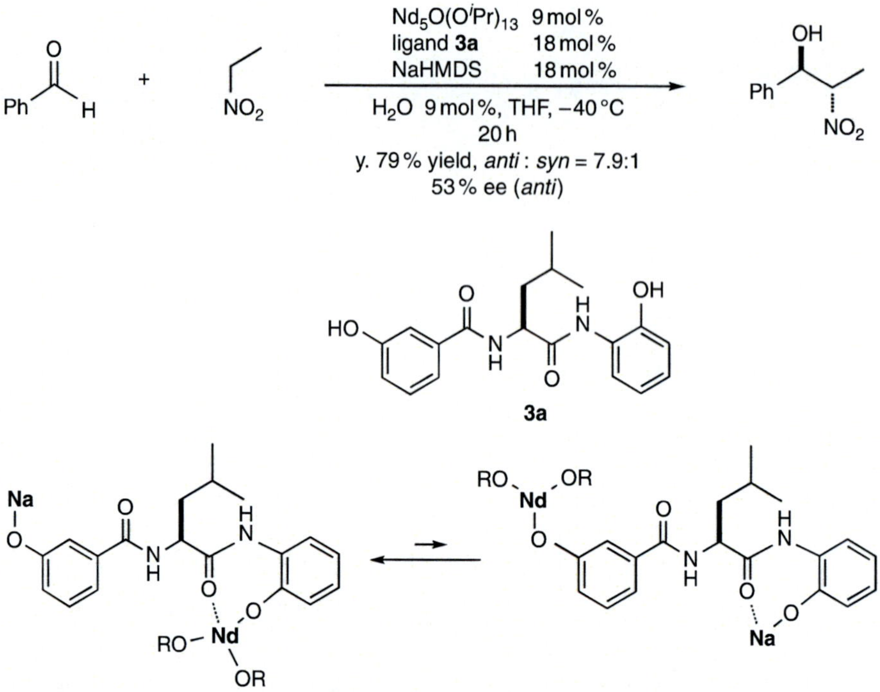

Fig. 11 Catalytic asymmetric nitroaldol reaction with Nd/Na heterobimetallic catalyst

$Nd_5O(O\text{-}iPr)_{13}$, and NaHMDS ($Na\{N(TMS)_2\}$), in which Nd^{3+} would be chelated at the *o*-aminophenol part and Na^+ would be located at the *m*-hydroxysalicylamide part, promoted the asymmetric nitroaldol reaction of benzaldehyde and nitroethane *anti*-selectively (*anti*:*syn* = 7.9:1) with moderate enantioselectivity (53% ee). Nd^{3+} aryloxide activated aldehyde as a Lewis acid, while Na aryloxide served as a Brønsted base to generate the Na-nitronate, allowing for the simultaneous activation of both reaction partners in an *anti*-parallel fashion. Systematic manipulation of the ligand structure identified that the difluoro-substituted ligand **3b** significantly improved both the *anti*-selectivity (*anti*:*syn* = 40:1) and enantioselectivity (84% ee) (Fig. 12) [40]. The likely basis for the enhanced stereoselectivity is the formation of the intramolecular hydrogen bonding between C–F–H–N, restricting the free rotation along the C–C single bond between *m*-hydroxybenzoyl group and benzamide moiety [41]. The orientation of *m*-hydroxybenzoyl would be key to achieve *anti*-parallel transition state to afford the nitroaldol product in a highly stereoselective manner.

The formation of white suspension was observed during the catalyst preparation process (Fig. 13). Whereas the mixture remained clear upon addition of $Nd_5O(O\text{-}iPr)_{13}$ to ligand **3b**, the subsequent addition of NaHMDS turned the mixture into a suspension. ESI TOF MS analysis of the catalyst at this stage revealed the formation of a heterobimetallic complex in which two metal cations Nd and Na are

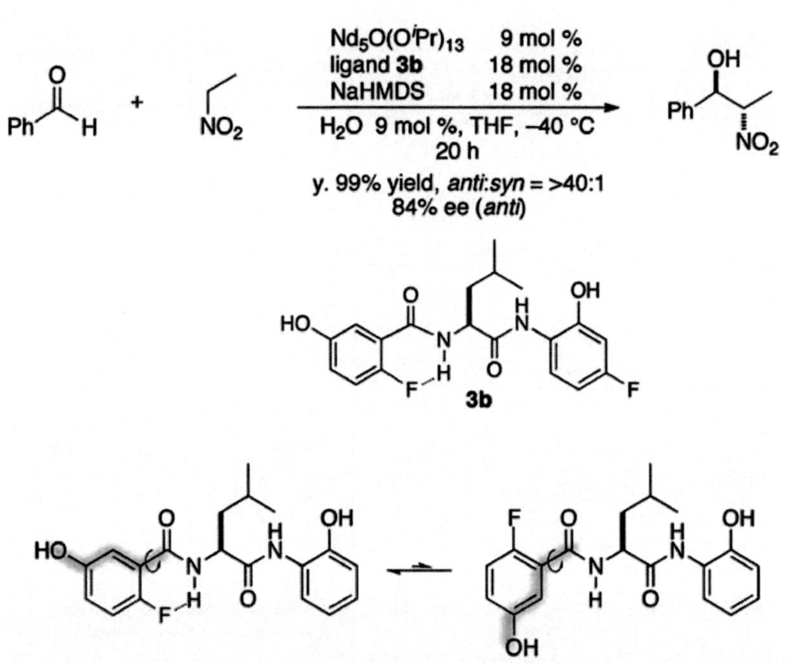

Fig. 12 Improvement of stereoselectivity with difluoro-substituted ligand 3b

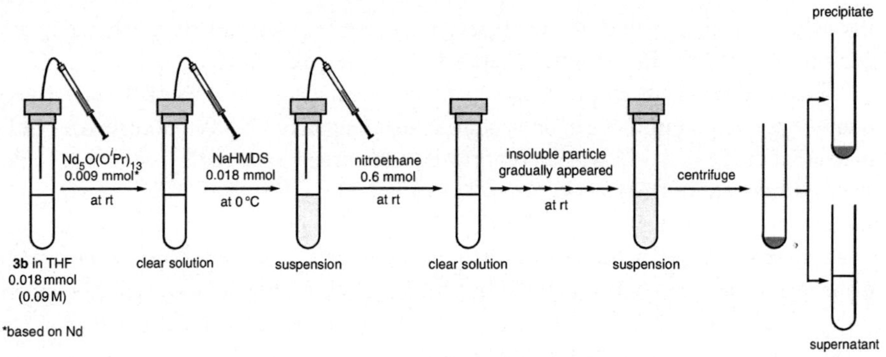

Fig. 13 Macroscopic state of the catalyst during preparation process

incorporated. Further addition of nitroethane to the suspension developed a clear solution, which gradually turned into a suspension again. To dissect the nature of the heterobimetallic catalyst in detail, the suspension was centrifuged to separate the supernatant and the precipitate, which were individually investigated in terms of catalytic activity and their composition. Scheme 7 detailed the individual evaluation of the supernatant and the precipitate as catalyst in the nitroaldol reaction of benzaldehyde and nitroethane. The difference of catalytic activity was not conclusive because the mole fraction of catalytically active components to the

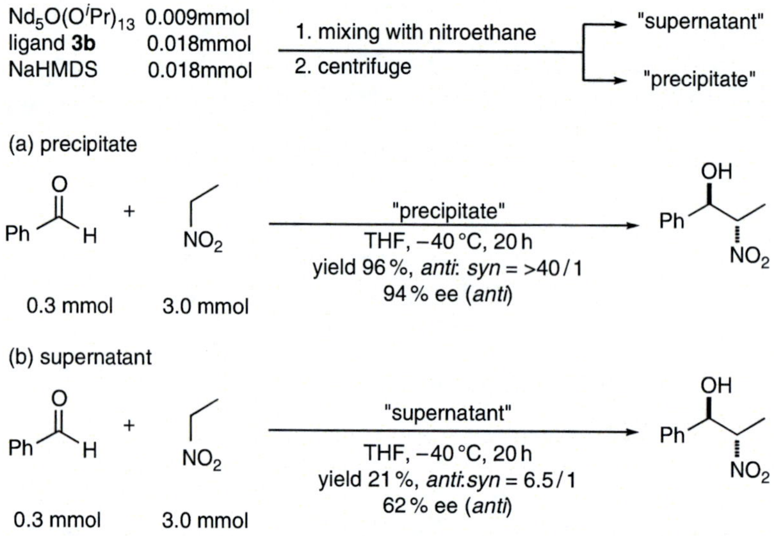

Scheme 7 Macroscopic state of the catalyst during the catalyst preparation process

supernatant and the precipitate was not determined. However, it is obvious that the precipitate, which was suspended in dry THF and used as catalyst, exhibited higher stereoselectivity than the supernatant, affording the desired *anti*-product in 99% yield, *anti*:*syn* = >40:1, and 94% ee. When the reaction was conducted using the whole catalyst solution, the reaction proceeded partially through the poor-performing catalyst in the supernatant to reduce the enantioselectivity (Fig. 12).

The detailed spectroscopic analysis of the precipitate using ICP-AES (inductively coupled plasma-atomic emission spectrometry) and XRF (X-ray fluorescence) indicated that ca. 85% of Nd was incorporated in the precipitate and approximate molar composition was Nd:ligand **3b**:Na = 1:0.96:1.8. In the ESI TOF MS analysis of the precipitate collected after the addition of nitromethane, a series of peaks derived from the Nd/ligand **3b**/Na heterobimetallic fragments including nitromethane were observed. It can be assumed that the (ligand **3b**)$_2$/Nd$_2$ fragments were self-assembled with the varying amount of Na cation and nitroethane to form the actual active catalyst as an insoluble heterogeneous entity.

Thus, developed heterogeneous heterobimetallic catalyst served as an efficient catalyst for the *anti*-selective catalytic asymmetric nitroaldol reaction using a range of aldehydes and nitroalkanes (Scheme 8) [40]. Aromatic aldehydes bearing various electron-withdrawing and -donating substituents were suitable substrate to provide the corresponding nitroaldol product in high *anti*- and enantioselectivity. The reaction with the aromatic aldehyde having highly electron-withdrawing substituent such as nitro and cyano afforded diminished stereoselectivity likely due to the competitive retro reaction. Moderate *anti*-selectivity was observed with aliphatic aldehydes. Of particular note is that nitropropane and functionalized nitroalkanes were also applicable in this catalytic system, affording the desired *anti*-1,2-nitro alkanols in a highly stereoselective manner.

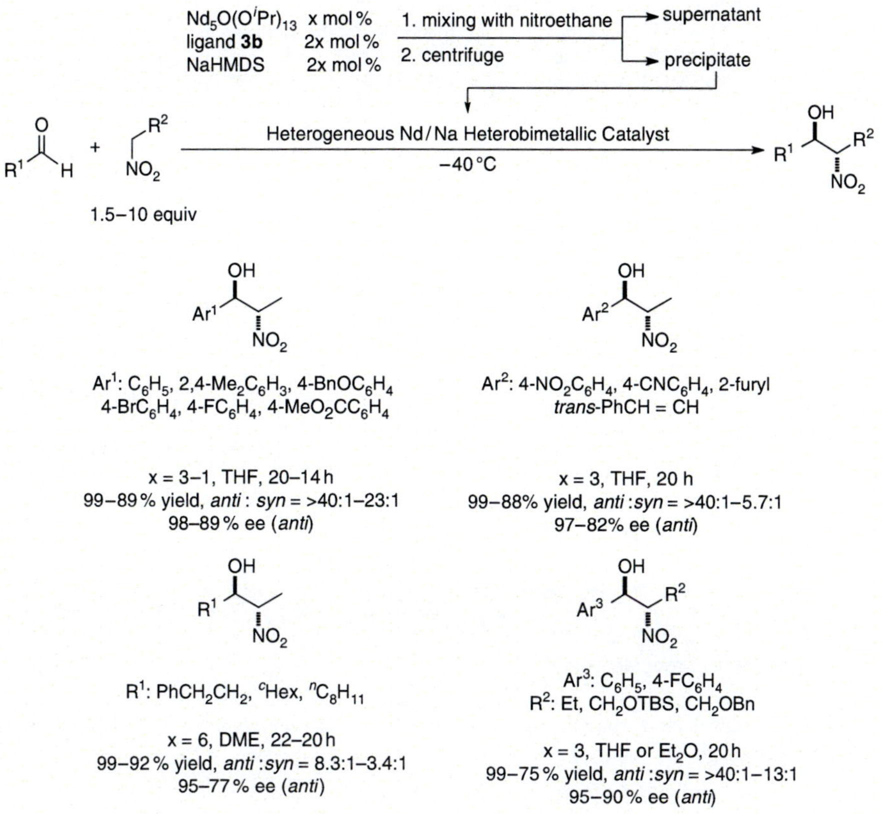

Scheme 8 *Anti*-selective catalytic asymmetric nitroaldol reaction promoted by the heterogeneous Nd/Na heterobimetallic catalyst

The utility of the *anti*-selective nitroaldol reaction was highlighted by the efficient large-scale synthesis of the key intermediate of a β_3-adrenoreceptor agonist **4**, which is under the clinical trial (Scheme 9). The *anti*-selective nitroaldol reaction of 4-benzyloxybenzaldehyde was conducted with 1 mol% of the heterogeneous heterobimetallic catalyst on 50 g scale, affording the desired product in 76% yield with *anti*:*syn* = >40:1 and 98% ee after recrystallization.

3.2 Anti-*Selective Catalytic Asymmetric Nitro-Mannich (aza-Henry) Reaction*

The heterobimetallic catalytic system using the amide-based ligand as a platform was extended to the diastereo- and enantioselective nitro-Mannich-type reaction. In an initial attempt using the heterogeneous heterobimetallic Nd/Na for the reaction of benzaldehyde-derived *N*-Boc imine and nitroethane exhibited poor stereoselectivity, likely because the coordination mode of *N*-Boc imine is different

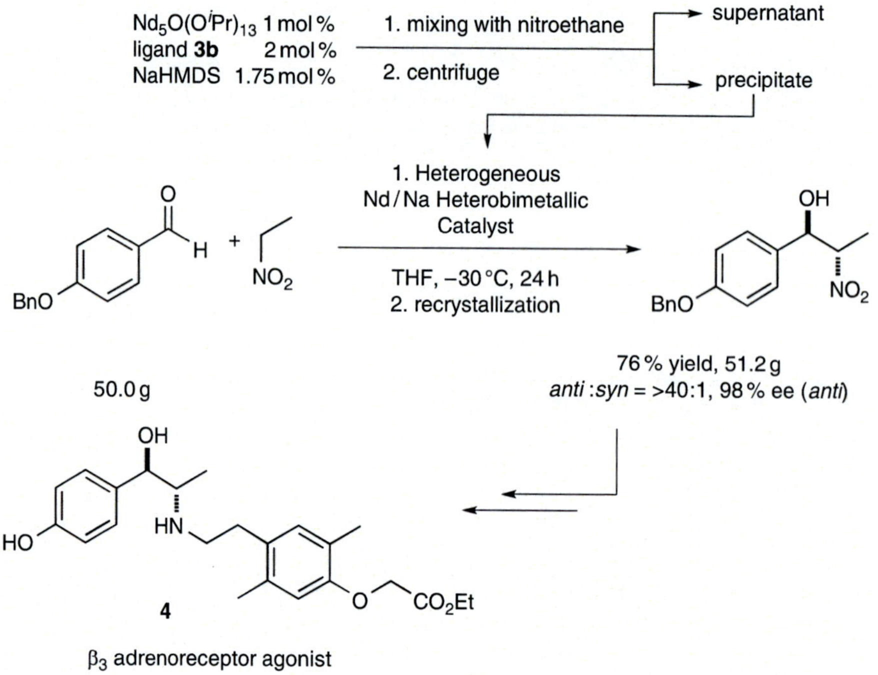

Scheme 9 *Anti*-selective catalytic asymmetric nitroaldol reaction promoted by the heterogeneous Nd/Na heterobimetallic catalyst

from that of aldehyde. A systematic metal screening identified that the optimized heterobimetallic combination of ytterbium and potassium for this specific transformation (Scheme 10) [42]. The heterobimetallic catalyst prepared from ligand **3b**, Yb(O-*i*Pr)$_3$, and KHMDS promoted the asymmetric nitro-Mannich-type reaction to afford *anti*-1,2-amino alkanols in moderate to high enantioselectivity. Imines bearing electron-withdrawing functionality gave poor stereoselectivity.

4 Enantioselective Homopolymetallic Catalysis

4.1 *Catalytic Enantioselective Protonation Using Chiral Poly Gadolinium Complexes*

We developed various catalytic enantioselective cyanation and azidation reactions using chiral poly rare earth metal (RE) complexes derived from ligands **5–8** and TMSCN or TMSN$_3$ as a stoichiometric nucleophile [43, 44]. General mechanism for those asymmetric catalyses is shown in Scheme 11. First, polymetallic complexes **9** containing defined higher order structures are generated via a reaction

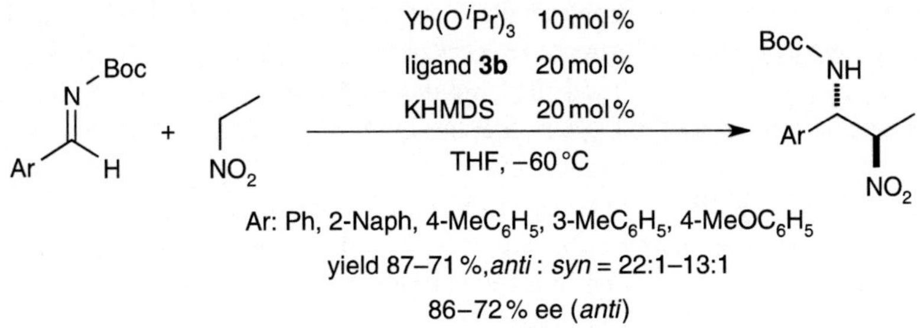

Scheme 10 *Anti*-selective catalytic asymmetric nitro-Mannich-type reaction with a Yb/K heterobimetallic catalyst

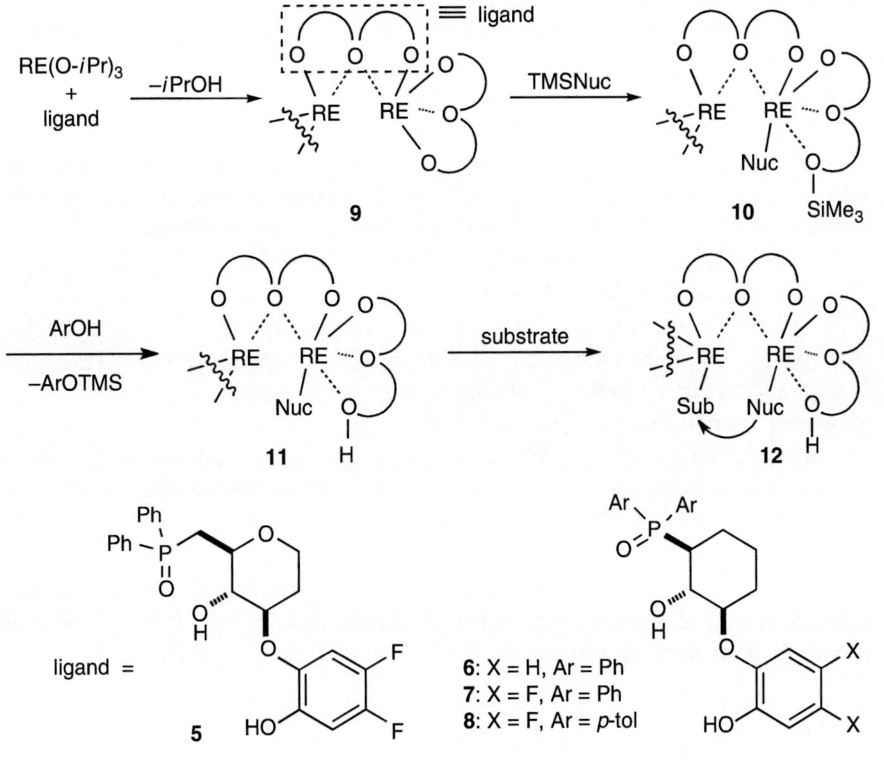

Scheme 11 General reaction mechanism for poly rare earth metal complex-catalyzed asymmetric cyanation and azidation

between RE(O-*i*Pr)$_3$ and a chiral ligand. Next, addition of TMSCN or TMSN$_3$ to **9** produces rare earth metal isonitrile or azide complexes **10** (Nuc = CN or N$_3$) through transmetalation. Protic additives, such as 2,6-dimethylphenol (DMP), often accelerate the reactions by incorporating a proton in the complex acting as a catalyst turnover accelerator (**11**). Substrates such as ketones, ketimines, α,β-unsaturated carbonyl compounds, and *N*-acyl aziridines coordinate to a Lewis acidic rare earth metal in a polymetallic complex, enhancing their electrophilicity. The enantioselective bond formation occurs through an intramolecular transfer of the activated rare earth metal-based nucleophiles to the activated electrophile, whose positions are defined by the higher order structure of the asymmetric catalyst (**12**).

Based on the general reaction mechanism shown in Scheme 11, we expected that polymetallic catalyst **9** would be also useful for enantioselective protonation of enol silyl ethers **13**. Thus, using enol silyl ethers as TMSNuc, transmetalation would produce **10** containing a rare earth metal enolate (Nuc = enolate). If protonolysis of the silyl ether moiety of **10** is faster than that of the rare earth metal enolate, **11** containing activated enolate and proton both at the defined positions in a chiral multimetallic complex would be generated. Enantioselective protonation should proceed from **11**, producing ketones **14** containing α-tertiary chiral center. The catalytic enantioselective protonation of prochiral enolates is a synthetically versatile reaction, but enantioselectivity is generally difficult to induce due to the extremely small size of proton [45–52].

Based on the above blueprint, we optimized the reaction. As a result, using a catalyst prepared from Gd(O-*i*Pr)$_3$ and ligand **7** or **8** mixed in a 1:1.2 ratio, high enantioselectivity was obtained from α-aliphatic-substituted tetralone- and indanone-derived substrates (Scheme 12) [53]. The enantioselectivity was not very sensitive to the size of the α-substituent. Ligands **7** and **8** afforded comparable enantioselectivity, but the reaction rate using a catalyst derived from **8** was higher than using the catalyst derived from **7**, which was particularly advantageous when catalyst loading was reduced. The enantioselective catalyst was determined to be a Gd:ligand = 5:6 complex by ESI-MS studies. Kinetic studies supported the initial hypothesis that this reaction proceeds through a gadolinium enolate, which was generated via rate-limiting transmetalation.

This chemistry was extended to a catalytic conjugate addition of cyanide to α-substituted α,β-unsaturated *N*-acylpyrroles **15** followed by enantioselective protonation (Scheme 13) [53]. In this reaction, gadolinium enolate was generated in situ via conjugate addition of a gadolinium isonitrile to **15**. Products **16** containing α-secondary alkyl- and α-aryl-substituted tertiary stereocenters, which are usually difficult to access via enantioselective alkylation or arylation, were successfully produced, in addition to α-methyl and ethyl-substituted *N*-acylpyrroles.

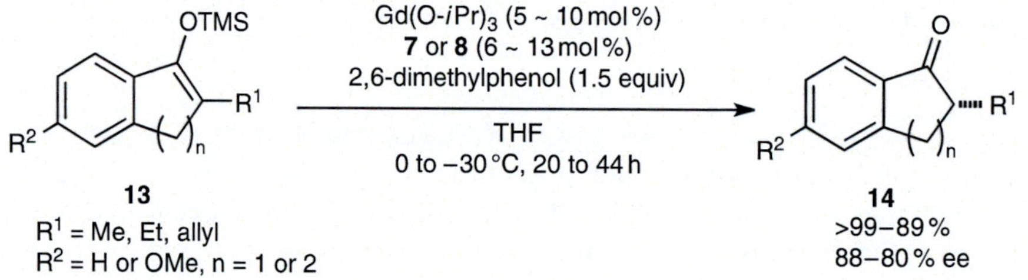

Scheme 12 Catalytic enantioselective protonation of enol silyl ethers

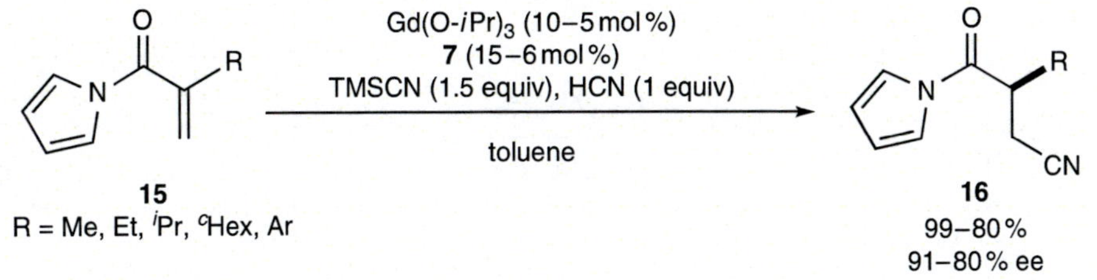

Scheme 13 Catalytic conjugate cyanation followed by enantioselective protonation of α-substituted α,β-unsaturated *N*-acylpyrroles

4.2 Catalytic Asymmetric Synthesis of R207910

Tuberculosis (TB) is one of the most serious worldwide infectious diseases. Despite the emergence of a multidrug resistant mutant, a new TB drug has not been introduced in the market for more than 40 years. R207910 (**17**), discovered by the Johnson & Johnson pharmaceutical company, is a promising potent anti-TB drug candidate that selectively inhibits the ATP synthase proton pump of both drug-sensitive and drug-resistant *Mycobacterium tuberculosis* [54, 55]. In contrast to its fascinating biological activity, however, the only existing synthetic route of **17** reported before our achievement is rather primitive, involving low-yielding stereo-random C–C bond formation, separation of the desired diastereomer as a minor isomer, and optical resolution using chiral HPLC [55]. We recently succeeded in developing the first catalytic asymmetric synthesis of R207910 (Scheme 14) [56].

Our initial synthetic plan was to construct α-chiral ketone **21** through a catalytic enantioselective protonation of the corresponding enol silyl ether, a methodology mentioned in the above section. This idea was not very successful even after intensive optimization (**21** with up to 63% ee was obtained). Therefore, we investigated an alternative method, enantioselective proton migration reaction from **20** to **21** through a chiral metal dienolate generated via deprotonation of **20**, taking advantage of the Brønsted basicity of metal complexes of **5–8** and related ligands.

Enone **20** was synthesized from *N*-protected quinolinone **18** and diketone **19** through site-selective aldol reaction followed by dehydration. Enone **20** was converted to α-chiral ketone **21** with 88% ee using 2.5 mol% yttrium catalyst prepared from YHMDS ($Y\{N(SiMe_3)_2\}_3$) and chiral ligand **22**, 1.25 mol% *p*-methoxypyridine *N*-oxide, and 0.5 mol% Bu_4NCl. Comprehensive metal screening of the catalyst led us to determine yttrium as the best metal for the enantioselective proton migration with regard to enantioselectivity. Nitrogen-containing ligand **22** produced higher enantioselectivity and catalyst activity than the previous ether-containing ligand **6**. Addition of a catalytic Lewis base improved the enantioselectivity, among which *p*-methoxypyridine *N*-oxide (MEPO) produced the best result. Finally, we found that addition of a catalytic amount of Bu_4NCl could accelerate the reaction without effecting the enantioselectivity. Because Bu_4NCl itself promoted racemic proton migration to some extent, the amount of Bu_4NCl must be much smaller (×1/5) than the amount of the asymmetric yttrium catalyst.

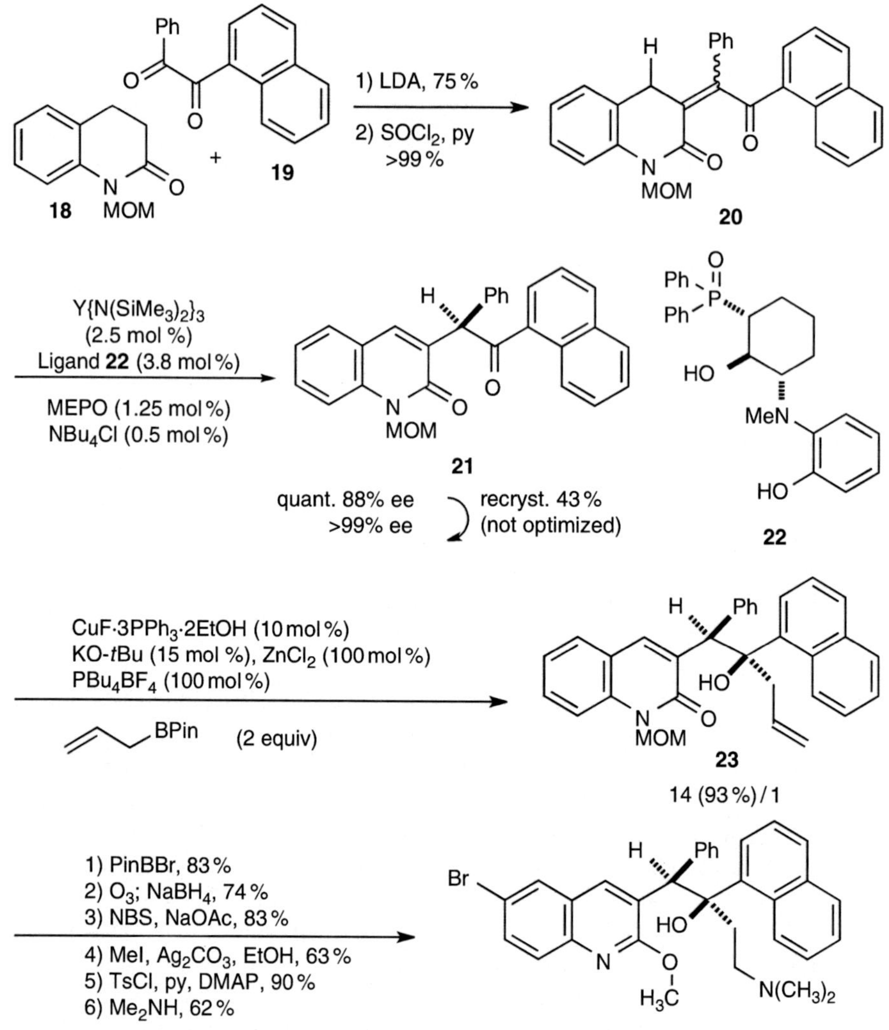

R207910 (**17**)

Scheme 14 Catalytic asymmetric synthesis of R207910

To gain insight into the reaction mechanism, we studied the catalyst composition using ESI-MS. An MS peak corresponding to Y:**22** = 2:3 complex **24** was predominantly observed in a solution prepared from $Y\{N(SiMe_3)_2\}_3$ and **22** mixed in a 2:3 ratio. By the addition of 0.5 equiv (to Y) of MEPO, MS peaks corresponding to **24** and a Y:**22**:MEPO = 2:3:1 complex were observed. Therefore, the Y:**22**:MEPO = 2:3:1 ternary complex would be the active enantioselective catalyst. This hypothesis was also supported by the observed dependency of the enantioselectivity on the Y:**22** ratio when the catalyst was prepared. Thus, enantiomeric excess of **21** gradually increased according to the increase of Y:**22** ratio from 1:1 to 1:1.5, and remained consistent from

1:1.5 to 1:2 ratio. This dependency again indicated that the active catalyst with optimum enantioselectivity was generated at Y:**22** = 1:1.5 ratio.

Based on the structural information and the experimental results that additive MEPO increased enantioselectivity while Bu_4NCl accelerated the reaction without affecting the enantioselectivity, we propose the working model shown in Scheme 15 for the catalytic cycle. MEPO coordinates to one of the yttrium metals of the bimetallic complex and favorably modifies the asymmetric environment of the catalyst. Meanwhile, Bu_4NCl might increase the Brønsted basicity of the catalyst, by generating an ammonium phenoxide (or alkoxide) moiety in **25**, which would deprotonate **20** via **26**. The thus-generated *pre*-transition state complex **27** contains an active Y-enolate and an acidic proton, both derived from substrate **20**, at defined positions in the complex. Asymmetric α-protonation would proceed in complex **27**, producing enantiomerically enriched **21** and regenerating catalyst **25**.

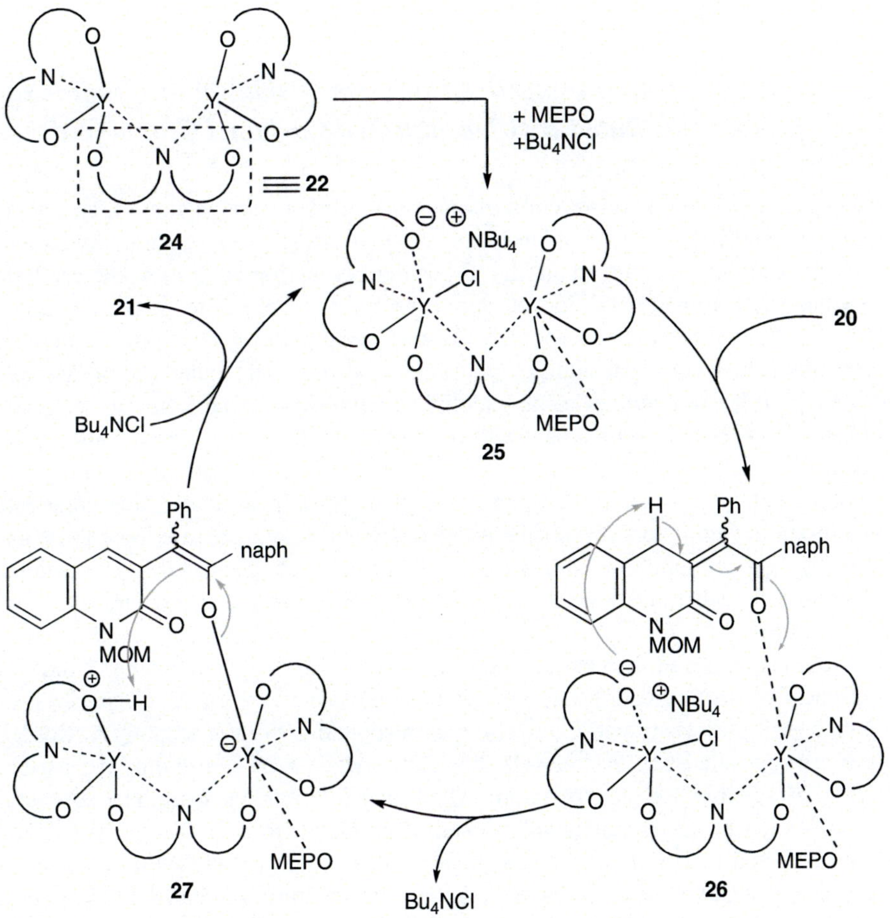

Scheme 15 Proposed catalytic cycle of the enantioselective proton migration

After enantiomerically pure **21** was produced via recrystallization, an allyl group was introduced with excellent diastereoselectivity (14:1) under the modified conditions using CuF catalysis [57, 58] developed in our group (Scheme 14, from **21** to **23**). Addition of 15 mol% KO-*t*Bu accelerated the reaction, and a combination of $ZnCl_2$ and Bu_4PBF_4 markedly improved the diastereoselectivity. Preliminary NMR studies suggested that ZnFCl and Bu_4PBF_3Cl might be generated from the additives, and a ZnFCl species might be beneficial for the improvement of the diastereoselectivity. It is noteworthy that addition of conventional nucleophiles, including lithium enolates, metal enolate equivalents derived from acetonitrile, vinyl Grignard reagents, allyl Grignard reagents ± $CeCl_3$, allylindium, allylzinc, and metal alkynides, failed due to both the high steric demand and the presence of the fairly acidic α-proton of the ketone carbonyl group in substrate **21**. This result highlights the utility of the CuF-catalyzed allylation reaction, which proceeds under nearly neutral conditions.

Enantiomerically and diastereomerically pure **17** was produced from **23** via six steps (Scheme 14).

4.3 *Catalytic Enantioselective Conjugate Addition of Cyanide to β,β-Disubstituted α,β-Unsaturated Carbonyl Compounds*

Catalytic asymmetric construction of quaternary carbon stereocenters is an important and challenging objective in chemical synthesis [59, 60]. Quaternary stereocenters can be constructed at the β-position of carbonyl groups by catalytic asymmetric conjugate addition of carbon-based nucleophiles to β,β-disubstituted α,β-unsaturated carbonyl compounds [61–64]. We previously developed a catalytic enantioselective conjugate addition of cyanide to β-monosubstituted α,β-unsaturated *N*-acylpyrroles and ketones using gadolinium catalysts derived from ligands **5** and **7** [65–71]. We planned to expand this reaction to a quaternary carbon construction using β,β-disubstituted α,β-unsaturated carbonyl compounds as substrates. Jacobsen's group recently reported the first example of catalytic enantioselective conjugate addition of cyanide to a β,β-disubstituted imide substrate, constructing a β-quaternary carbon containing a synthetically versatile cyanide group [70]. The catalyst activity and substrate generality of this isolated example, however, were not satisfactory.

Although rare earth metal complexes derived from ligands **6** and **7** did not produce satisfactory catalyst activity and enantioselectivity for the conjugate cyanation of β,β-disubstituted enones, the corresponding strontium catalyst afforded promising results. Use of other alkaline earth metals such as calcium and barium was not suitable for this reaction. After intensive studies including structural modification of the chiral ligand, a catalyst prepared from $Sr(O\text{-}i\text{Pr})_2$ and ligand **28** mixed in a 1:1.7 ratio and the use of TBSCN + 2,6-dimethylphenol as a cyanide source (gradually generating HCN in the reaction mixture) produced the optimized result. The catalyst loading was reduced to 0.5 mol% in ideal cases. Substrate scope of this reaction was very broad (Scheme 16) [72]. Excellent enantioselectivity

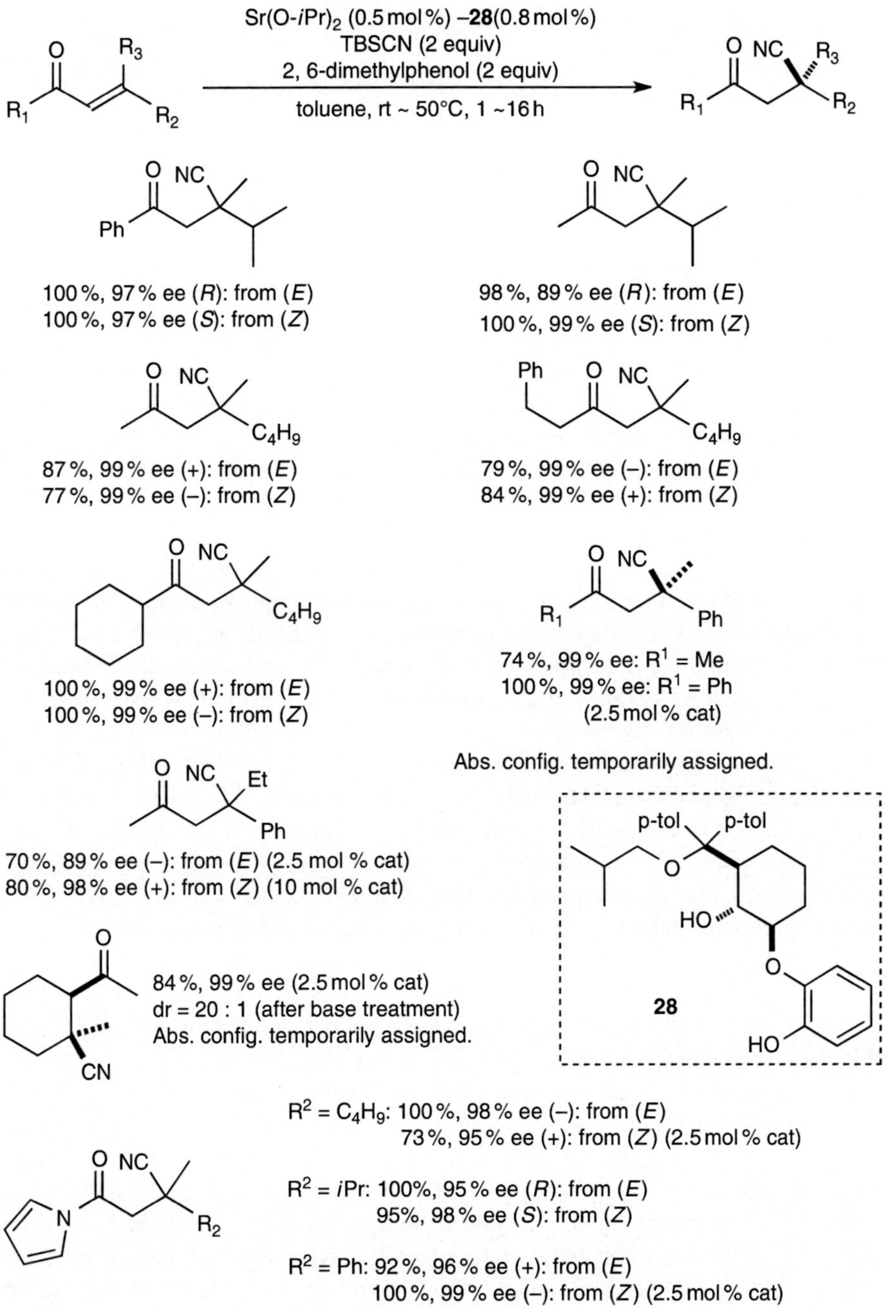

Scheme 16 Catalytic enantioselective conjugate addition of cyanide to β,β-disubstituted α,β-unsaturated carbonyl compounds

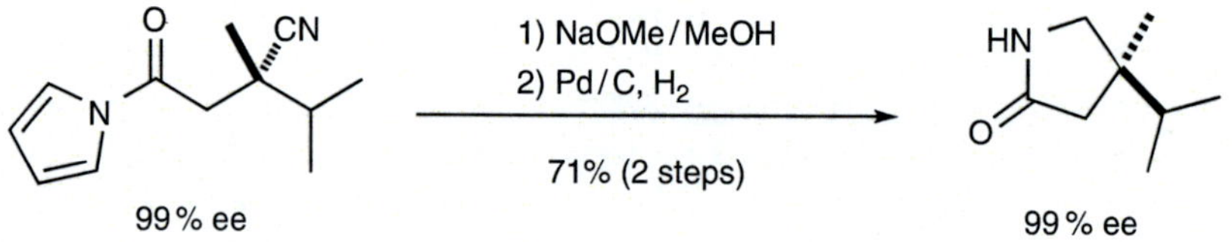

Scheme 17 A typical synthetically useful conversion of the product

was realized from a wide range of β,β-disubstituted enones including aromatic- and aliphatic-substituted substrates. (*E*) and (*Z*) substrates produced opposing enantiomers. The reaction also proceeded from α,β,β-trisubstituted enone with high enantioselectivity. Although the product of asymmetric cyanation was a 1:1 mixture of diastereomers in this case, the diastereoselectivity was enriched, via epimerization of the α-stereocenter with base treatment of the crude mixture to 20:1 in high yield without affecting the excellent enantioselectivity. In all entries, the reactions were completely 1,4-selective. The reaction was also applicable to synthetically useful ester equivalents, *N*-acylpyrroles [73, 74].

Due to the versatility of cyanide and *N*-acylpyrroles, various synthetically useful conversions of the products were possible. A typical example is shown in Scheme 17.

ESI-MS studies revealed that the active catalyst for this enantioselective conjugate cyanation is a 3:5 complex of a strontium atom and ligand **28**. In addition, the use of HCN or TMSCN + 2,6-dimethylphenol as a cyanide source, instead of TBSCN + 2,6-dimethylphenol, produced consistent enantioselectivity, indicating that the actual stoichiometric cyanide source is HCN. These results led us to propose a working hypothesis for the catalytic cycle as shown in Scheme 18. Catalyst 3:5 complex **29** deprotonates HCN, generating a strontium isocyanide complex **30**. After coordination of a β,β-disubstituted α,β-unsaturated carbonyl substrate to complex **30**, enantioselective cyanation proceeds through intramolecular cyanide transfer (**31**), resulting in **32** containing a strontium enolate. The enolate is protonated by a proton in the complex, liberating the enantiomerically enriched product with regenerating active catalyst **29**.

In addition to the excellent enantioselectivity and broad substrate scope, complete 1,4-regioselectivity is an important characteristic of this reaction. Previous studies suggested that the conjugate cyanation products were derived from kinetically formed cyanohydrins through catalyzed cyanide migration to thermodynamically more stable 1,4-adducts [75]. To assess the ability of the strontium complex in promotion of cyanide migration reaction from 1,2-adducts to 1,4-adducts, we subjected racemic cyanohydrin **33** under the reaction conditions in the presence of 10 mol% of the strontium complex (Scheme 19). As a result, conjugate cyanation product **34** was obtained in quantitative yield with >99% ee. Moreover, this cyanide migration reaction was determined to be an intermolecular process, according to a crossover experiment. Therefore, even if 1,2-addition of cyanide proceeded, the catalyst promoted retro-cyanation from the resulting cyanohydrin, and the subsequent irreversible asymmetric 1,4-cyanation produced the desired 1,4-product. This ability of strontium complex **29** would be attributed to the complete 1,4-regioselectivity observed in the current reaction.

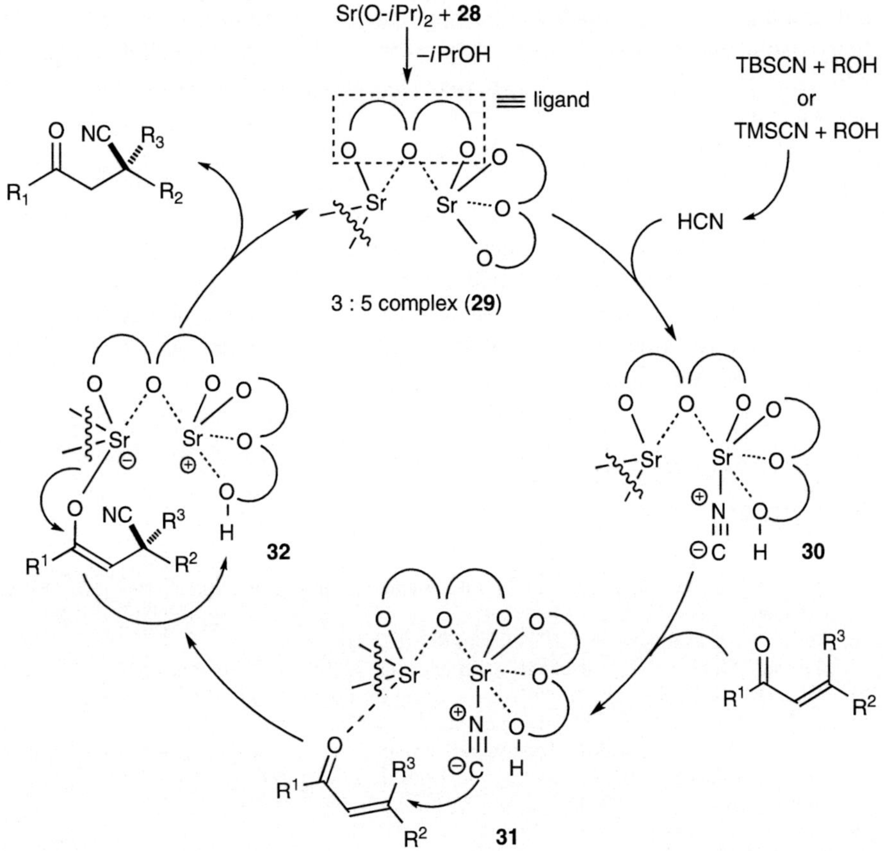

Scheme 18 Proposed catalytic cycle of enantioselective conjugate cyanation promoted by a strontium complex

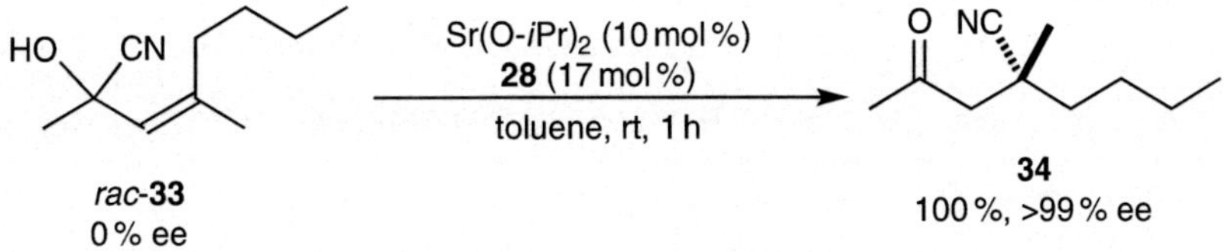

Scheme 19 Catalytic enantioselective rearrangement of cyanide: a possible origin for the complete 1,4-regioselectivity

5 Conclusion

In this chapter, recent achievements in asymmetric cooperative catalysis exploiting the power of multimetallic catalysts are described. The three types of multidentate ligands, Schiff base ligands, amide-based ligands, and sugar-derived ligands, serve as a suitable platform for the construction of homobimetallic, heterobimetallic, and polymetallic catalysts, which exhibit high catalytic performance and stereoselectivity. Simultaneous activation of substrate pair at the multiple catalytically active sites enables the catalytic asymmetric construction of quaternary stereogenic centers via intermolecular C–C bond-forming reactions. Spatial arrangement of distinct metal cations in chiral environment allows the reaction to proceed through a thoroughly different transition state, attaining different stereoselectivity that cannot be observed in a monometallic catalytic system. Changing the combination of metal cations gave rise to a dramatic change of catalytic activity, catalytic specificity, and stereoselectivity, indicating that wider range of multimetallic catalysts can be developed than the monometallic counterparts. There would be no doubt to expect the emergence of a plethora of multimetallic asymmetric catalysts in near future and their application to the practical organic synthesis.

References

1. Shibasaki M, Yamamoto Y (2004) Multimetallic catalysts in organic synthesis. Wiley, Weinheim
2. Shibasaki M, Yoshikawa N (2002) Chem Rev 102:2187
3. Shibasaki M, Kanai M, Matsunaga S, Kumagai N (2009) Acc Chem Res 42:1187
4. Yamamoto H, Futatsugi K (2005) Angew Chem Int Ed 44:1924
5. Ma J-A, Cahard D (2004) Angew Chem Int Ed 43:4566
6. Houk KN, List B (ed) (2004) Acc Chem Res issue 8
7. Taylor MS, Jacobsen EN (2005) Angew Chem Int Ed 45:1250
8. Berkessel A, Gröger H (2005) Asymmetric organocatalysis. Wiley, New York
9. Jacobsen EN (2000) Acc Chem Res 33:421, review
10. Katsuki T (2004) Chem Soc Rev 33:437, review
11. Haak RM, Wezenberg SJ, Kleij AW (2010) Chem Commun 46:2713, review on bimetallic salens
12. Handa S, Gnanadesikan V, Matsunaga S, Shibasaki M (2007) J Am Chem Soc 129:4900
13. Handa S, Gnanadesikan V, Matsunaga S, Shibasaki M (2010) J Am Chem Soc 132:4925
14. Handa S, Nagawa K, Sohtome Y, Matsunaga S, Shibasaki M (2008) Angew Chem Int Ed 47:3230
15. Kobayashi S, Sugiura M, Kitagawa H, Lam WWL (2002) Chem Rev 102:2227, review on rare earth metal triflates in organic synthesis
16. Wang SX, Wang MX, Wang DX, Zhu J (2007) Org Lett 9:3615
17. Tsuruta H, Yamaguchi K, Imamoto T (1999) Chem Commun 1703
18. Mihara H, Xu Y, Shepherd NE, Matsunaga S, Shibasaki M (2009) J Am Chem Soc 131:83840
19. Furutachi M, Chen Z, Matsunaga S, Shibasaki M (2010) Molecules 15:532
20. Chen Z, Furutachi M, Kato Y, Matsunaga S, Shibasaki M (2009) Angew Chem Int Ed 48:2218
21. Kato Y, Furutachi M, Chen Z, Mitsunuma H, Matsunaga S, Shibasaki M (2009) J Am Chem Soc 131:9168
22. Chen Z, Morimoto H, Matsunaga S, Shibasaki M (2008) J Am Chem Soc 130:2170
23. Chen Z, Yakura K, Matsunaga S, Shibasaki M (2008) Org Lett 10:3239
24. Xu Y, Lu G, Matsunaga S, Shibasaki M (2009) Angew Chem Int Ed 48:3353
25. Shepherd NE, Tanabe H, Xu Y, Matsunaga S, Shibasaki M (2010) J Am Chem Soc 132:3666
26. Mouri S, Chen Z, Matsunaga S, Shibasaki M (2009) Chem Commun 5138
27. Kato Y, Chen Z, Matsunaga S, Shibasaki M (2009) Synlett 1635
28. Xu Y, Lu G, Matsunaga S, Shibasaki M (2010) Org Lett 12: 3246

29. Mouri S, Chen Z, Mitsunuma H, Furutachi M, Matsunaga S, Shibasaki M (2010) J Am Chem Soc 132:1255
30. Yoshino T, Morimoto H, Lu G, Matsunaga S, Shibasaki M (2009) J Am Chem Soc 131:17068
31. Henry C (1895) C R Hebd Seances Acad Sci 120:1265
32. Shibasaki M, Gröger H (1999) Nitroaldol reaction. In: Jacobsen EN, Pfaltz A, Yamamoto H (eds) Comprehensive asymmetric catalysis. Springer, Berlin, pp 1075–1090
33. Palomo C, Oiarbide M, Laso A (2007) Eur J Org Chem 2561
34. Sasai H, Tokunaga T, Watanabe S, Suzuki T, Itoh N, Shibasaki M (1995) J Org Chem 60:7338
35. Sohtome Y, Takemura N, Takada K, Takagi R, Iguchi T, Nagasawa K (2007) Chem Asian J 2:1150
36. Arai T, Takashita R, Endo Y, Watanabe M, Yanagisawa A (2008) J Org Chem 73:4903
37. Uraguchi D, Sakaki S, Ooi T (2007) J Am Chem Soc 129:12392
38. Purkarthofer T, Gruber K, Gruber-Khadjawi M, Waich K, Skranc W, Mink D, Griengl H (2006) Angew Chem Int Ed 45:3454
39. Nitabaru T, Kumagai N, Shibasaki M (2008) Tetrahedron Lett 49:272
40. Nitabaru T, Nojiri A, Kobayashi M, Kumagai N, Shibasaki M (2009) J Am Chem Soc 131:13860
41. Li C, Ren S-F, Hou J-L, Yi H-P, Zhu S-Z, Jiang XK, Li Z-T (2005) Angew Chem Int Ed 44:5725
42. Nitabaru T, Kumagai N, Shibasaki M (2010) Molecules 15:1280
43. Shibasaki M, Kanai M (2007) Org Biomol Chem 5:2027
44. Kanai M, Kato N, Ichikawa E, Shibasaki M (2005) Synlett 1491
45. Cheon CH, Yamamoto H (2008) J Am Chem Soc 130:9246
46. Marinescu S, Nishimata T, Mohr J, Stoltz BM (2008) Org Lett 10:1039
47. Poisson T, Dalla V, Marsais F, Dupas G, Oudeyer S, Levacher V (2007) Angew Chem Int Ed 46:7090
48. Mitsuhashi K, Ito R, Arai T, Yanagisawa A (2006) Org Lett 8:1721
49. Yanagisawa A, Touge T, Arai T (2005) Angew Chem Int Ed 44:1546
50. Sugiura M, Nakai T (1997) Angew Chem Int Ed Engl 36:2366
51. Duhamel L, Duhamel P, Plaquevent JC (2004) Tetrahedron: Asymmetry 15:3653
52. Yanagisawa A, Ishihara K, Yamamoto H (1997) Synlett 411
53. Morita M, Drouin L, Motoki R, Kimura Y, Fujimori I, Kanai M, Shibasaki M (2009) J Am Chem Soc 131:3858
54. Andries K, Verhasselt P, Guillemont J, Göhlmann HWH, Neefs JM, Winkler H, Gestel JV, Timmerman P, Zhu M, Lee E, Williams P, de Chaffoy D, Huitric E, Hoffner S, Cambau E, Truffot-Pernot C, Lounis N, Jarlier V (2005) Science 307:223
55. van Gestel JFE, Guillemont JEG, Venet MG, Poignet HJJ, Decrane LFB, Odds FC. US Patent 2005/0148581
56. Saga Y, Motoki R, Makino S, Shimizu Y, Kanai M, Shibasaki M (2010) J Am Chem Soc 132:7905
57. Wada R, Oisaki K, Kanai M, Shibasaki M (2004) J Am Chem Soc 126:8910
58. Wada R, Shibuguchi T, Makino S, Oisaki K, Kanai M, Shibasaki M (2006) J Am Chem Soc 128:7687
59. Trost BM, Jiang C (2006) Synthesis 369
60. Cozzi PG, Hilgraf R, Zimmermann N (2007) Eur J Org Chem 36:5969
61. Hird AW, Hoveyda AH (2005) J Am Chem Soc 127:14988
62. d'Augustin M, Palais L, Alexakis AJ (2005) Angew Chem Int Ed 44:1376
63. Fillion E, Wilsily A (2006) J Am Chem Soc 128:2774
64. Shintani R, Tsutsumi Y, Nagaosa M, Nishimura T, Hayashi T (2009) J Am Chem Soc 131:13588
65. Mita T, Sasaki K, Kanai M, Shibasaki M (2005) J Am Chem Soc 127:514
66. Fujimori I, Mita T, Maki K, Shiro M, Sato A, Furusho S, Kanai M, Shibasaki M (2007) Tetrahedron 63:5820

67. Tanaka Y, Kanai M, Shibasaki M (2008) J Am Chem Soc 130:6072
68. Sammis GM, Jacobsen EN (2003) J Am Chem Soc 125:4442
69. Sammis GM, Danjo H, Jacobsen EN (2004) J Am Chem Soc 126:9928
70. Mazet C, Jacobsen EN (2008) Angew Chem Int Ed 47:1762
71. Wang J, Li W, Liu Y, Chu Y, Lin L, Liu X, Feng X (2010) Org Lett 12:1280
72. Tanaka Y, Kanai M, Shibasaki M (2010) J Am Chem Soc 132:8862. doi:10.1021/ja1035286
73. Matsunaga S, Kinoshita T, Okada S, Harada S, Shibasaki M (2004) J Am Chem Soc 126:7559
74. Evans DA, Borg G, Scheidt KA (2002) Angew Chem Int Ed 41:3188
75. Nagata W, Yoshioka M (1997) Org React 25:255, see also [63]

Top Organomet Chem (2011) 37: 31–53
DOI: 10.1007/3418_2011_2

Published online: 12 March 2011

Bifunctional Transition Metal-Based Molecular Catalysts for Asymmetric Syntheses

Takao Ikariya

Abstract The discovery and development of conceptually new chiral bifunctional molecular catalysts based on the metal/NH acid–base synergy effect are described. The chiral bifunctional molecular catalysis originally developed for asymmetric transfer hydrogenation of ketones is applicable in enantioselective hydrogenation of polar functionalities as well as practical oxidative reactions including aerobic oxidation of alcohols. The structural modification and electronic fine-tuning of the protic amine chelating ligands are crucial to develop unprecedented catalytic reactions. The present bifunctional transition metal-based molecular catalyst offers a great opportunity to open up new fundamentals for stereoselective molecular transformations.

Keywords Aerobic oxidation · Asymmetric reduction · Bifunctional molecular catalyst · Concerto catalysis · Cooperating ligand

Contents

T. Ikariya
Department of Applied Chemistry, Graduate School of Science and Engineering, Tokyo Institute of Technology, 2-12-1 O-okayama, Meguro-ku, Tokyo 152-8552, Japan
e-mail: tikariya@apc.titech.ac.jp

1 Introduction

A deep understanding of the mechanism of catalytic transformations as well as a great insight into the architecture of molecular catalysts has enabled rational design of sophisticated metal-based molecular catalysts [1–4]. Particularly, much effort has been recently paid to the development of bifunctional molecular catalysts based on the combination of two or more active sites working in concert, to attain highly efficient molecular transformations for organic synthesis. We have recently developed metal–ligand cooperating bifunctional catalysts (concerto catalyst), in which the non-innocent ligands directly participate in the substrate activation and in the bond formation. This bifunctional molecular catalysis is now an attractive and general strategy to highly realize effective molecular transformation [5–9].

The first metal-cooperating ligand bifunctional catalyst for reductive transformation of carbonyl compounds was reported by Shvo in 1985 [10, 11]. In 1995, Noyori and Ikariya found a prototype of the bifunctional ruthenium catalysts based on the M/NH acid base synergy effect, RuH(Tsdpen)(η^6-arene), [TsDPEN: *N*-(*p*-toluenesulfonyl)-1,2-diphenylethylenediamine] [12–14] as shown in Fig. 1, for asymmetric transfer hydrogenation of ketones and imines. Milstein's pincer-type bifunctional catalysts based on the aromatization–dearomatization of the pyridyl ligand facilitate hydrogenation and transfer hydrogenation [15, 16]. Grützmacher's diolefinic amido Rh catalysts based on the Rh/NH bifunctionality promote hydrogenation and transfer hydrogenation of olefinic compounds [17, 18].

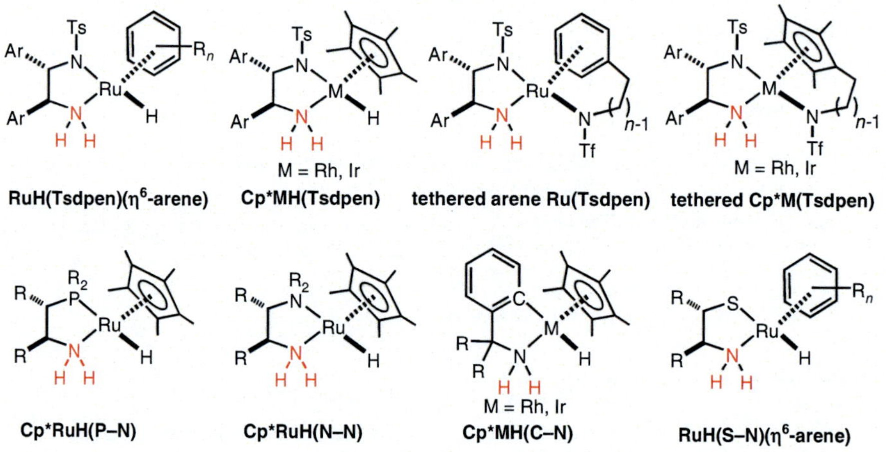

Fig. 1 A prototype and modified bifunctional catalysts bearing chelating amine ligands

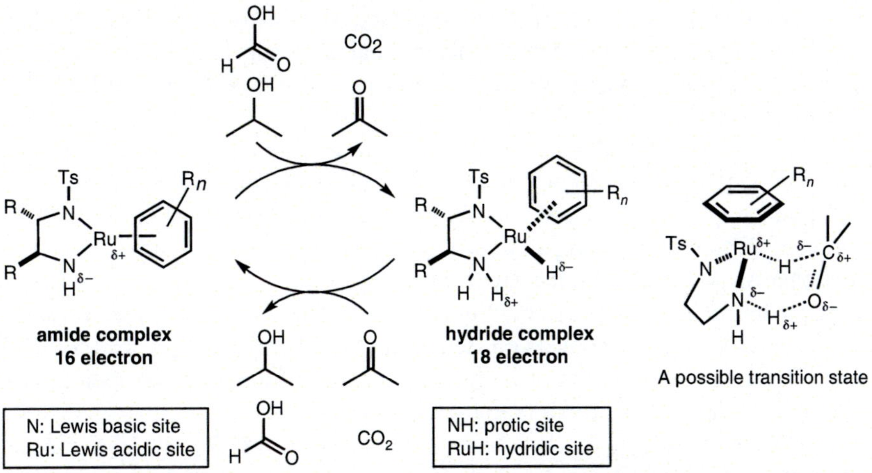

Fig. 2 Concept of bifunctional molecular catalysis based on M/NH acid–base synergy effect

Detailed structural studies on the active catalysts and kinetic studies as well as computational analysis for this hydrogen transfer revealed that an excellent catalytic performance is attributable to the metal–ligand cooperation based on a metal/NH bifunctionality [19]. This conceptually new bifunctional catalysis is unique and simple. Only the chiral amido and chiral amine hydrido ruthenium complexes are involved as catalysts and intermediates. The hydrogen transfer between an alcohol and a ketone takes place reversibly through a six-membered pericyclic transition state as shown in Fig. 2. An important and unprecedented aspect is that the carbonyl compound can be activated via outer-sphere mechanism, in which the metal and the amine ligand work in concert for its efficient activation, leading to high reaction rate and excellent stereoselectivities. The presence of an NH moiety in the ligands is crucially important to determine the catalytic performance in terms of reactivity and selectivity.

In this chapter, I outline our recent progress in asymmetric reductive and oxidative transformations with bifunctional molecular catalysts based on ruthenium, rhodium, and iridium complexes bearing chiral chelating amine ligands.

2 Asymmetric Transfer Hydrogenation of Simple and Functionalized Aromatic Ketones with the Bifunctional Catalysts

The concept of the chiral bifunctional η^6-arene–Ru complexes, RuCl(Tsdpen)-(η^6-arene), has been successfully extended to analogous Cp*Rh and Ir complexes, Cp*MCl(Tsdpen) (Cp* = η^5-C(CH$_3$)$_5$, M = Rh, Ir) [20, 21], as well as a new family of Cp*Ru complexes bearing the protic amine chelating ligands (Fig. 1) for hydrogenation [22], which will be discussed later. The chiral N-sulfonylated diamine complexes serve as highly efficient catalysts for asymmetric transfer hydrogenation of simple alkyl aryl ketones and imines [13]. Chiral N-tosylated diamines, β-amino alcohols (O–N) [23–25], diamines (N–N) [26], amino phosphines (P–N) [22], and aminoethanethiols (S–N) [27] serve as excellent cooperating ligands and lead to high reactivity and enantioselectivity in these asymmetric reactions.

In general, 2-propanol and formic acid can be used as very cheap hydrogen sources. Particularly, 2-propanol is a safe, nontoxic, environmentally friendly hydrogen source. Although the asymmetric reduction in 2-propanol gives satisfactory results, an inherent problem of the reaction using 2-propanol is the reversibility, leading to limited conversion determined by thermodynamic factors of the system and the deterioration of enantiomeric purity of the products upon the long exposure of the reaction mixture to the catalyst.

On the other hand, the asymmetric reduction using formic acid proceeds irreversibly with kinetic enantioselection and, in principle, 100% conversion. However, this bifunctional Ru catalyst also efficiently promotes hydrogenation of CO_2 to give formic acid and its derivatives [28]. Therefore, effective removal of CO_2 with inert gas allow complete conversion, in particular, in a large-scale reaction.

Asymmetric reduction of simple aromatic ketones with a mixture of formic acid and triethylamine containing the bifunctional catalyst is characterized by high efficiency in terms of activity, selectivity, wide applicability, and practicability. As a result of the coordinatively saturated nature of the bifunctional amine hydrido Ru complex, the reduction proceeds chemoselectively without the interference of amino, ester, hydroxyl, carbonyl, sulfido, sulfone, nitro, azide, and chloride groups, neither furan, thiophene, and quinoline rings, nor the olefinic linkage. For example, a reaction of the keto esters with a mixture of $HCO_2H/N(C_2H_5)_3$ containing Ru–TsDPEN complexes gives the corresponding chiral alcohols with moderate to excellent ee's (Fig. 3) [8, 29].

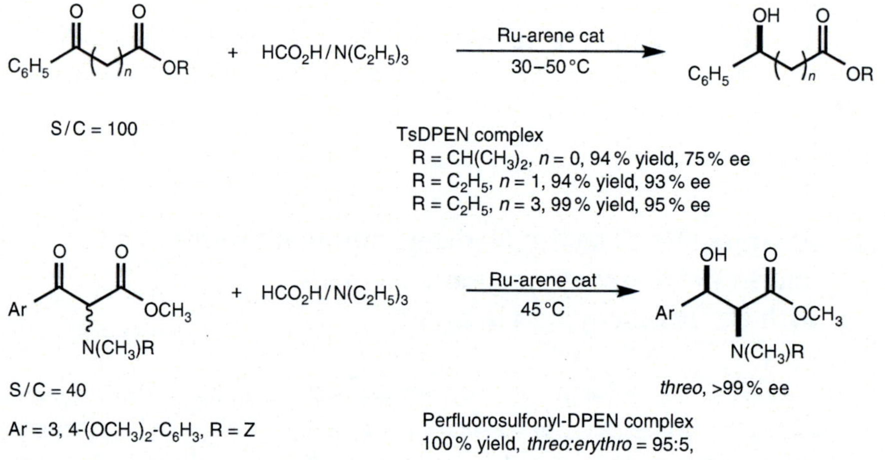

Fig. 3 Asymmetric reduction of benzoylacetate esters and β-keto esters

β-(3,4-Dimethoxyphenyl)serine methyl ester is obtainable in high diastereomeric and enantiomeric excesses from similar stereoselective transfer hydrogenation of β-keto-α-methylamino acid ester with a mixture of $HCO_2H/N(C_2H_5)_3$ and chiral arene–Ru catalysts bearing *N*-**perfluorobutanesulfonyl-1,2-diamine ligand** [29].

1,2-Diaryldiketones are stereoselectively reducible with the chiral Ru–*p*-cymene complex in a mixture of $HCO_2H/N(C_2H_5)_3$ to give the chiral 1,2-diols with excellent ee's (Fig. 4) [30, 31]. Notably, the outcome of the asymmetric reduction of benzils relies strongly on the property of the benzoin intermediates and the reactivity of the chiral Ru complex. For example, a reaction of racemic benzoin with the (*S*,*S*)-Ru catalyst gives (*R*,*R*)-diol with >99% ee at the early stage of the reaction (4% yield), while after 24 h, a chiral diol with the same de's and ee's as observed at the initial stage of the reaction is quantitatively obtainable, indicating that the reaction proceeds through a DKR of the intermediary benzoin [31]. This chiral diol can be readily converted to the corresponding chiral diphenylethylenediamine by the conventional procedure. Similarly, the reaction of 1,3-diphenylpropane-1,3-dione produces the corresponding chiral diol with 99% ee and in 99% yield (*dl*:*meso* = 94:6). The reduction of 1,3-pentanedione gives no reduction product under the same conditions [32, 33].

Asymmetric reduction of unsymmetrically substituted 1,2-diketones with the chiral Ru catalyst gives a partly reduced chiral α-hydroxy ketone at 10°C, while at higher temperature, 40°C, chiral *anti*-1,2-diols with an excellent ee is obtainable (Fig. 5) [34]. This method can be applied to access (1*R*,2*S*)-1-(4′-methoxyphenyl)-1,2-propanediol (98% ee), which is a major metabolite of *trans*-anethole in the rat.

Another valuable class of functionalized ketones is acetophenones bearing a functional group at the α-position. The reactions of acetophenones bearing CN, N_3,

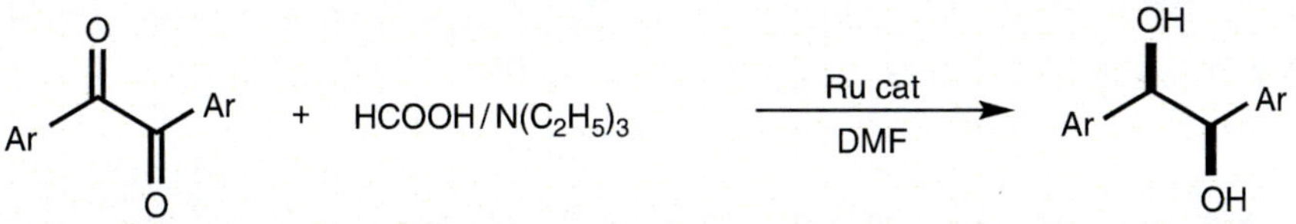

Ar	temp, °C	time, h	yield, %	*dl*:*meso*	ee, %
C_6H_5	40	24	100	98.4:1.6	>99
p-CH_3-C_6H_4	40	48	67	96.7:3.3	>99
p-CH_3O-C_6H_4[a]	35	48	75	94.4:5.6	>99
p-F-C_6H_4	40	24	100	94.2:5.8	>99

Conditions: Ru cat = RuCl[(*S*, *S*)-Tsdpen](*p*-cymene), S / C = 1000, $HCOOH/N(C_2H_5)_3$ = 4.4 / 2.6. [a] S / C = 200, $HCOOH/N(C_2H_5)_3$ = 4.4 / 4.4 in 1.2 M DMF.

Fig. 4 Asymmetric reduction of benzils with chiral Ru catalyst

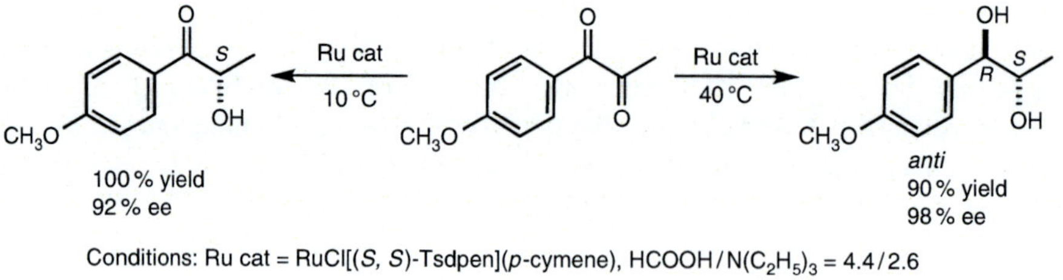

Fig. 5 Asymmetric reduction of unsymmetrically substituted 1,2-diketone

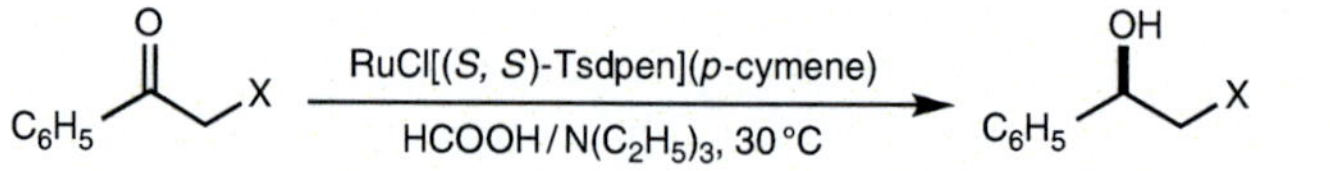

X	S/C	HCOOH: $N(C_2H_5)_3$	h	yield, %	% ee
CN	1000	3.1:2.6	24	100	98
N_3	100	3.1:5.2	24	65	92
NO_2	200	6.0:2.4 (DMF)	16	90	98

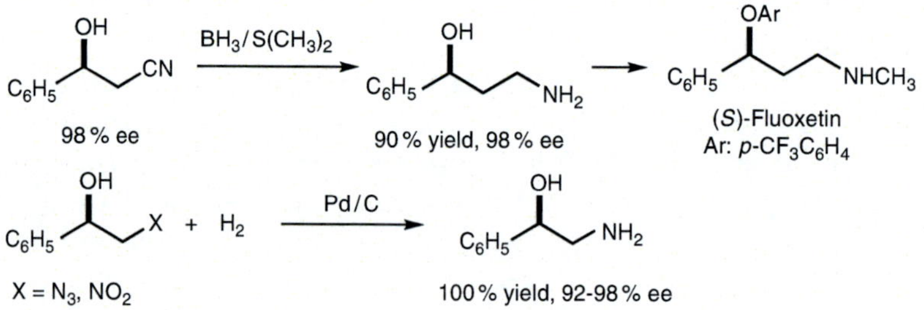

Fig. 6 Asymmetric reduction of acetophenones bearing CN, N_3, and NO_2 groups

and NO_2 with a mixture of $HCO_2H/N(C_2H_5)_3$ containing the chiral Ru catalysts smoothly proceed to give the corresponding chiral alcohols with an excellent ee (Fig. 6) [32]. These alcohols can be easily transformed by the conventional reduction of the functional groups to chiral β- and γ-amino alcohols with high ee.

A chiral Cp*Rh complex, Cp*RhCl[(*R*,*R*)-Tsdpen], is the most reactive catalyst for the asymmetric reduction of a variety of ring-substituted α-chloroacetophenones. The reduction with an azeotropic mixture of $HCO_2H/N(C_2H_5)_3$ and the Rh catalyst proceeds rapidly to give almost quantitatively the corresponding chiral alcohol with 96% ee and an initial turnover frequency (TOF) exceeding 2,500 h^{-1} (0.7 s^{-1}) [35].

The use of a Wills' tethered Ru catalyst gives significant improvement in the reactivity; the reaction with an S/C = 200 gives the reduction product quantitatively with 95% ee after 1.5 h [36]. The resulting chiral 2-chlorophenylethanol is easily convertible to chiral styrene oxide with NaOH in water without loss of ee.

A more appealing feature is that one-pot synthesis of chiral styrene oxides can be performed by sequential asymmetric reduction of chloroacetophenones with the chiral Rh in 2-propanol followed by treatment of the reaction mixture with NaOH aqueous solution, leading to the desired products in isolated yields of 80–90% with 96–98% ee in a single reactor (Fig. 7) [37]. For example, (*S*)-*m*-chlorostyrene oxide, which is a key intermediate for the preparation of several β3-adrenergic receptor agonist compounds, is easily obtained from a one-pot procedure.

Diastereoselective reduction of enantiomerically enriched aliphatic chlorinated ketones bearing another stereogenic center, N-substituted (3*S*)-3-amino-1-chloro-4-phenyl-2-butanones with the chiral Rh catalyst gives the corresponding chiral alcohols in excellent yields with high de's. Using Cp*RhCl[(*S*,*S*)-Tsdpen] as a catalyst, the (2*S*,3*S*)-alcohol can be obtained with an excellent de, while the antipodal (*R*,*R*)-Rh catalyst gives rise to the (2*R*,3*S*)-alcohol (Fig. 8) [38]. A sequential asymmetric reduction of *N*-(*tert*-butoxycarbonyl)-(3*S*)-3-amino-1-chloro-4-phenyl-2-butanone with a mixture of $HCO_2H/N(C_2H_5)_3$ in 2-propanol containing the (*S*,*S*)-Rh catalyst followed by treatment of the reaction mixture with 1 M NaOH aqueous solution at 0°C gives (2*S*,3*S*)-*N*-(*tert*-butoxycarbonyl)-3-amino-1,2-epoxy-4-phenylbutane with 90% de, as crystals after the addition of water (Fig. 8). These chiral epoxides serve as potential chiral building blocks for the synthesis of pharmaceuticals such as inhibitors of HIV protease and of β-secretase in Alzheimer's disease.

Sulfur- or oxygen-containing ketones are also reducible with a mixture of formic acid and triethylamine containing chiral Ru catalyst to the corresponding chiral alcohols with 97–99% ee (Fig. 9) [8]. The resulting chiral alcohols are key intermediates for the synthesis of MK-0417, a carbonic anhydrase inhibitor. In a similar manner, asymmetric reduction of acetylpyridine and its derivatives bearing an electron-withdrawing group at 10°C gives chiral pyridylethanols in almost quantitative yield and with up to 92% ee [39], one of which is an intermediate of

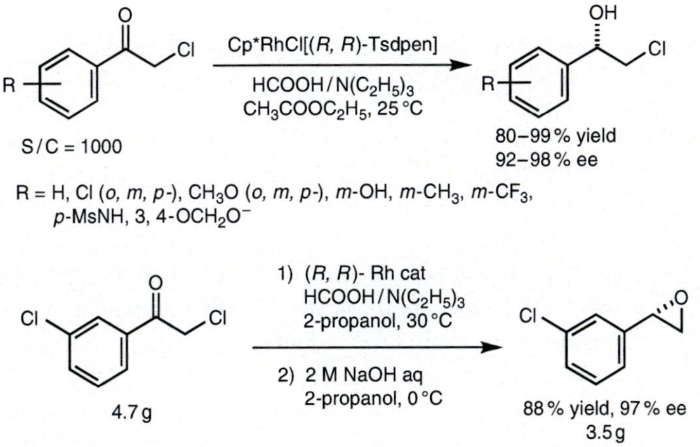

Fig. 7 Asymmetric reduction of α-chloroacetophenones and one-pot procedure for the synthesis of chiral epoxide

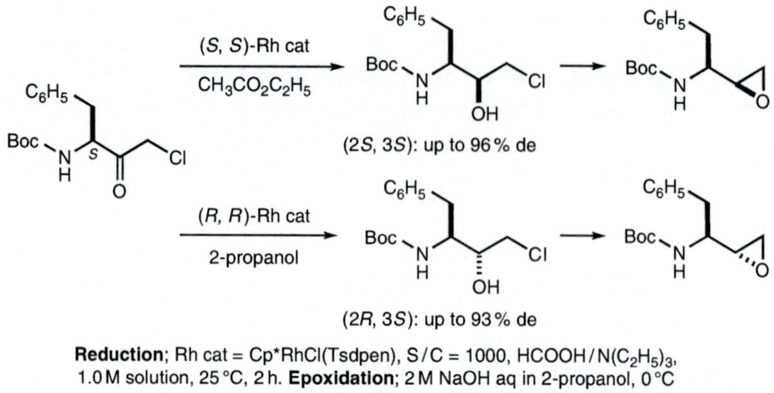

Fig. 8 Asymmetric reduction of *N*-(*tert*-butoxycarbonyl)-(3*S*)-3-amino-1-chloro-4-phenyl-2-butanone and one-pot procedure for the synthesis of chiral epoxide

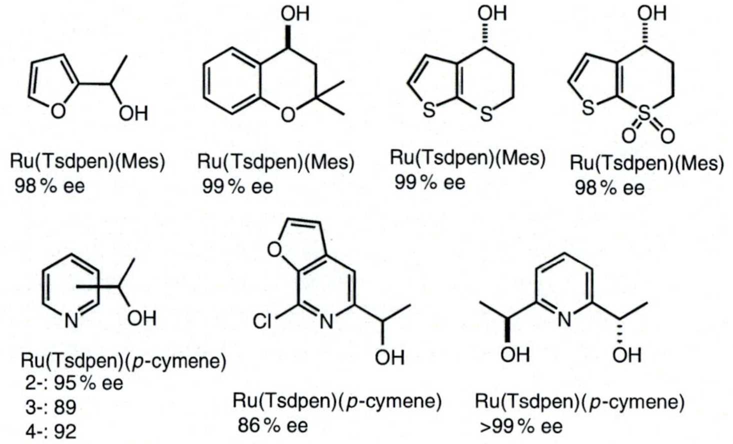

Fig. 9 Asymmetric reduction products of hetero atom containing ketones

PNU-142721, a potent *anti*-HIV medicine. Wills reported that his tethered version of the Ru(TsDPEN) catalyst is also highly effective for asymmetric reduction of 2,6-diacetylpyridine with a mixture of $HCO_2H/N(C_2H_5)_3$ to a chiral diol with 99.6% ee in 91% yield [36].

Both aryl- and alkylethynyl ketones are also reducible with 2-propanol containing the (*S*,*S*)-Ru catalyst to give chiral propargylic alcohols with an excellent ee and in good yield (Fig. 10). The asymmetric reduction of chiral acetylenic ketones with a pre-existing stereogenic center leads to diastereomeric propargylic alcohols. Using (*R*,*R*)- or (*S*,*S*)-catalyst for the reduction of the (*S*)-ketone with 98% ee leads to (3*S*,4*S*)-alcohol in 97% yield and with >99% ee and (3*R*,4*S*)-isomer in 97% yield and with >99% ee [40]. Similarly, functionalized acetylenic ketones bearing chiral centers can be reduced with excellent diastereoselectivity [41].

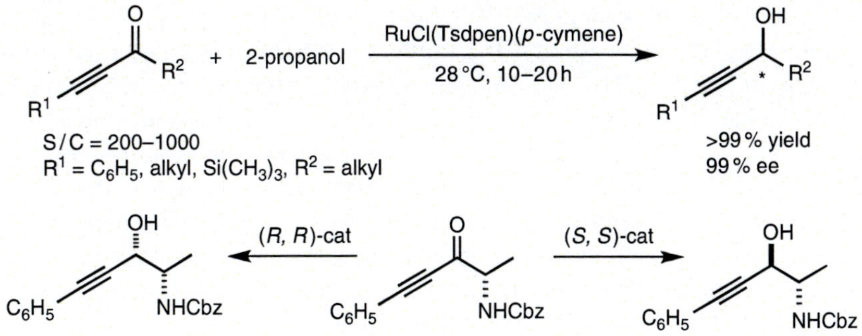

Fig. 10 Asymmetric reduction of α,β-acetylenic ketones

3 Catalytic Hydrogenation of Carboxylic Acid Derivatives Containing C=O Double Bonds with Bifunctional Molecular Catalysts

3.1 Heterolytic Cleavage of Molecular Hydrogen with Bifunctional Catalysts

Direct hydrogenation using molecular hydrogen with bifunctional molecular catalyst is a straightforward, practical, and environmentally benign process. In 1995, Noyori and coworkers discovered that a combined catalyst system including *trans*-[$RuCl_2$(diphosphine)(diamine)] and a base serves as a highly efficient catalyst for the hydrogenation of ketones and aldehydes, in which excellent chemo- and stereoselectivity originates from the Ru/NH bifunctional property [5, 42, 43]. The active catalyst, the amido Ru complex, undergoes heterolytic H_2 cleavage to generate the hydrido Ru complex. NMR studies of *trans*-[RuH(η^2-H_2)(diphosphine)(diamine)]$^+$ in 2-propanol-d_8 have revealed that subtle interplay between ligated η^2-H_2 and solvated 2-propoxide is responsible for the heterolytic cleavage of H_2 in 2-propanol [44, 45].

In 2001, we found that H_2 bound to the coordinatively unsaturated Cp*Ru[$(CH_3)_2N(CH_2)_2NH_2$]$^+$(OR)$^-$ complex (*N,N*-dimethylaminoethylamine, N–N) undergoes heterolytic cleavage through a hydrogen-bonding network in 2-propanol to give coordinatively saturated Cp*RuH[$(CH_3)_2N(CH_2)_2NH_2$], on the basis of very careful isotope labeling experiments [26]. The alcohol-assisted H_2 activation on the amido Ru complex is a key step for hydrogenation with half-sandwich type of bifunctional catalysts (Fig. 11). Later, this reaction pathway was confirmed by Brandt and Andersson using theoretical calculations [46]. Similarly, Cp*Ru(P–N) complexes (2-phosphinoethylamines: P–N) can also facilitate heterolytic cleavage of H_2 with the help of conjugate base of acidic compounds under mild conditions [22].

Noyori [47], our group [48, 49], and others [50–53] reported that some isolable cationic amine Ru and Ir complexes, Ru(OTf)[Tsdpen](η^6-*p*-cymene) and $[Cp^*Ir(Tscydn)(CH_3CN)]^+SbF_6^-$, are highly effective for the asymmetric hydrogenation of ketones or imines in methanol or 2-propanol containing no base. A key step is the heterolytic H_2 cleavage, which is promoted by the Ru (OTf) or Ir(SbF_6) system. It generates an active hydride complex, RuH[Tsdpen] (η^6-*p*-cymene) with simultaneous release of HOTf or $HSbF_6$ in highly polarized media like methanol. After the concerted hydride/proton transfer from the hydride complex substrates, the 16-electron amido Ru complex is regenerated and then reacts with HOTf to complete the catalytic cycle.

These results led us to design a new type of bifunctional molecular catalysts having a triflylamide tethers linked to η^6-arene or Cp* ring for asymmetric hydrogenation of ketones, in which tethered anion might work as a conjugate base [54]. The tethered arene–Ru complex, Ru[Tsdpen][η^6:η^1-$C_6H_5(CH_2)_n$NTf] (Tf = CF_3SO_2) (Figs. 1 and 11), and tethered Cp* bound Rh complex, [η^5:η^1-$(CH_3)_4C_5(CH_2)_n$NTf]M[Msdpen] (M = Rh, Ir, Ms = CH_3SO_2), which have been structurally modified based on the hydrogen transfer catalyst, Ru[(*S*,*S*)-Tsdpen] (η^6-arene) and Cp*M[Msdpen] can effect efficiently asymmetric hydrogenation of aromatic ketones to give chiral alcohols with an excellent ee [55].

Thus, the ligand modification by changing the amine chelating ligands and by linking the triflylamide ligand to the innocent ligands causes a drastic change in the catalyst performance. Both Cp*Ru(N–N) and Cp*Ru(P–N) complexes as well as the tethered ones can facilitate heterolytic cleavage of H_2 with help of conjugate base of acidic compounds under mild conditions and can efficiently effect hydrogenation of a variety of carbonyl compounds.

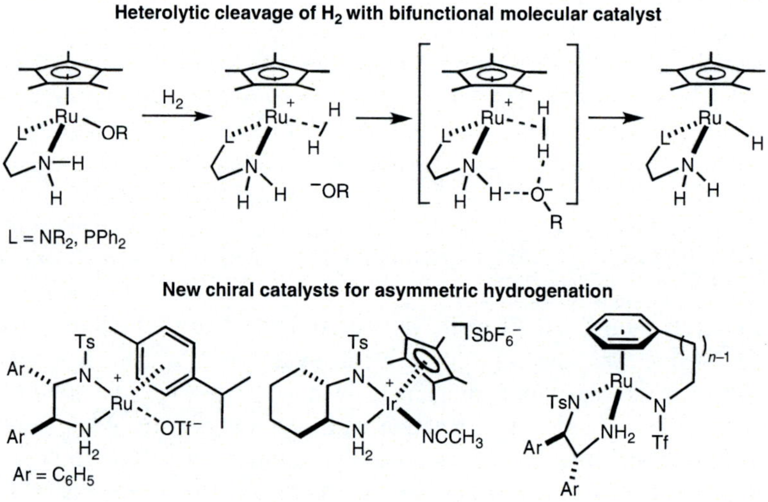

Fig. 11 H_2 activation with bifunctional molecular catalysts

3.2 Hydrogenation of Imides

The newly developed half-sandwich-type bifunctional catalysts Cp*Ru(N–N), Cp*RuCl[$(CH_3)_2N(CH_2)_2NH_2$], Cp*Ru(P–N), and Cp*Ru[$(C_6H_5)_2P(CH_2)_2NH_2$] serve as efficient catalysts for the chemoselective hydrogenation of ketones, as a result of its ability to perform alcohol-assisted heterolytic cleavage of H_2 and the Ru/NH bifunctionality. The use of the electron-donating N–N and the P–N ligand causes a positive effect on H_2 activation on the amido complexes. Since the electronic effect of the chelating amine ligands on both amido and amine complexes might conflict each other, even a slight change in the electronic properties of the amine ligands causes a significant change in the catalytic performance [22, 56]. In fact, Cp*Ru(P–N) complex efficiently catalyzes the hydrogenation of imides, while Cp*Ru(N–N) is totally inactive [57]. The electron-withdrawing nature of the P–N ligand might enhance the Lewis acidity of the NH proton in the hydrido complex Cp*RuH[$(C_6H_5)_2P(CH_2)_2NH_2$], leading to facile activation of polar functionalities imides with the hydride complex.

A variety of imides are chemoselectively convertible to the corresponding alcohols and carboxamides in 2-propanol containing Cp*RuCl(P–N) and KO*t*-Bu as the catalyst under mild conditions as shown in Fig. 12 [57]. This hydrogenation method is characterized by its excellent chemoselectivity, substrate scope, and controllable stereoselectivity by the chiral modification of the ligand structures. In fact, it is applicable to the deprotection of primary amines from *N*-phthaloyl-protected amino acid ester derivatives. *N*-phthaloyl-L-Phe methyl ester undergoes hydrogenation to generate *N*-(*o*-hydroxymethylbenzoyl)-L-Phe methyl ester, whose acid-promoted cyclization liberates the HCl salt of L-Phe methyl ester with concomitant formation of phthalide in high yields (Fig. 12) [57].

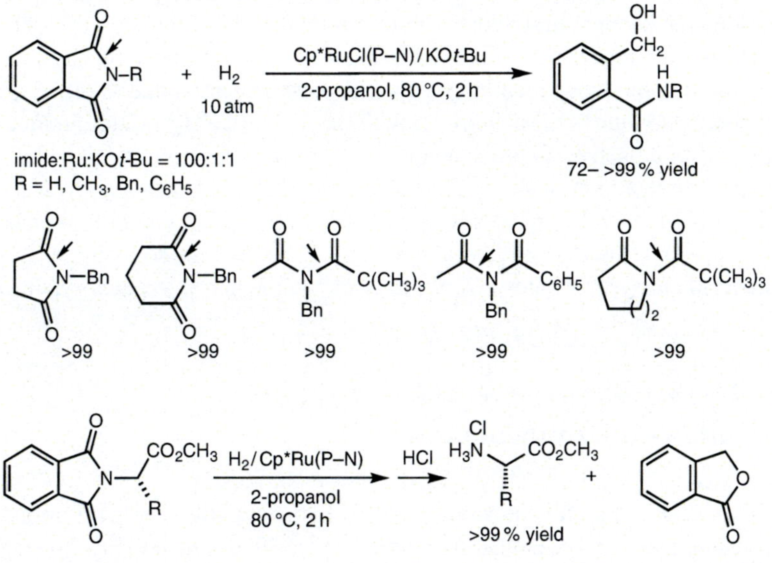

Fig. 12 Chemoselective hydrogenation of imides with bifunctional Cp*Ru(P–N) catalyst

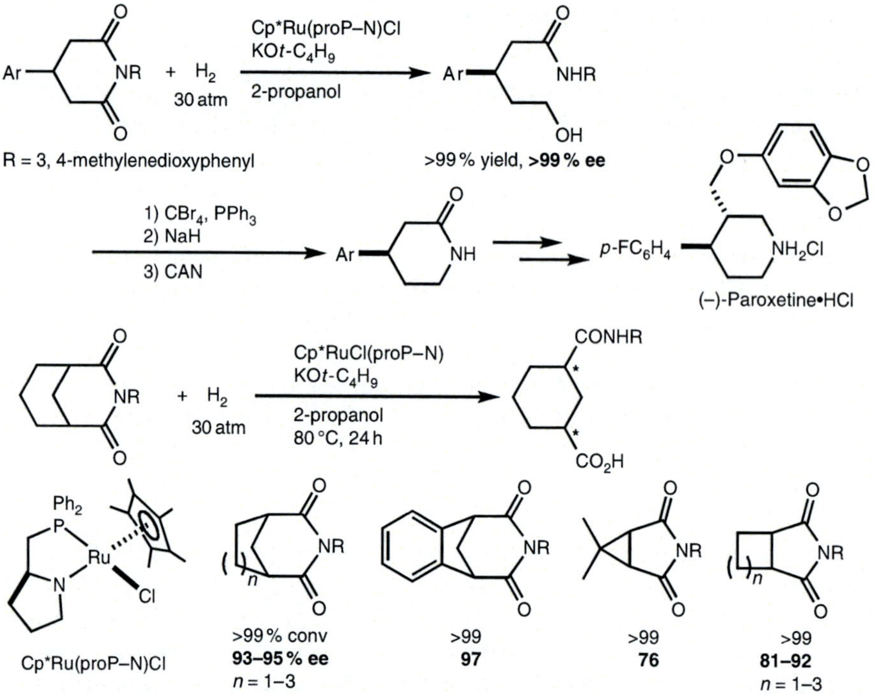

Fig. 13 Enantioselective hydrogenation of prochiral cyclic imides via desymmetrization

The chiral version of the Cp*Ru(P–N) catalyst bearing the chiral P–N ligand derived from L-proline promotes the enantioselective hydrogenation of prochiral 4-arylglutarimides via desymmetrization to provide the corresponding hydroxyamides with excellent ees and in high yields as shown in Fig. 13 [57]. Further synthetic transformation of the chiral hydroxyamides provides chiral piperidinone derivatives which serve as important synthetic intermediates for a number of physiologically active chiral compounds including the antidepressant paroxetine. Similarly, readily accessible *sym*-glutar- or succinimides with the *N*-3,4-$(OCH_2O)C_6H_3$ group undergo highly enantioselective hydrogenation via desymmetrization to give the corresponding hydroxyamides with excellent ee's (Fig. 13) [58]. The present hydrogenation method would provide highly functionalized chiral hydroxyamides, which would otherwise require tedious multistep synthesis. Thus, this imide hydrogenation with bifunctional catalyst provides a versatile synthetic method in organic synthesis.

3.3 Hydrogenation of N-Acylcarbamates and N-Acylsulfonamides

The Cp*Ru(P–N) catalyst promotes the hydrogenation of acyl moieties in *N*-acylcarbamates and *N*-acylsulfonamides under 30 atm of H_2 (Fig. 14) [59]. It is

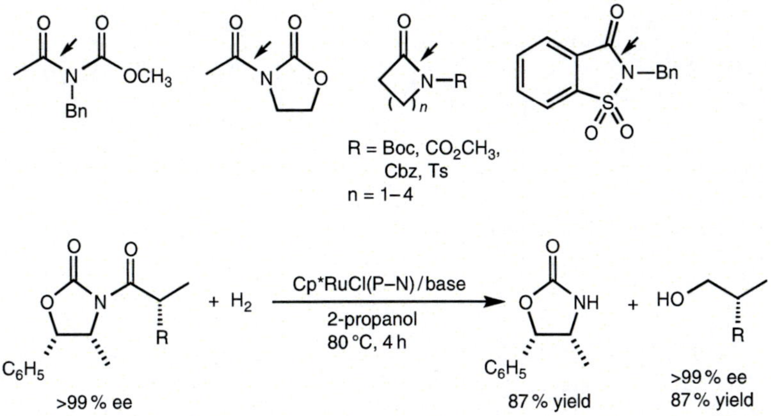

Fig. 14 Hydrogenation of *N*-acylcarbamates and related compounds

preferable to use *t*-BuOH for the hydrogenation of substrates with an adequately electrophilic acyl group, because sterically small alcoholic solvents sometimes cause undesired alcoholysis of the substrates. Moreover, the rates of the reaction strongly rely on the electron-withdrawing nature of substituents on nitrogen in the substrates. In fact, the acceleration effect of the *N*-substituents of pyrrolidinone derivatives has been found to increase in the following order: Cbz $<$ Boc $<$ CO_2CH_3 $<$ SO_2CH_3 $\approx$ $SO_2C_6H_4$-*p*-CH_3. The present hydrogenation is applicable to the reductive treatment of chiral *N*-acyloxazolidinones, which are useful synthetic intermediates in the asymmetric synthesis developed by Evans (Fig. 14). For example, *N*-acyloxazolidinone undergoes selective hydrogenation in the presence of Cp*RuCl(P–N) and KO*t*-Bu, to furnish the corresponding chiral alcohol without any damage to the stereochemistry, along with the original chiral auxiliary in high yields. This method may be an environmentally benign catalytic alternative to the method using $LiAlH_4$, which sometimes causes difficulty in the recovery of the chiral auxiliaries.

3.4 Hydrogenation of Esters and Lactones

In 1997, Teunissen and Elsevier reported the first homogeneous hydrogenation of unactivated esters to their corresponding alcohols catalyzed by a $Ru(acac)_3/CH_3C[CH_2P(C_6H_5)_2]_3$ system in dry methanol [60]. In 2006, Milstein and coworkers reported that a new mode of bifunctional catalysts based on the metal–ligand cooperation via aromatization–dearomatization of pyridine-based pincer-type ligands in the RuH(PNN)(CO) pincer catalyst can facilitate hydrogenation of esters to alcohols under mild conditions (Fig. 15) [61]. Since then, a few groups reported that some combined bifunctional catalysts based on the metal–NH cooperating effect, $RuCl_2(P–N)_2$, PNNP-tetradentate Ru catalyst with a base [62], and RuH_2(diphosphine)(diamine) [63] promoted efficiently hydrogenation of esters in reasonably good reactivity.

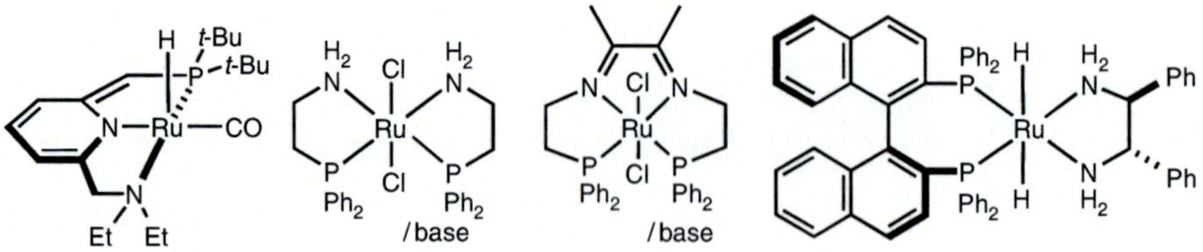

Fig. 15 Examples of efficient bifunctional catalysts for ester hydrogenation

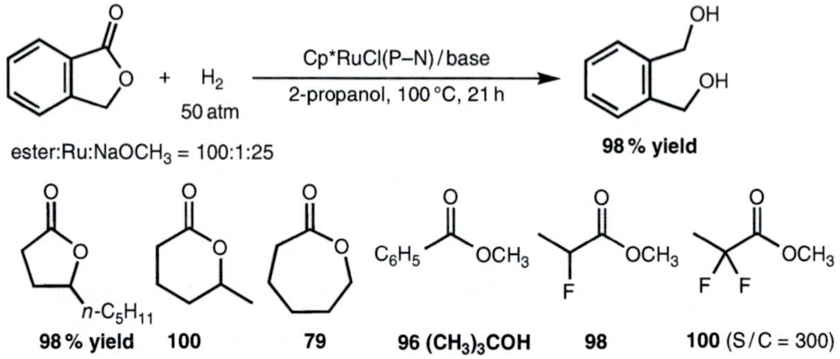

Fig. 16 Hydrogenation of esters and lactones with the Cp*Ru(P–N) catalyst

We have also found that our bifunctional catalyst Cp*Ru(P–N) and a large amount of base promote the hydrogenation of esters under more forcing conditions [22, 56]. For example, phthalide is cleanly reduced to *o*-xylyleneglycol under 50 atm of H_2 at 100°C in the presence of Cp*Ru(P–N) catalysts (Fig. 16). A variety of esters and lactones undergo hydrogenation to give the corresponding alcohols and diols, respectively.

3.5 Hydrogenolysis of Epoxides

The Cp*Ru(P–N) catalyst systems also efficiently facilitate hydrogenation of a variety of terminal epoxides, leading to the corresponding secondary alcohols preferentially in high yields for the steric reasons (Fig. 17). Alkenyl epoxides gave secondary alkenyl alcohols in quantitative yield without formation of saturated alcohols or epoxides. Terminal epoxides bearing another oxygen functionality on the side chain also undergo hydrogenolysis to afford the corresponding secondary alcohols in good yields, indicating that groups next to the epoxide group do not interact with the metal center because of the Ru/NH bifunctionality of the Cp*Ru catalyst [64]. Although stereospecific hydrogenolysis of chiral terminal epoxides was hampered by the competing racemization of the chiral product alcohols (vide infra), this catalytic hydrogenolysis provides a new alternative to stoichiometric metal hydride reduction.

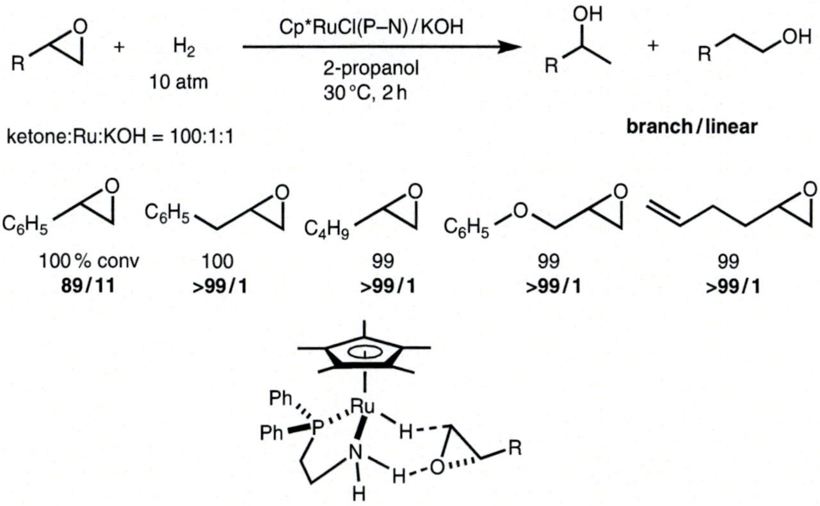

Fig. 17 Hydrogenation of terminal epoxides with the Cp*Ru(P–N) catalyst

4 Oxidative Transformations with Bifunctional Molecular Catalysts

As mentioned in the introduction, the hydrogen transfer between alcohols and ketones with the bifunctional molecular catalyst occurs reversibly through a six-membered pericyclic transition state. When suitable hydrogen acceptors such as acetone and oxygen molecule can be used, the reverse reaction, dehydrogenative oxidation of alcohols would attract much attention in organic synthesis (Fig. 18). The ligand modification is also crucial to determine the catalytic performance for oxidative transformation. Noticeably, the bifunctional Cp*Ru(P–N) catalyst exhibits excellent catalytic activity toward both hydrogenation and transfer hydrogenation, while the Cp*Ru(N–N) complex serves as only hydrogenation activity. Therefore, the (P–N) complex is a choice of the catalyst for oxidative transformation.

*4.1 Dehydrogenative Oxidation of Alcohols with Cp*Ru(P–N) Catalysts*

The bifunctional catalyst Cp*Ru(P–N) but not with Cp*Ru(N–N) is an excellent catalyst for the racemization of chiral non-racemic *sec*-alcohols [65]. In fact, asymmetric hydrogenation of prochiral ketones becomes possible with chiral Cp*Ru(N–N) catalyst system, thanks to its reluctance to dehydrogenate alcohols, but not with chiral Cp*Ru(P–N) systems since concurrent racemization of the product alcohols deteriorates the ee value of the alcoholic products regardless of whether the hydrogen source is H_2 or 2-propanol [64, 65].

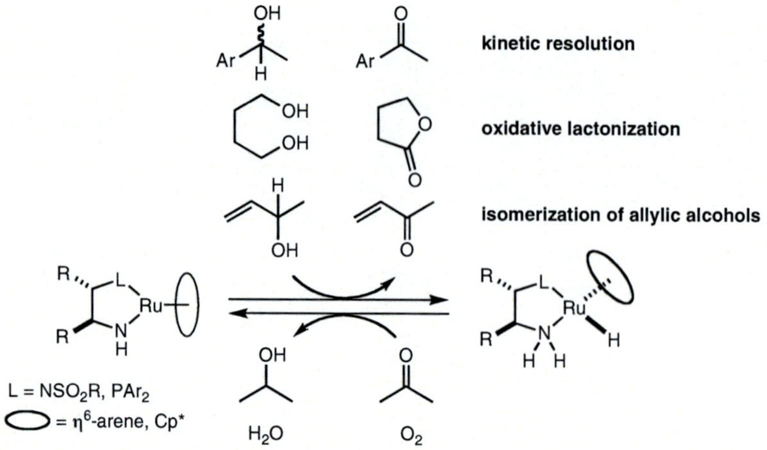

Fig. 18 Bifunctional catalyst promoted oxidative transformations

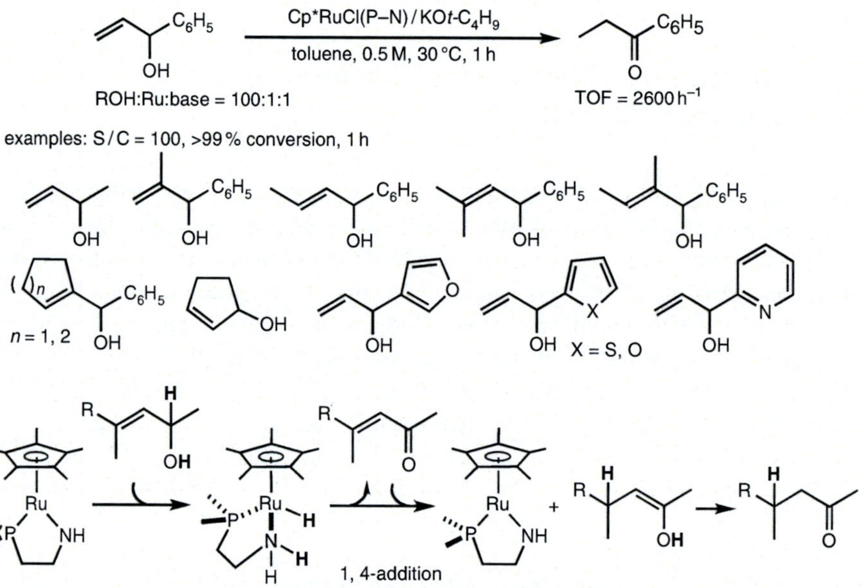

Fig. 19 Isomerization of allylic alcohols with bifunctional catalysts

Two synthetically viable, oxidative transformations of alcohols have been successfully developed. First, allylic alcohols undergo intramolecular hydrogen transfer to give saturated carbonyl compounds in aprotic media containing Cp*Ru(P–N) catalysts; the TOF of the reaction at 30°C exceeds 2,500 h^{-1} (Fig. 19) [66].

Unlike conventional catalysts, this catalyst system discriminates olefins with an allylic hydroxy group from other olefinic groups as a result of Ru/NH bifunctionality as shown in Fig. 19. This unique chemoselectivity is applicable in the preparation of

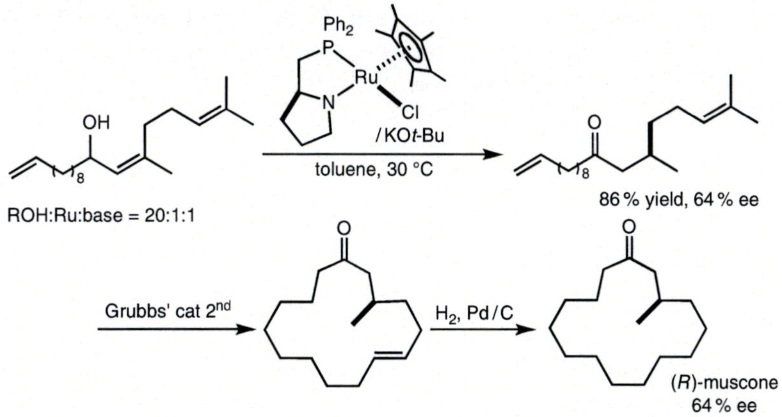

Fig. 20 Asymmetric isomerization of allylic alcohol and its application to muscone synthesis

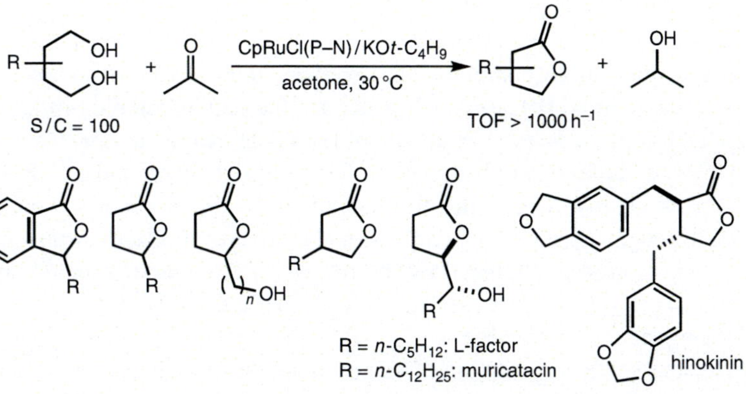

Fig. 21 Dehydrogenative oxidation of diols with bifunctional Cp*Ru(P–N) catalysts

macrocyclic ketones starting from readily available acyclic allylic alcohols equipped with two isolated C=C double bonds. For example, the present asymmetric isomerization provides a chiral ketone having two olefinic units suitable for ring-closing metathesis as shown in Fig. 20. Muscone can be conveniently prepared using asymmetric isomerization with chiral Cp*Ru(P–N) catalysts as a key step [66].

Second, 1,4-diols undergo intermolecular hydrogen transfer, giving lactones efficiently in acetone containing Cp*Ru(P–N) catalysts; the TOF of this reaction at 30°C exceeds 1,000 h^{-1} (Fig. 21) [67]. The catalytic oxidative lactonization of diols is characterized by its unique chemo- and regioselectivity. The significant rate difference between primary and secondary alcohols in dehydrogenation, and the rate difference between 1,4-diols and 1,5- or 1,6-diols enable unique oxidative lactonization of triols.

The oxidation of triols provides exclusively γ-butyrolactones including L-factor and muricatacin, where the remote OH groups remain intact regardless of whether they are primary or secondary. Due to its high efficiency and experimental simplicity, the present catalytic oxidative transformation provides a powerful and environmentally benign alternative for Fétizon oxidation.

4.2 Aerobic Oxidation of Alcohols with Bifunctional Molecular Catalysts

When molecular oxygen can be used as a hydrogen acceptor for the dehydrogenative oxidation of alcohols with the bifunctional catalyst, aerobic oxidation becomes a simple and minimal organic waste process. We found that a new series of bifunctional hydrido(amine)-iridium complexes with electron-donating C–N chelate amine ligands Cp*IrH[κ^2(*N*,*C*)-{NH_2CR_2-2-C_6H_4}] (R = C_6H_5, CH_3) rapidly react with molecular oxygen under mild conditions to generate the corresponding amido Ir complexes Cp*Ir[κ^2(*N*,*C*)-{NH_2CR_2-2-C_6H_4}] as shown in Fig. 22 [68].

Other oxidants like hydroperoxides also effect the transformation of the hydrido complex to the amido complex. The reaction of the hydrido complex with an equimolar amount of H_2O_2 in THF generates the amido complex in an excellent yield in addition to a detectable amount of H_2O. The treatment of *tert*-BuOOH with the hydrido complex also clearly gives the amido complex and *tert*-BuOH [69]. These results clearly indicate that the reaction of the hydrido complex with O_2 may proceed through O_2 insertion into the metal hydride bond to form an amine–hydroperoxo complex, followed by the release of the amido complex and H_2O_2. The resulting H_2O_2 product then reacts with the hydrido complex to provide the amido complex and water (Fig. 22).

On the basis of a combination of the reactions shown in Fig. 22, we successfully developed the aerobic oxidation of alcohols with bifunctional Cp*Ir, Cp*Rh, and

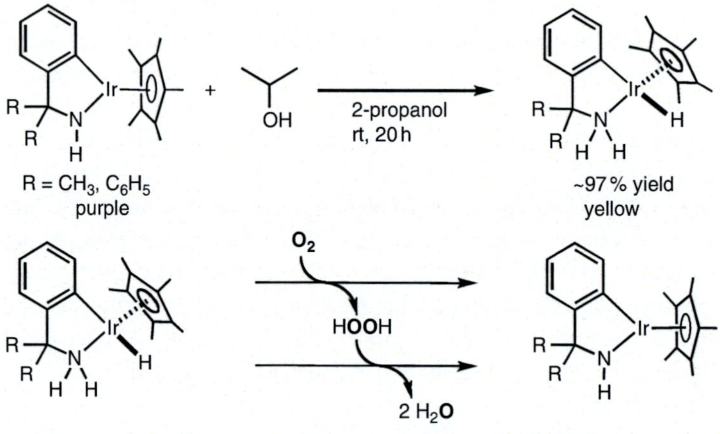

Fig. 22 Stoichiometric reactions of bifunctional CpIrH(C–N) complexes with O_2 and H_2O_2

(η^6-arene)Ru catalysts bearing the C–N ligands [68, 70]. Although a wide variety of homogeneous and heterogeneous systems based on transition metals have been explored, limited examples of Rh- and Ir-catalyzed reaction have been reported (Fig. 23). For example, the reaction of 1-phenylethanol proceeded smoothly under atmospheric pressure of air at 30°C in THF containing amido Ir complex gives acetophenone in 72% yield. Notably, an analogous hydrido complex bearing an *N,N*-dimethylamino group did not exhibit catalytic activity under otherwise identical conditions, indicating that the M/NH bifunctionality is also crucial for O_2 activation and that the aerobic oxidation proceeds through the interconversion between the amine/amido catalyst intermediates. Binary catalyst systems, including the chloro(amine)-Rh and -Ru complexes and $KOC(CH_3)_3$, are applicable to the aerobic oxidation.

Other 1-phenylethanols with substituents on the arene ring, sterically congested diphenylmethanol, and an aliphatic secondary alcohol are convertible into the corresponding ketones using the amido Ir catalyst as shown in Fig. 24.

When primary alcohols are used for aerobic oxidation under identical conditions, the oxidative dimerization product, esters are obtainable as shown in Fig. 25. In fact, the reaction of benzyl alcohols gives the corresponding benzyl benzoate derivatives in a range of 62–64% yields. The oxidation of 1,2-benzenedimethanol

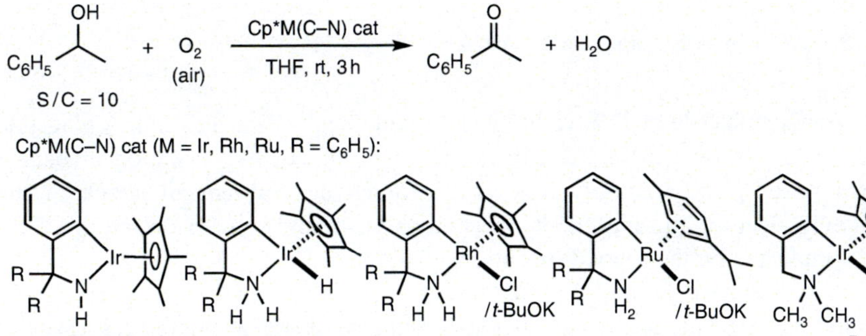

72% yield **63** **72** **42** **5**

Fig. 23 Aerobic oxidation of 1-phenylethanol with bifunctional catalysts

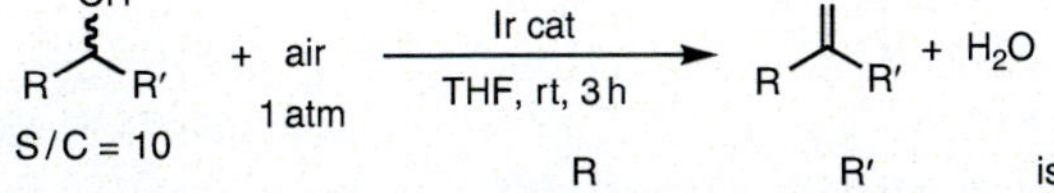

Ir cat

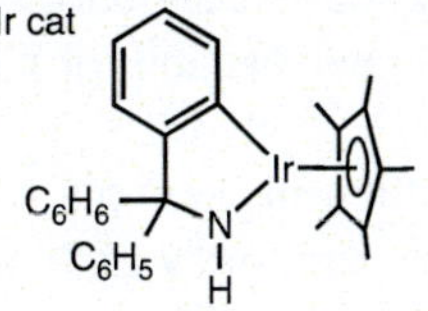

R	R′	isolated yield, %
4-$CH_3OC_6H_4$	CH_3	62
4-ClC_6H_4	CH_3	55
C_6H_5	C_6H_5	62
$C_6H_5CH_2CH_2$	CH_3	87
2-cyclohexen-1-ol		47

Fig. 24 Aerobic oxidation of secondary alcohols with bifunctional amido Ir catalyst

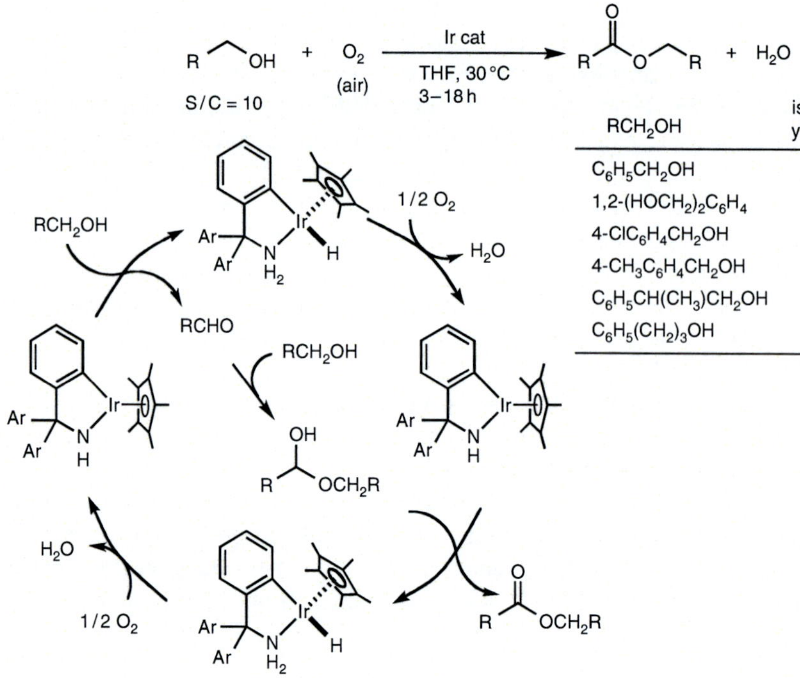

RCH_2OH	isolated yield, %
$C_6H_5CH_2OH$	64
1,2-$(HOCH_2)_2C_6H_4$	72
4-$ClC_6H_4CH_2OH$	62
4-$CH_3C_6H_4CH_2OH$	64
$C_6H_5CH(CH_3)CH_2OH$	31
$C_6H_5(CH_2)_3OH$	46

Fig. 25 Aerobic oxidation of primary alcohols with the Cp*Ir(C–N) catalyst

also affords phthalide in 72% yield by an intramolecular esterification. A plausible mechanism is shown in Fig. 25. In the presence of O_2, the oxidation of benzyl alcohol takes place smoothly to give benzaldehyde. Subsequent attack of the remaining alcohol affording the hemiacetal and its ready conversion into the ester is accomplished by the second oxidation.

This aerobic oxidation of alcohols is more appealing when applied to the kinetic resolution of racemic secondary alcohols with chiral amido complexes [69–76]. When a THF (1.0 M) solution of racemic 1-phenylethanol with the chiral Cp*Ir (C–N) was treated with air at 30°C for 4 h, (*R*)-1-phenylethanol was recovered with a 48% yield and 12% ee catalysts (Fig. 26). Noticeably, the use of the chiral amido Ir complex bearing the DPEN ligand, Cp*Ir[(*S*,*S*)-Msdpen] (Ms = methanesulfonyl), significantly improves the enantiomer discrimination ability, and desired (*R*)-1-phenylethanol is obtainable in 48% yield and 98% ee with a k_f/k_s ratio of up to 90. A 1-phenylethanol derivative having an electron-donating CH_3O group at the para position is efficiently resolved with catalyst. Similarly, the *R*-enantiomers with >99% ee and with 46–50% yields are readily obtainable from the reactions of 1-indanol and 1-tetralol at ambient temperature [69].

In contrast to previously reported kinetic resolution of alcohols with isolable chiral amido Ru using acetone [19], aerobic kinetic resolution with binary catalyst systems including chiral Ir and Rh complexes Cp*MCl[(*S*,*S*)-Tsdpen] (M = Ir, Rh) and a base proceeds smoothly to give the desired chiral alcohols.

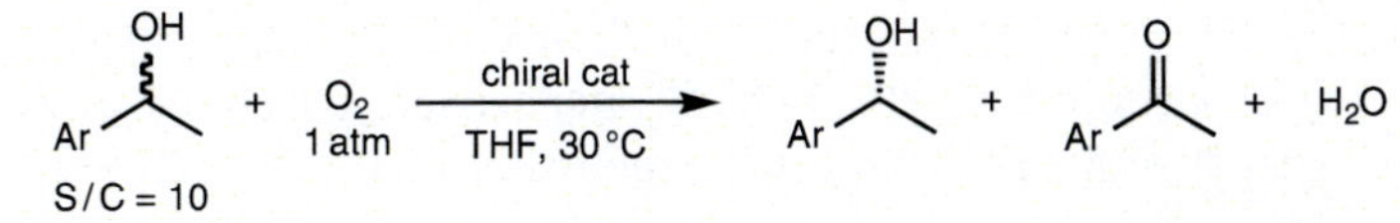

chiral cat:	alcohol			unreacted alcohol		
	Ar	[M]	time, h	recover %	ee, %	k_f/k_s
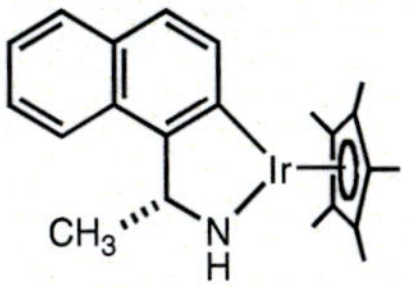	C_6H_5	1	4	48	12	1.5
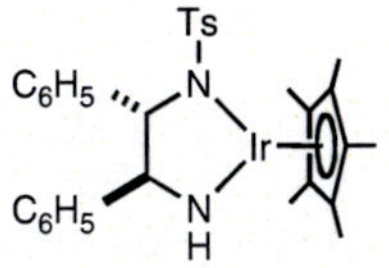	C_6H_5	1	24	39	86	9
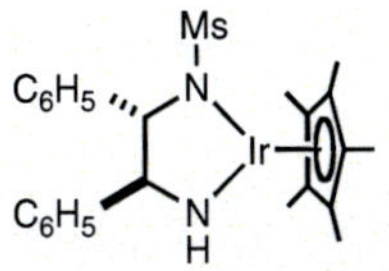	C_6H_5	1	22	44	84	13
	C_6H_5	0.2	38	48	98	91
	4-$CH_3C_6H_4$	0.2	19	38	98	17
	1-indanol	1	2	42	98	18
	1-indanol	0.1	6	50	>99	>100
	1-tetralol	0.1	6.5	46	>99	78
Cp*RhCl(Tsdpen)/*t*-BuOK	C_6H_5	1	3.5	46	93	29

Fig. 26 Aerobic oxidative kinetic resolution of racemic secondary alcohols

5 Conclusions

This chapter has described mainly recent advances in chemistry of our chiral bifunctional transition metal molecular catalyst based on metal/NH acid–base synergy effect for stereoselective reductive and oxidative transformations. The concept of the bifunctional catalyst design is now successfully applicable to a new family for asymmetric hydrogenation of polar functionalities by the ligand modification and electronic fine-tuning. A key step in the hydrogenation is the alcohol-assisted heterolytic cleavages of molecular H_2 with the chiral amido complexes, leading to the amine hydrido Ru complexes. The acidic amine proton and the metal hydride activate polarized C–O bond through outer-sphere mechanism, in which the reacting substrate is not bonded directly to the central metal. Another unprecedented aspect is that the oxygen molecule can be readily activated by the bifunctional system leading to aerobic oxidation of alcohols. Thus, the present bifunctional molecular catalyses can provide a wide substrate scope and applicability in organic synthetic chemistry. Thus, the rational design of the amine ligand that adjusts the balance of the electronic factors on the M/NH units in the bifunctional catalysts is crucially important to exploit further unprecedented catalyst performance [9, 77–80]. Finally, the industrial outlook for asymmetric reduction with bifunctional catalysts is bright, because of their excellent catalyst performance, wide substrate scope, operational simplicity, and economic viability as well as the growing awareness of the need for green chemistry [6].

Acknowledgments This work was supported by a Grant-in-Aid from the Ministry of Education, Culture, Sports, Science, and Technology, Japan (Nos. 18065007 "Chemistry of *Concerto* Catalysis" and 2225004) and partially supported by The G-COE Program.

References

1. Noyori R (1994) Asymmetric catalysis in organic synthesis. Wiley, New York
2. Ojima I (ed) (2010) Catalytic asymmetric synthesis, 3rd edn. Wiley, New York
3. Jacobsen EN, Pfaltz A, Yamamoto H (eds) (1999) Comprehensive asymmetric catalysis. Springer-Verlag, Heidelberg
4. Hartwig JF (2010) Organotransition metal chemistry from bonding to catalysis. University Science Books, Mill Valley
5. Noyori R (2002) Angew Chem Int Ed 41:2008
6. Ikariya T, Blacker AJ (2007) Acc Chem Res 40:1300
7. Ikariya T, Murata K, Noyori R (2006) Org Biomol Chem 4:393
8. Noyori R, Hashiguchi S (1997) Acc Chem Res 30:97
9. Ikariya T, Gridnev ID (2009) Chem Rec 9:106
10. Blum Y, Czarkie D, Rahamie Y, Shvo Y (1985) Organometallics 4:1459
11. Shvo Y, Czarkie D, Rahamim Y (1986) J Am Chem Soc 108:7400
12. Hashiguchi S, Fujii A, Takehara J, Ikariya T, Noyori R (1995) J Am Chem Soc 117:7562
13. Fujii A, Hashiguchi S, Uematsu N, Ikariya T, Noyori R (1996) J Am Chem Soc 118:2521
14. Uematsu N, Fujii A, Hashiguchi S, Ikariya T, Noyori R (1996) J Am Chem Soc 118:4916
15. Zhang J, Leitus G, Ben-David Y, Milstein D (2005) J Am Chem Soc 127:10840
16. Kohl S, Weiner L, Schwartsburd L, Konstantinovski L, Shimon LJW, Ben-David Y, Iron M, Milstein D (2009) Science 324:74
17. Läng F, Breher F, Stein D, Grützmacher H (2005) Organometallics 24:2997
18. Maire P, Breher F, Schönberg H, Grützmacher H (2005) Organometallics 24:3207
19. Haack KJ, Hashiguchi S, Fujii A, Ikariya T, Noyori R (1997) Angew Chem Int Ed Engl 36:285
20. Murata K, Ikariya T, Noyori R (1999) J Org Chem 64:2186
21. Mashima K, Abe T, Tani, K (1998) Chem Lett:1199
22. Ito M, Ikariya T (2007) Chem Commun:5134
23. Takehara J, Hashiguchi S, Fujii A, Inoue S, Ikariya T, Noyori R (1996) J Chem Soc Chem Commun:233
24. Alonso DA, Nordin SJM, Roth P, Tarnai T, Andersson PG, Thommen M, Pittelkow U (2000) J Org Chem 65:3116
25. Petra DGI, Kamer PCJ, van Leeuwen PWNM, Goubitz K, van Loon AM, de Vries JG, Schoemaker HE (1999) Eur J Inorg Chem:2335
26. Ito M, Hirakawa M, Murata K, Ikariya T (2001) Organometallics 20:379
27. Ito M, Shibata Y, Watanabe A, Ikariya T (2009) Synlett 10:1621
28. Koike T, Ikariya T (2004) Adv Synth Catal 346:37
29. Mohar B, Valleix A, Desmurs JR, Felemez M, Wagner A, Mioskowski C (2001) Chem Commun:2572
30. Schwink L, Ireland T, Püntener K, Knochel P (1998) Tetrahedron Asymmetr 9:1143
31. Murata K, Okano K, Miyagi M, Iwane H, Noyori R, Ikariya T (1999) Org Lett 1:1119
32. Watanabe M, Murata K, Ikariya T (2002) J Org Chem 67:1712

33. Cossy J, Eustache F, Dalko PI (2001) Tetrahedron Lett 42:5005
34. Koike T, Murata K, Ikariya T (2000) Org Lett 2:3833
35. Hamada T, Torii T, Izawa K, Noyori R, Ikariya T (2002) Org Lett 4:4373
36. Morris DJ, Hayes AM, Wills M (2006) J Org Chem 71:7035
37. Hamada T, Torii T, Izawa K, Ikariya T (2004) Tetrahedron 60:7411
38. Hamada T, Torii T, Onishi T, Izawa K, Ikariya T (2004) J Org Chem 69:7391
39. Okano K, Murata K, Ikariya T (2000) Tetrahedron Lett 41:9277
40. Matsumura K, Hashiguchi S, Ikariya T, Noyori R (1997) J Am Chem Soc 119:8738
41. Marshall JA, Bourbeau MP (2003) Org Lett 5:3197
42. Ohkuma T, Ooka H, Hashiguchi S, Ikariya T, Noyori R (1995) J Am Chem Soc 117:2675
43. Ohkuma T, Noyori R (2001) Angew Chem Int Ed 40:40
44. Hamilton RJ, Leong CG, Bigam G, Miskolzie M, Bergens SH (2005) J Am Chem Soc 127:4152
45. Sandoval CA, Yamaguchi Y, Ohkuma T, Kato K, Noyori R (2006) Mag Reson Chem 44:66
46. Hedberg C, Källström K, Arvidsson PI, Brandt P, Andersson PG (2005) J Am Chem Soc 127:15083
47. Ohkuma T, Utsumi N, Tsutsumi K, Murata K, Sandoval C, Noyori R (2006) J Am Chem Soc 128:8724
48. Shirai S, Nara H, Ikariya T (2005) Abstr Symp Organomet Chem 52:320
49. Shirai S, Nara H, Kayaki Y, Ikariya T (2009) Organometallics 28:802
50. Ohkuma T, Utsumi N, Watanabe M, Tsutsumi K, Arai N, Murata K (2007) Org Lett 9:2565
51. Heiden ZM, Rauchfuss TB (2006) J Am Chem Soc 128:13048
52. Heiden ZM, Rauchfuss TB (2009) J Am Chem Soc 131:3593
53. Li C, Wang C, Villa-Marcos B, Xiao J (2008) J Am Chem Soc 130:14450
54. Ito M, Endo Y, Ikariya T (2008) Organometallics 27:6053
55. Ito M, Endo Y, Tejima N, Ikariya T (2010) Organometallics 29:2397
56. Ito M, Ikariya T (2008) J Synth Org Chem Jpn 66:1042
57. Ito M, Sakaguchi A, Kobayashi C, Ikariya T (2007) J Am Chem Soc 129:290
58. Ito M, Kobayashi C, Himizu A, Ikariya T (2010) J Am Chem Soc 132:11414
59. Ito M, Koo LH, Himizu A, Kobayashi C, Sakaguchi A, Ikariya T (2009) Angew Chem Int Ed 48:1324
60. Teunissen HT, Elsevier CJ (1997) Chem Commun:667
61. Zhang J, Leitus G, Ben-David Y, Milstein D (2006) Angew Chem Int Ed 45:1113
62. Saudan LA, Saudan CM, Dabiéux C, Wyss P (2007) Angew Chem Int Ed 46:7473
63. Takebayashi S, Bergens SH (2009) Organometallics 28:2349
64. Ito M, Hirakawa M, Osaku A, Ikariya T (2003) Organometallics 22:4190
65. Ito M, Osaku A, Kitahara S, Hirakawa M, Ikariya T (2003) Tetrahedron Lett 44:7521
66. Ito M, Kitahara S, Ikariya T (2005) J Am Chem Soc 127:6172
67. Ito M, Osaku A, Shiibashi A, Ikariya T (2007) Org Lett 9:1821
68. Arita S, Koike T, Kayaki Y, Ikariya T (2008) Organometallics 27:2795
69. Arita S, Koike T, Kayaki Y, Ikariya T (2008) Angew Chem Int Ed 47:2447
70. Heiden ZM, Rauchfuss TB (2007) J Am Chem Soc 129:14303
71. Arita S, Koike T, Kayaki Y, Ikariya T (2008) Chem Asian J 3:1479
72. Jensen DR, Pugsley JS, Sigman MS (2001) J Am Chem Soc 123:7475
73. Sigman MS, Jensen DR (2006) Acc Chem Res 39:221
74. Ferreira EM, Stoltz BM (2001) J Am Chem Soc 123:7725
75. Radosevich AT, Musich C, Toste FD (2005) J Am Chem Soc 127:1090
76. Shimizu H, Onitsuka S, Egami H, Katsuki T (2005) J Am Chem Soc 127:5396
77. Ikariya T, Kuwata S, Kayaki Y (2010) Pure Appl Chem 82:1471
78. Ikariya T, Gridnev ID (2010) Top Catal 53:894
79. Kuwata S, Ikariya T (2010) Dalton Trans 39:2984
80. Ishiwata K, Kuwata S, Ikariya T (2009) J Am Chem Soc 131:5001

Top Organomet Chem (2011) 37: 55–84
DOI: 10.1007/3418_2011_6

Published online: 21 April 2011

Bond Activation by Metal–Ligand Cooperation: Design of "Green" Catalytic Reactions Based on Aromatization–Dearomatization of Pincer Complexes

Chidambaram Gunanathan and David Milstein

Abstract We have developed a new mode of bifunctional catalysis based on metal–ligand cooperation, involving aromatization–dearomatization of pyridine- and acridine-derived pincer complexes. This type of metal–ligand cooperation is involved in the recently discovered environmentally benign reactions of alcohols, catalyzed by PNP and PNN pincer complexes of ruthenium, including: (a) dehydrogenation of secondary alcohols to ketones, (b) dehydrogenative coupling of primary alcohols to form esters and H_2, (c) unprecedented amide synthesis: catalytic coupling of amines with alcohols, with liberation of H_2, (d) direct synthesis of imines from alcohols and amines with H_2 liberation, (e) direct conversion of alcohols to acetals with H_2 liberation, (f) selective synthesis of primary amines from alcohols and ammonia, and (g) hydrogenation of esters to alcohols under mild conditions. These reactions are very efficient, proceed under neutral conditions, and produce no waste.

Keywords Alcohols · Amides · Amines · Carbon dioxide · Catalysis · Dehydrogenative coupling · Esters · Hydrogen · Imines · Iridium · Metal–ligand cooperation · O–H activation · Pincer complexes · PNN · PNP · Ruthenium

Contents

C. Gunanathan and D. Milstein (✉)
Department of Organic Chemistry, The Weizmann Institute of Science, 76100 Rehovot, Israel
e-mail: david.milstein@weizmann.ac.il

1 Introduction and Background

Transition metal-catalyzed advanced synthetic methods and strategies have enabled synthesis of molecular targets with extraordinary complexity. An important challenge for catalysis today is the discovery of environmentally benign, "green" syntheses with heightened levels of efficiency. In conventional metal-based homogeneous catalysts, catalysis is essentially performed by the metal center, with the ligands not actively participating in bond making and breaking of the reacting substrates [1]. However, there are unique classes of catalysts that operate by bifunctional modes, involving cooperation between the metal center and ligands. Such bifunctional catalysis is widely found in nature; for example, the Cu(II)-containing enzyme galactose oxidase chemoselectively oxidizes primary alcohols to aldehydes by a mechanism in which the tyrosinyl radical coordinated to Cu(II) center participates in the reaction [2]. The two electron reduction of two protons to yield dihydrogen catalyzed by FeFe-hydrogenase also uses bifunctional catalysis involving the metal center and a pendant amine group [3] (for proposed mechanism (DFT), see [4–6]. Recently, bifunctional catalysis was shown to play an important role in the development of several homogeneous catalytic reactions [7–11]. For example, the Ru(II)-catalyzed hydrogenation of unsaturated polar bonds, in which H_2 activation involves both the metal center and a polar (N- or O-based) ligand, followed by a concerted transfer of acidic and hydridic hydrogens to an unsaturated substrate to provide hydrogenated products [12–15].

Pincer complexes have found important applications in synthesis, bond activation, and catalysis [16–26]. Among these, pincer complexes of ^{i}Pr-PNP (2,6-bis-(di-*iso*-propylphosphinomethyl)pyridine), ^{t}Bu-PNP (2,6-bis-(di-*tert*-butylphosphinomethyl)pyridine), and PNN (2-(di-*tert*-butylphosphinomethyl)-6-diethylaminomethyl)pyridine) ligands exhibited diverse reactivity [27–60]. These bulky, electron-rich pincer ligands can stabilize coordinatively unsaturated complexes and participate in unusual bond activation and catalytic processes.

Recently, we have discovered a new mode of bond activation by metal–ligand cooperation based on aromatization–dearomatization of pyridine- and acridine-based heteroaromatic pincer complexes [50–60]. Deprotonation of a pyridinylmethylenic proton of a pyridine-based pincer complex can lead to dearomatization. The dearomatized complex can then activate a chemical bond (H–Y, Y = H, OR, NR_2, C) by cooperation between the metal and the ligand, thereby regaining aromatization (Fig. 1). The overall process does not involve a change in the metal oxidation state. In this chapter, we describe the novel, environmentally benign catalytic transformations that operate via this new metal–ligand cooperation based on aromatization–dearomatization processes.

2 Metal–Ligand Cooperation: Facile Ligand Dearomatization–Aromatization

Pincer ruthenium complexes composed of ^{i}Pr-PNP and PNN ligands undergo deprotonation at the phosphine arm upon reaction with a base and resulted in dearomatization [50, 51] of the pyridine ring (Schemes 1 and 2). This is made possible by (a) the relatively low resonance energy of pyridine (28 kcal/mol; compared to benzene, 36 kcal/mol), (b) the acidity of pyridinylmethylenic protons in the pyridine-based pincer complexes, and (c) stabilization of the dearomatized ligand by the metal center. For example, the pyridine-based pincer complexes **1** and **4** undergo smooth deprotonation to quantitatively yield complexes **2** and **5** (Schemes 1 and 2). NMR studies of **2** and **5** indicate dearomatization, as the pyridine protons are shifted to lower frequency (olefinic region). The structure of complex **2** is unequivocally corroborated by a single-crystal X-ray diffraction study (Fig. 2), which shows that the bond length of C1 – C2 (1.450 Å) is much shorter than that of C7 – C6 (1.552 Å), while the bond length of C1 – P1 (1.803 Å) is shorter than C7 – P2 (1.843 Å) to a smaller extent, indicating that the dearomatized resonance structure of **2** contributes much more than the aromatized resonance structure (Fig. 1) of phosphor-ylide form [51]. Importantly, the dearomatized complexes of **2** and **5** activate dihydrogen by cooperation between the ruthenium center and the deprotonated phosphine arm resulting in aromatization to quantitatively yield the ruthenium *trans*-dihydride complexes of **3** and **6**, respectively (Schemes 1 and 2). The magnetically equivalent *trans*-dihydrides resonate as a triplet in complex **2** at −4.96 ppm (J_{PH} = 20.0 Hz) and as a doublet at −4.06 ppm (J_{PH} = 17.0 Hz) in complex **5**. Both complexes **3** and **6** slowly lose H_2 at room temperature to regenerate complexes **2** and **5**, respectively.

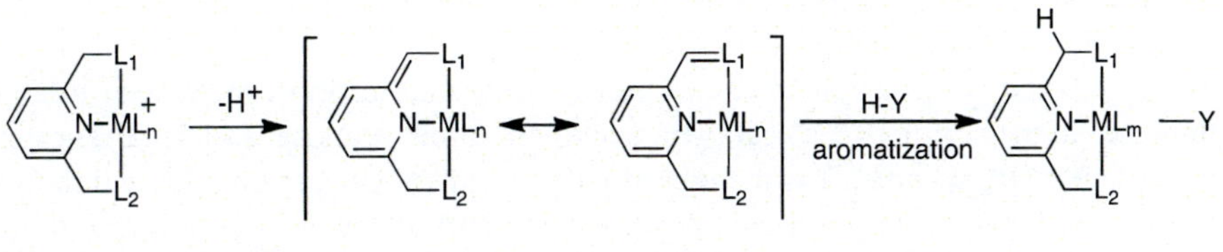

Fig. 1 Metal–ligand cooperation based on aromatization–dearomatization

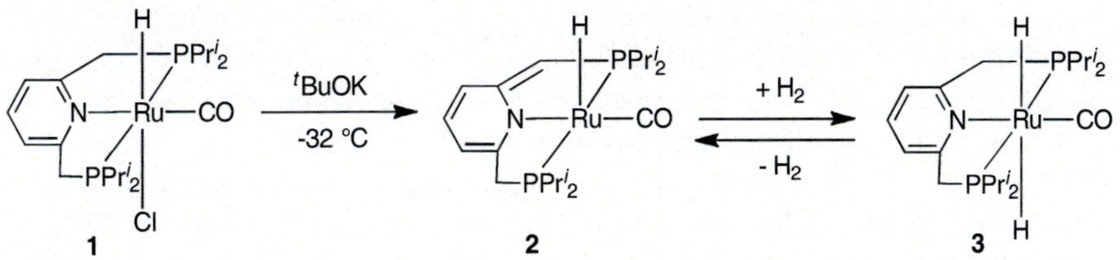

Scheme 1 Preparation of dearomatized PNP pincer complex **2** and its reversible reaction with H_2

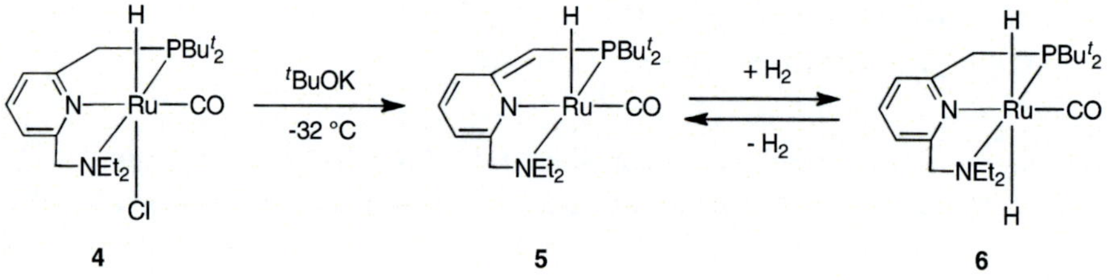

Scheme 2 Preparation of dearomatized PNN pincer complex **5** and its reversible reaction with H_2

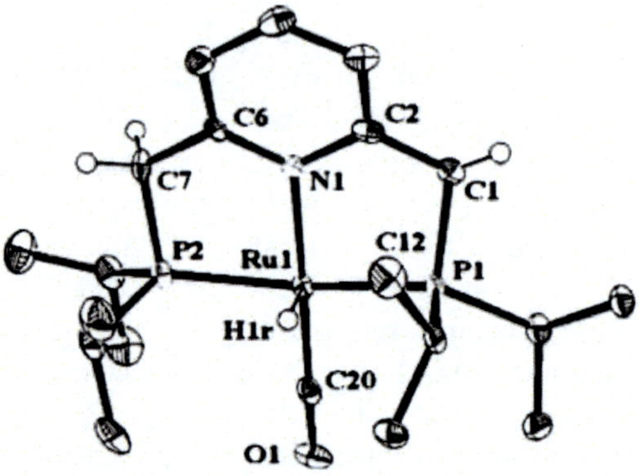

Fig. 2 X-ray crystal structures of complex **2**. Selected bond lengths (Å): C1 − C2 = 1.450, C7 − C6 = 1.552, C1 − P1 = 1.803, C7 − P2 = 1.843

3 Direct Dehydrogenative Functionalization of Alcohols

3.1 Facile Conversion of Alcohols into Esters and Dihydrogen

Metal–Ligand cooperation by aromatization–dearomatization can play an important role in the design of new catalytic reactions. A few years ago, we have reported that the PNP Ru complex **1** in the presence of an equivalent amount of base (relative to Ru) catalyzes the acceptorless dehydrogenation of secondary alcohols to ketones with the liberation of molecular hydrogen [40]. Interestingly, when primary alcohols were reacted with complex **1**, in the presence of a base, an unusual reaction took place, namely, the dehydrogenative coupling to form esters with formation of H_2 [50]. Thus, refluxing a hexanol (bp 157°C) solution containing 0.1 mol% complex **1** with 0.1 mol% KOH under argon in an open system for 24 h resulted in the formation of hexyl-hexanoate in 67% yield. Assuming that an alkoxy complex is an intermediate in the reaction, formed by alkoxide substitution of the chloride ligand, which undergoes β-H elimination, we have realized that this process is likely to be hampered by the lack of a *cis* coordination site. Thus, we have prepared the PNN complex **4**, which possesses a hemilabile amine "arm". Indeed, this complex, in the presence of an equivalent of base, was a significantly

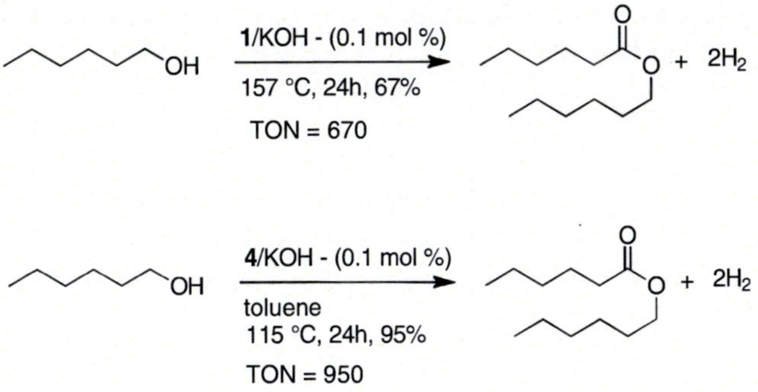

Scheme 3 Catalytic conversion of alcohols to ester by PNP (**1**) and PNN (**4**) complexes

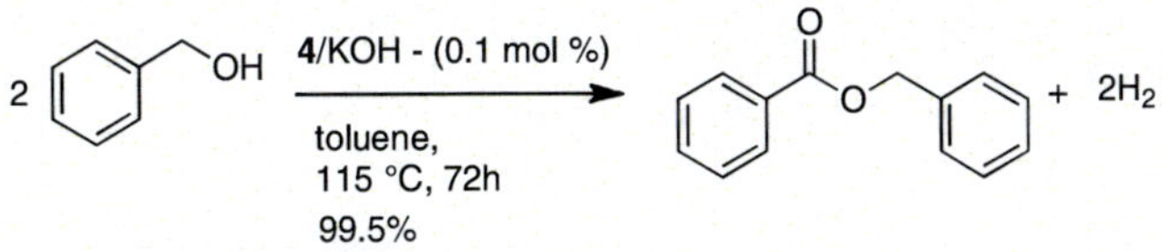

Scheme 4 Catalytic conversion of benzyl alcohol to benzyl benzoate by PNN complex **4**

better catalyst [50] than the corresponding PNP complex **1**, leading to 95% (950 turnovers) hexyl-hexanoate after 24 h at 115°C (Scheme 3).

Complex **4** in the presence of 1 equivalent of base is an efficient dehydrogenative esterification catalyst, resulting in good turnovers and yields of esters with various alcohols at 115°C. Reaction follow-up with benzyl alcohol indicated that 91% benzyl benzoate was formed after 6 h, with TOF reaching 333 h^{-1} at the level of 50% benzyl benzoate. Formation of the ester became very slow after 6 h, perhaps because of retardation of the reaction by the high ester concentration. However, almost quantitative formation of benzyl benzoate was obtained (Scheme 4).

In an attempt to totally eliminate the need for added base, which could further improve the reaction, complex **4** was reacted with ^{t}BuOK to obtain the coordinatively unsaturated 16 e- Ru(II) neutral complex **5**. Indeed, **5** is an excellent catalyst for the dehydrogenative coupling of alcohols to esters with liberation of H_2 under neutral conditions [50]. Table 1 provides a few examples. Only traces of aldehydes are formed. This catalytic reaction provides a new "green" pathway for the synthesis of esters. Prior examples of coupling of alcohols to esters are considerably less efficient: see [50] and [61, 62].

There are two potential pathways by which an ester can be formed, both operating via an intermediate aldehyde, namely a Tischenko type condensation, or hemiacetal formation followed by its dehydrogenation. Our results establish that the second pathway is operative [50]. Thus, using benzaldehyde (in the absence of alcohol) did not yield any ester. On the other hand, reaction of benzaldehyde with one equivalent of benzyl alcohol led to quantitative formation of benzyl benzoate (1).

Table 1 Dehydrogenative coupling of alcohols to form esters and dihydrogen catalyzed by dearomatized Ru PNN complex **5**

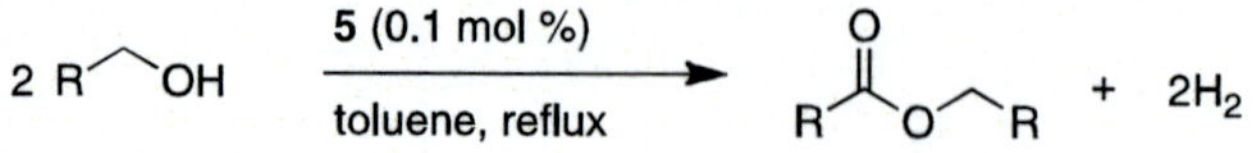

Entry	Alcohol	Ester	Time (h)	Ester (%)	Aldehyde (%)
1[a]	1-butanol	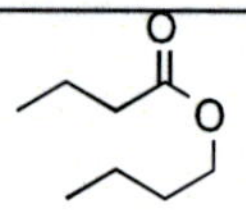	5	90	0.5
2	1-hexanol	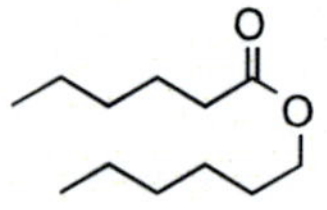	6	99	0
3	Benzyl alcohol	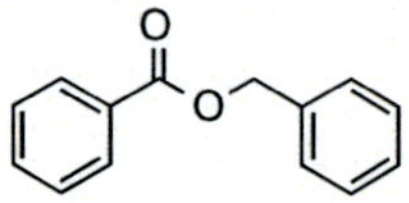	4	92	1

Conditions: 0.01 mmol catalyst **5**, 10 mmol alcohol, and toluene (2 ml) were refluxed (115 °C) under Ar flow

[a]Neat reaction, heated at 117°C

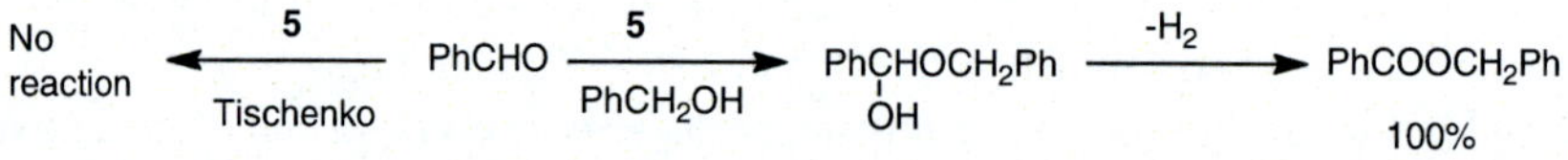

(eq 1)

Figure 3 outlines the proposed catalytic cycle for the dehydrogenative coupling of alcohols to form esters. Facile O–H activation of the alcohol by complex **5** results in aromatization with formation of the saturated hydrido-alkoxy complex **7**. Amine "arm" opening provides the required *cis* coordination site for β-H elimination, generating an intermediate aldehyde (which may stay coordinated) and the *trans*-dihydride complex **6**. The coordinatively saturated complex **6** liberates H_2, as we have described above (Scheme 2), thus regenerating the unsaturated 16 e^- complex **5**. The generated aldehyde is in equilibrium with the corresponding hemiacetal, which undergoes a similar catalytic cycle, providing the product ester.

3.2 Direct Synthesis of Amides from Alcohols and Amines

Next, the coupling of alcohols and amines was studied. In principle, three reaction pathways can be envisioned, proceeding via either intermediate hemiaminal (Fig. 4) or hemiacetal. Hemiaminals are generally very unstable, undergoing facile water liberation to form an imine, which can undergo hydrogenation (with the H_2 liberated in the first stage) to yield a secondary amine. Such reactions are known [63]. A competing pathway proceeding via the hemiacetal to yield esters is also

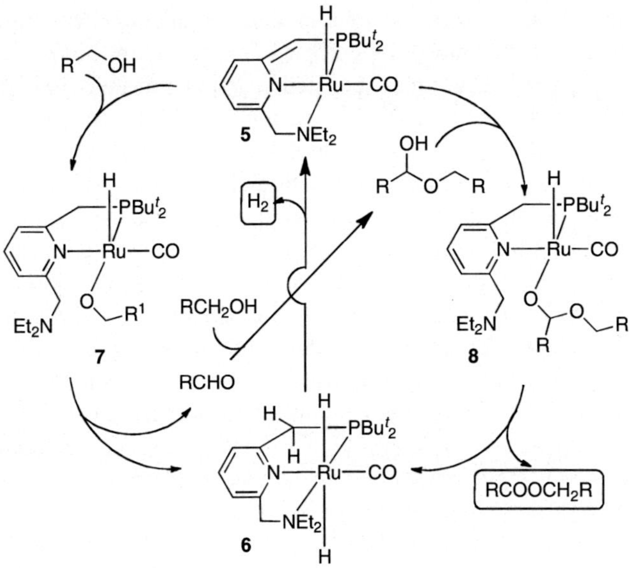

Fig. 3 Proposed mechanism for the direct conversion of alcohols to esters catalyzed by complex **5** via metal–ligand cooperation

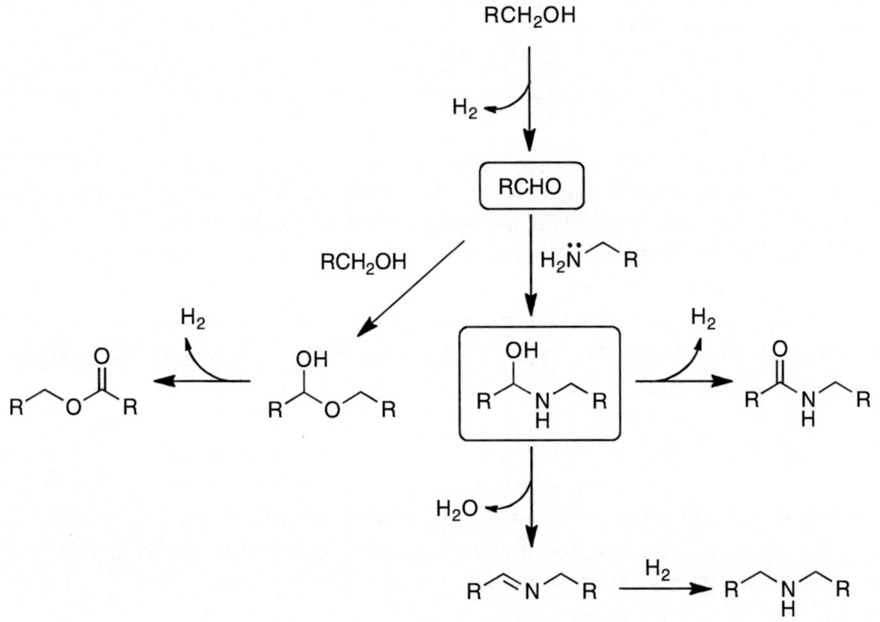

Fig. 4 Possible dehydrogenative pathways for the reactions of alcohols with amines

possible in case of amines of relatively low nucleophilicity. We were particularly interested in the possibility of H_2 liberation from the hemiaminal intermediate to form an amide.

Gratifyingly, heating a toluene solution containing equivalent amounts of alcohol and amine with 0.1 mol% of complex **5** resulted in amide formation with liberation of H_2 [52]. This unprecedented reaction proceeds in high yields and high turnover numbers under neutral conditions, with no generation of waste, the only byproduct being H_2 (valuable by itself), thus providing an attractive, "green" route to the synthesis of amides. The reaction can also be conducted using the air-stable hydridochloride complex **4** as a catalyst, in the presence of an equivalent (relative to Ru) of a base, such as isopropoxide. Examples are listed in Table 2.

While yields are good to excellent in most cases, lower yield is obtained as a result of steric hindrance (entries 4, 9). Also, as a result of the lower nucleophilicity of aniline, the self-dehydrogenative coupling of the alcohol becomes competitive, leading to some ester formation. Noteworthy, no reaction took place with dibenzylamine, indicating that the reaction is limited to primary amines. Diamines react normally, leading to bis-amides (Table 3). Diethylenetriamine reacts only at the primary sites, requiring no protection of the secondary amine functionality.

Mechanistic studies revealed that the reaction does not involve ester intermediacy and proceed via a hemiaminal-type mechanism (Scheme 5). When the benzyl benzoate and benzylamine were reacted under similar experimental conditions of the amidation reactions, either in the presence or absence of catalyst **5**, no amide formation took place and the starting materials were recovered. These studies preclude the involvement of a hemiacetal pathway in the reactions. Further, the dehydrogenation of hemiaminal prevails over the anticipated water elimination to result in an imine, which upon H_2 addition would provide the secondary amine. Secondary amines were observed only in trace amounts when linear alcohols were used.

The postulated catalytic cycle for the direct amidation by reaction of alcohols and amines could be envisioned as similar to the catalytic cycle for esterification reaction of alcohols with intermediates being the hemiaminal instead of hemiacetal (Fig. 5). The intermediate aldehyde, generated in the alcohol dehydrogenation cycle, reacts with the primary amine to form an intermediate hemiaminal, which undergoes dehydrogenation to form an amide.

However, the observed remarkable selectivity for amide formation suggests that free hemiaminal intermediate in solution is not involved, since the latter would likely liberate water faster than H_2. We believe that the amine reacts with the intermediate aldehyde while it is still coordinated to the metal center, leading to formation of the quaternary salt **12** (Fig. 6), which may transfer a proton to the dearomatized phosphine arm (to give intermediate **10**). Subsequent β-hydride elimination and ring closure by the amine arm would lead to the *trans*-dihydride complex **6** and product amide. Liberation of dihydrogen from **6** would complete the catalytic cycle with regeneration of catalyst **5**.

Very recently, interesting catalytic conversion of alcohols to esters, acids, and amides by hydrogen transfer to a sacrificial acceptor was reported [64, 65]. Low turnover catalytic dehydrogenative coupling reaction of amines with alcohols to form amides with H_2 liberation under basic conditions was also reported very recently [66–71].

Table 2 Direct dehydrogenative coupling of amines with alcohols to form amides and H_2

$$R^1CH_2OH + R^2NH_2 \xrightarrow[\text{toluene, reflux}]{\mathbf{5}\ (0.1\text{mol }\%)\ \Delta} R^2NHCOR^1 + 2H_2$$

Entry	R^1CH_2OH	R^2NH_2	Time (h)	Amides	Yield (%)[a]
1	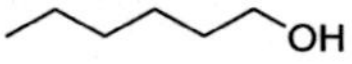		7	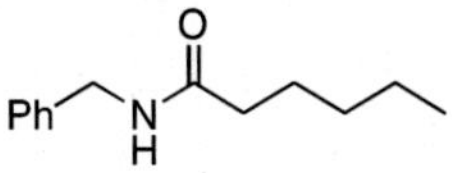	96
2		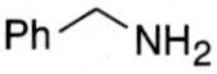	7	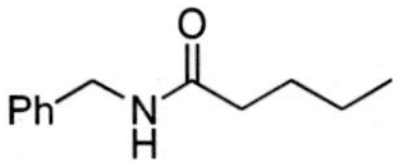	97
3	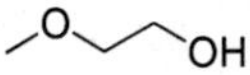	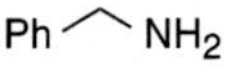	9	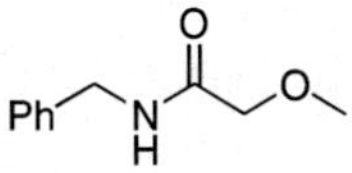	99
4	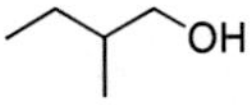	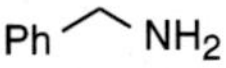	12	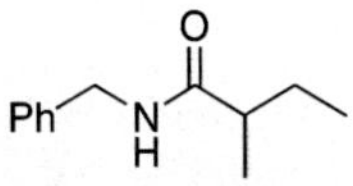	70[b]
5		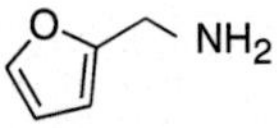	8	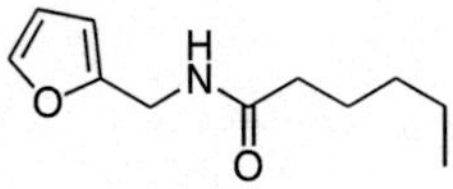	78[b]
6	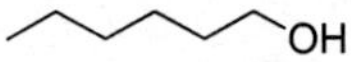	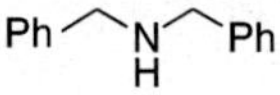	8	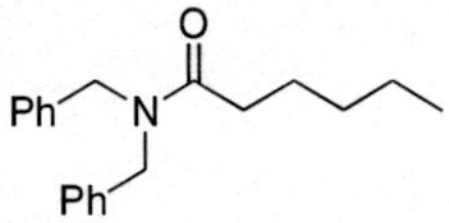	0[b]
7		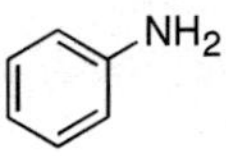	8	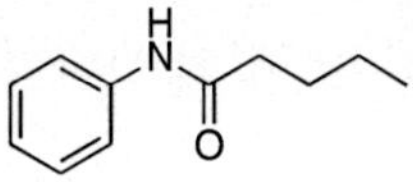	58[b]
8	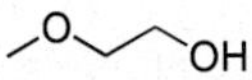	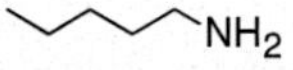	8	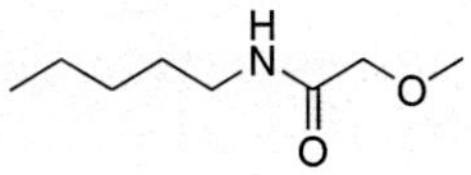	99
9		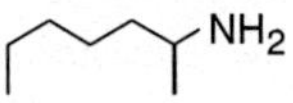	8	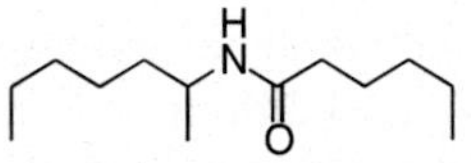	72[b]
10	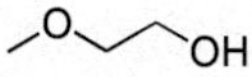	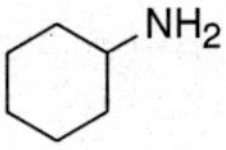	8	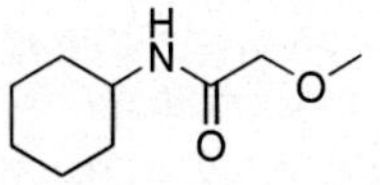	99

Conditions: Catalyst **5** (0.01 mmol), alcohol (10 mmol), amine (10 mmol), and toluene (3 mL) were refluxed under Ar flow

[a]Isolated yields

[b]The remaining alcohol was converted into the corresponding ester. In the reactions involving hexanol and pentanol, trace amounts of the corresponding secondary amines were detected (GC–MS)

Table 3 Dehydrogenative coupling of di- and tri-amines with alcohols to form bis-amides, catalyzed by complex **5**

Entry	Diamine	Time (h)	Bis-amide	Yield (%)
1	Ethylenediamine	9	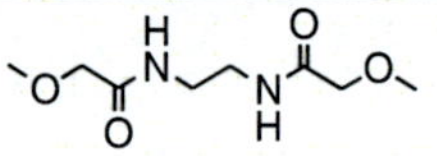	99
2	Diethylenetriamine	8	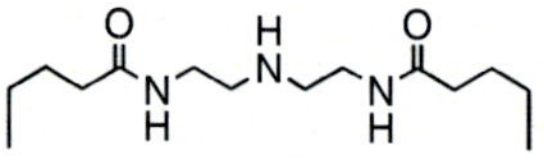	88
3	1,6-Diaminohexane	9	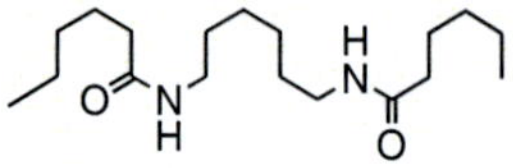	95

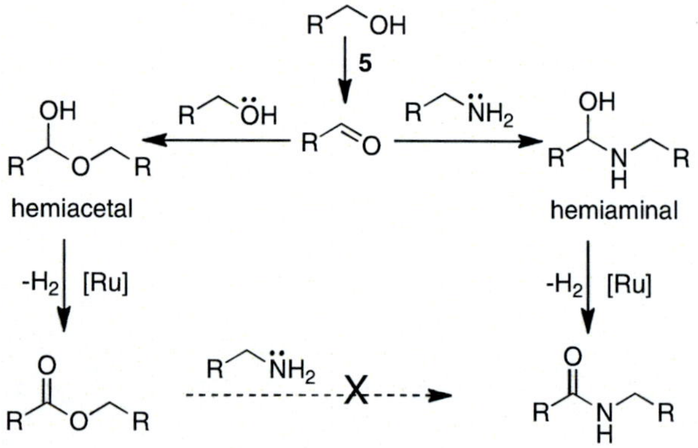

Scheme 5 Pathways in the alcohol amidation reaction: hemiacetal is not an intermediate

3.3 *Coupling of Alcohols with Amines to Form Imines and H_2*

When the PNP complex **2** or **2**-^{t}Bu (Fig. 7) was used in the coupling of alcohols and amines, a different course of reaction took place that provided an efficient method for the synthesis of imines [72]. This environmentally benign process for imines directly from alcohols and amines occurs with liberation of H_2 gas and water, gives high turnover numbers, and no waste products. Furthermore, the reaction proceeds under neutral conditions and no hydrogen acceptor is needed. Remarkably, hydrogenation of the imine does not take place.

A variety of alcohols undergo efficient dehydrogenative coupling with amines (Table 4). A range of alcohols and amines containing either electron-donating or -withdrawing substituents reacted to yield the corresponding imines in very good yields. Steric hindrance in the amines is also tolerated (entries 7, 9). This new catalytic reaction works very well in the challenging synthesis of unstable aliphatic imines (entries 9). However, competing esterification and amidation occurred leading to the formation of esters and amides in minor yields [72]. Convenient to practical applications, the reaction can be carried out in air (entry 10).

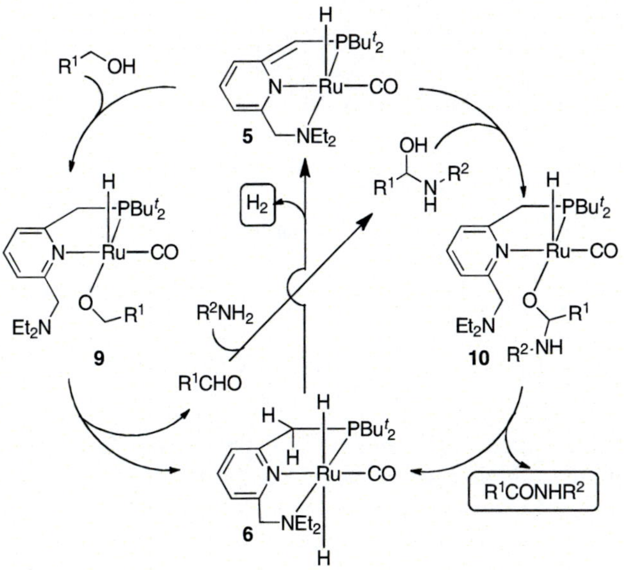

Fig. 5 Possible catalytic cycle for the direct synthesis of amides from alcohols and amines, catalyzed by the PNN ruthenium pincer complex **5**

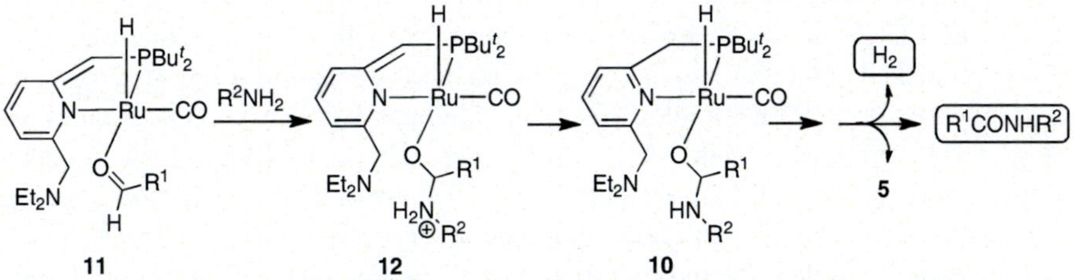

Fig. 6 Formation of amides via intermediate quaternary ammonium salt

The proposed mechanism for direct imination of alcohols by amines is delineated in Fig. 8. Activation of the O–H bond of the alcohol by complexes **2** or **2**-^{t}Bu likely results in the aromatized intermediate **13**, which upon β-H elimination (perhaps involving "arm" opening) yields an aldehyde-coordinated intermediate **14**. Dissociation of the aldehyde leads to the known dihydride **3**, which liberates H_2 to regenerate complex **2** or **2**-^{t}Bu. Reaction of the aldehyde with the amine in solution generates an unstable hemiaminal, which loses water to produce the product imine.

The striking difference in the catalytic activity of the PNN complex **5**, which leads to amide formation [52], and PNP **2** or **2**-^{t}Bu complexes, which catalyze imine formation under similar conditions, can be rationalized as follows. In the case of the PNN complex, which bears a hemilabile amine arm, the aldehyde stays coordinated to the metal center long enough to be attacked by the amine (Fig. 6), while in the case of the PNP complex closure of the phosphine "arm" occurs more rapidly, resulting in liberation of the aldehyde and its attack by the amine takes place in solution, generating a hemiaminal intermediate which liberates water (Fig. 8).

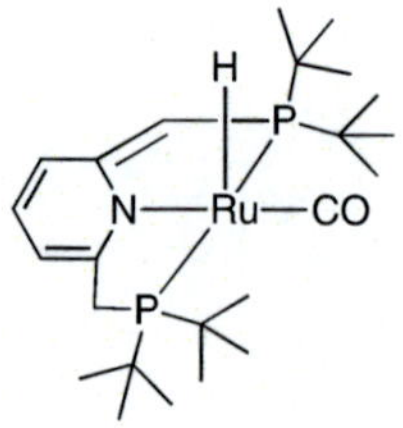

2-Bu^t

Fig. 7 Dearomatized PNP complex with bulky *tert*-butyl phosphine substituents

3.4 Catalytic Conversion of Alcohols to Acetals and H_2

While continuing studies on pyridine-based pincer complexes, we extended our investigations to acridine-derived pincer complexes [73, 74]. We designed and synthesized the acridine-based new PNP complex **16** (PNP = 4,5-bis(di-*iso*-propyl-phosphinomethyl)acridine) in three steps from acridine (Scheme 6). The structure of complex **16**, determined by X-ray diffraction, reveals a distorted octahedral geometry around the ruthenium center (Fig. 9), with an unusually long Ru–N bond (2.479 Å; compared to 2.103 Å in complex **4** [50]) and the CO ligand coordinated *trans* to the acridinyl nitrogen atom. Upon complexation, the acridine ligand becomes bent at the middle aryl ring to adopt a boat-shaped structure with a dihedral angle of 167.6° [74]. This diminished aromaticity upon coordination likely persists in solution as indicated by ^{1}H-NMR of the PNP complex **16** which exhibits the C9*H* proton of the acridine ring at 8.15 ppm, an upfield shift of 0.46 ppm relative to C9*H* of uncomplexed PNP ligand (8.61 ppm) (an acridine-based PNP ligand lacking benzylic groups forms complexes with planar coordination geometry [75]).

Interestingly, unlike the pyridine-based PNP and PNN pincer complexes **1** and **4** which did not react with alcohols in neutral media, the acridine-derived air-stable PNP complex **16** catalyzes the conversion of primary alcohols to acetals [75] under neutral conditions (Scheme 7). Small amounts of esters and aldehydes were also formed. Murahashi et al. reported the first acetal formation directly from the alcohols [62]. Efficient direct catalytic transformations of alcohols to acetals are of interest as this circumvents the need for aldehydes or aldehyde derivatives.

Further studies revealed that temperatures of 130°C and above are necessary for the formation of acetals from alcohols and added HCl has no influence. We believe that the reaction proceeds via enolether intermediates (Fig. 10, **17**), which are generated by water elimination from the initially formed hemiacetals. This is supported by the observation of *E, Z*-mixtures of enolether intermediates **17** in the reaction mixture. In line with this postulate, reaction of complex **16** with benzyl alcohol (under neutral conditions), which lacks β-hydrogens and cannot form an enolether, did not yield benzylacetal; rather, benzylbenzoate was quantitatively formed. Thus, formation of the acetal product likely takes place by the addition of alcohols to the C=C bond of the enolether (route b, Fig. 10). It is not clear at this

Table 4 Direct synthesis of imines from alcohols and amines catalyzed by the dearomatized ruthenium complex **2**-Bu^t

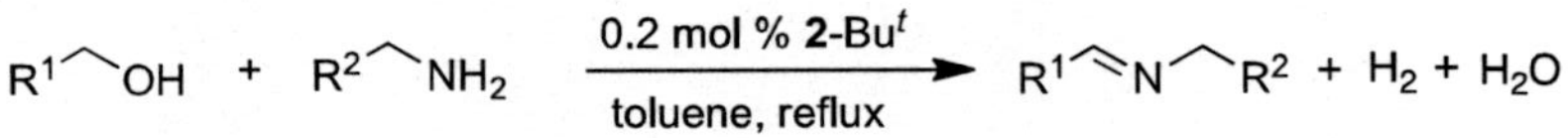

Entry	R^1CH_2OH	R^2NH_2	Time (h)	Conv. of alcohol	R^1=N–R^2	Yield of imine (%)[a]
1	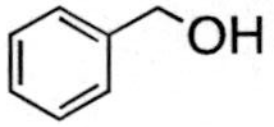	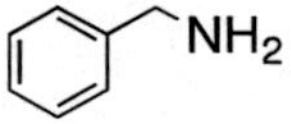	56	>94	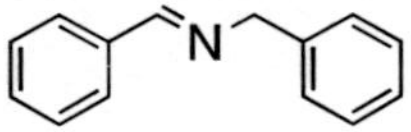	79
2	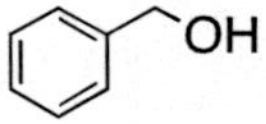		52	100	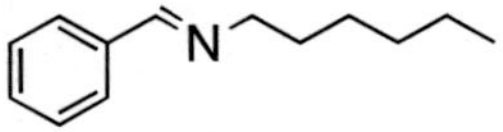	82
3	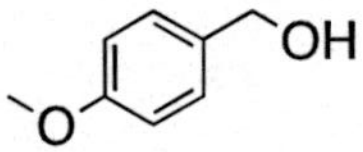		48	100	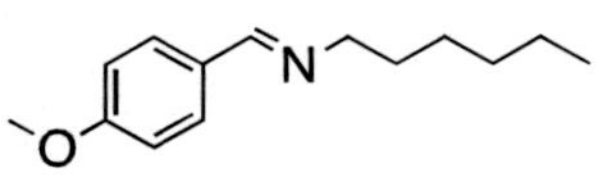	89
4	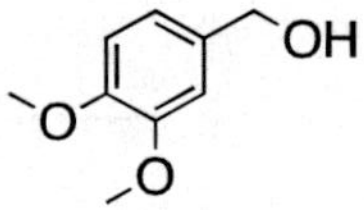	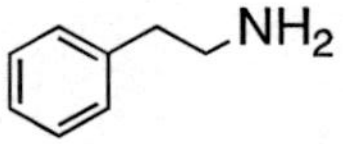	48	100	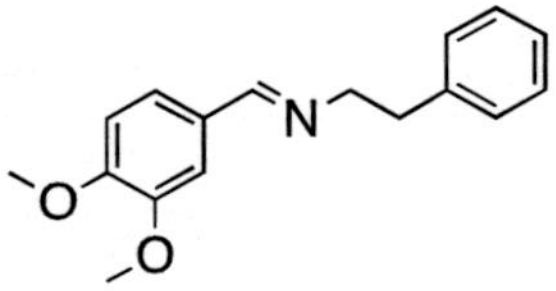	92
5	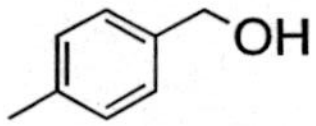		48	97	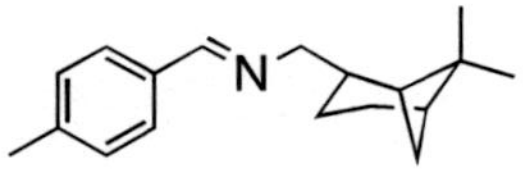	88
6	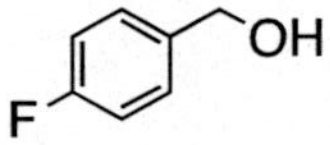	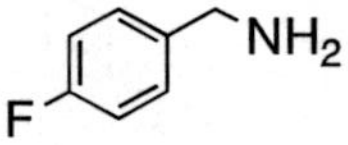	56	90	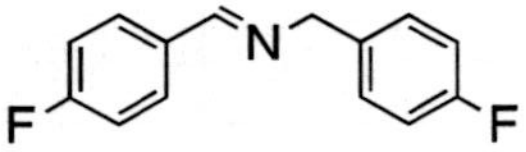	77
7	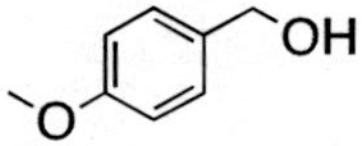	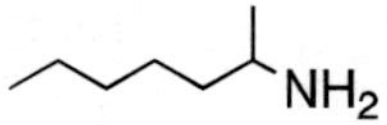	48	97	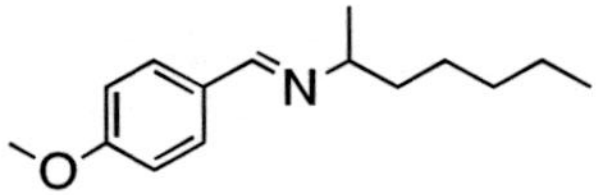	84

(*continued*)

Table 4 (*continued*)

Entry	R^1CH_2OH	R^2NH_2	Time (h)	Conv. of alcohol	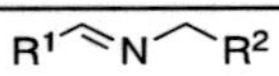	Yield of imine (%)[a]
8	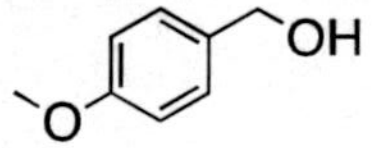	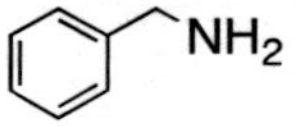	48	100	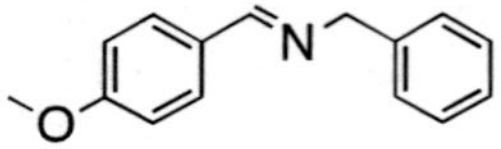	90
9	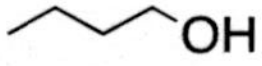	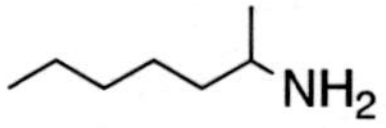	56	–	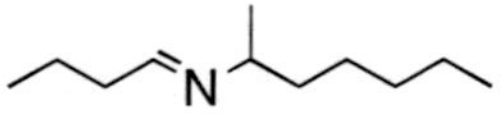	86
10[b]	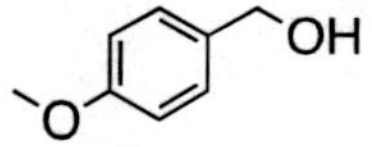	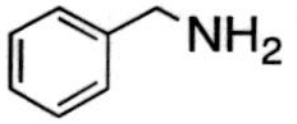	24	100	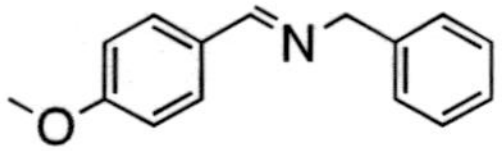	89

Conditions: Complex **2**-Bu^t (0.02 mmol), alcohol (10 mmol), amine (10.1 mmol), *m*-xylene (1 mmol, internal standard), and toluene (3 mL) were heated at reflux in a Schlenk tube. Conversion of alcohol was determined by GC

[a]Isolated yields

[b]Under air

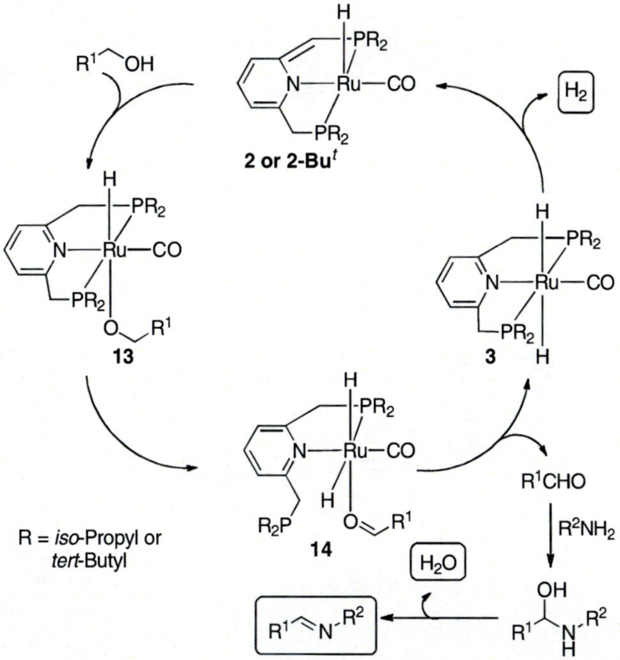

Fig. 8 Proposed catalytic cycle for the synthesis of imines from alcohols and amines, catalyzed by PNP ruthenium pincer complexes

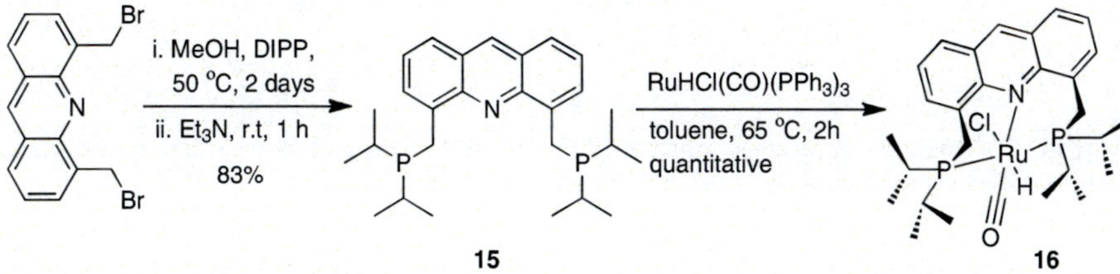

Scheme 6 Synthesis of new PNP ligand and ruthenium(II) pincer complex based on acridine

stage why the acridine-based complex **16** catalyzes acetal formation in the absence of base, while no reaction was observed under the same conditions with the pyridine-based complexes [50] (*vide supra*). Perhaps the much longer Ru–N bond and bent middle acridine ring result in "hemilability" of the acridine moiety, affording a potential vacant site and a localized "internal base". Alcohol coordination, followed by its deprotonation by the adjacent acridinyl nitrogen, can lead after β-H elimination to a hemiacetal (via aldehyde), which undergoes dehydration.

However, when complex **16** was reacted with alcohol in the presence of one equivalent of base, esters were obtained (Scheme 8) as observed for the pyridine-based pincer complexes (e.g., **1** and **4**). An Ru(0) intermediate which might have formed upon reaction of complex **16** with KOH may be the actual

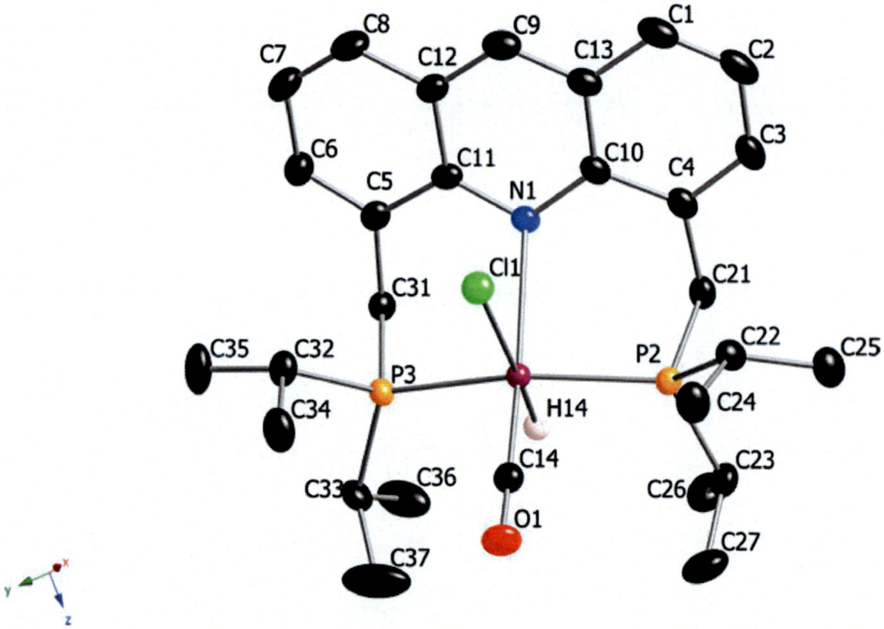

Fig. 9 X-ray structure of acridine-derived PNP pincer complex **16**

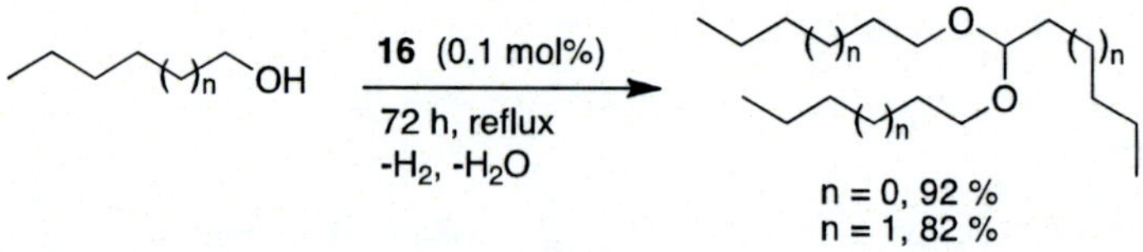

Scheme 7 Synthesis of acetals directly from alcohols catalyzed by acridine-based pincer complex **16**

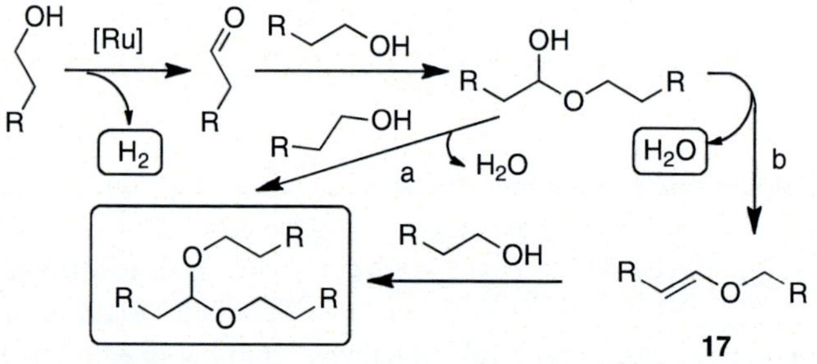

Fig. 10 Possible mechanistic pathways for the direct acetalization of alcohols. Involvement of enolether intermediate **17**

catalyst in the ester formation. Further studies indicate that the reactions likely proceeds via the hemiacetal intermediate [62] rather than by a Tischenko disproportionation [76] of an aldehyde intermediate (eq 1).

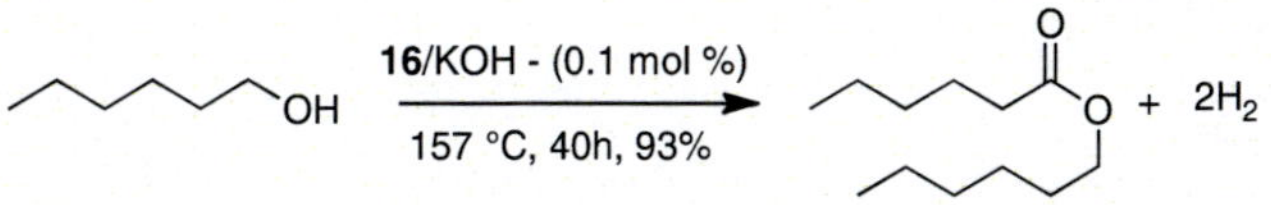

Scheme 8 Esterification of alcohols catalyzed by the acridine-based pincer complex **16** in the presence of base

3.5 Selective Synthesis of Primary Amines from Alcohols and Ammonia

The acridine-based PNP pincer complex **16** catalyzes the selective formation of primary amines [73] directly from alcohols and ammonia (Table 6). Among the amines, the terminal primary amines are the most useful, but in general their selective synthesis is challenging. Primary amines are more nucleophilic than ammonia and compete with it in reaction with electrophiles such as alkyl halides or aldehydes, producing secondary amines, which can also react, leading to the formation of mixtures of products.

Suppression of the secondary amine formation is a primary challenge in these reactions. Optimization studies involving aliphatic alcohols indicated that in the presence of ammonia, the formation of secondary amines could occur in two pathways: (1) reaction of the product primary amine (in competition with ammonia) with the in situ-generated aldehyde to form a hemiaminal followed by water elimination to give the imine (Fig. 11, route a), or (2) N–H activation of the initially formed primary amine leading to the imine followed by nucleophilic attack by another molecule of the primary amine on the terminal-imine and elimination of ammonia from intermediates **18** (Fig. 11, route b), and the subsequent hydrogenation delivers secondary amines. Because of route b, it is important to stop the reaction as soon as the alcohol has been consumed, to avoid the self degradation of the primary amine.

An assortment of primary alcohols was directly converted to primary amines upon reaction with ammonia (Table 5). Aryl methanols underwent facile reaction to provide benzylamines in good yields. Benzyl alcohols bearing electron-donating substituents reacted faster than benzyl alcohols with an electron-withdrawing substituents (entries 1–3). The electron-rich heteroaryl methanols exhibited excellent selectivity for primary amines. Pyridine-2-yl-methanol and 2-furylmethanol were converted to the corresponding primary amines in 96% and 94.8% yields, respectively (entries 4–5). Increasing the steric hindrance at the β-position of alkyl alcohols diminished the formation of imines and the corresponding secondary amines, and hence increased the selectivity and yields of primary amines (entries 9–11). It is noteworthy that the strained four-membered ring in the oxetane alcohol remained intact, resulting in high yield of the primary amine (entry 11). The reaction took place effectively also in neat alcohols, requiring no added solvent.

The fact that the formation of water byproduct in the reaction system does not adversely affect catalysis inspired us to investigate the use of water as the reaction

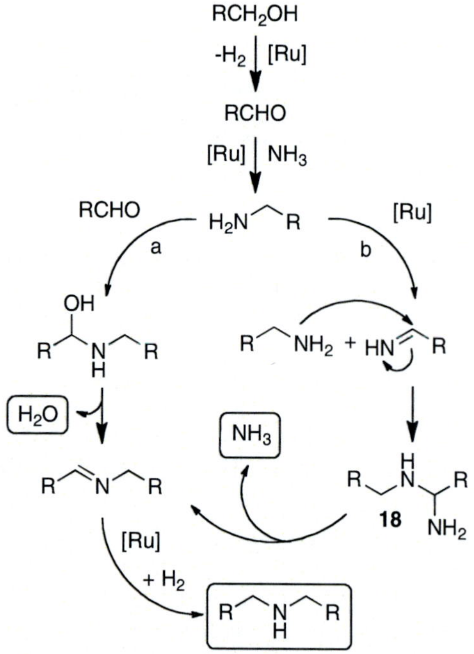

Fig. 11 Possible pathways for the formation of secondary amines

medium. The direct amination of alcohols with ammonia by complex **16** proceeded "on water" with excellent selectivity for primary amines (Table 6). The presence of water in large excess was advantageous since it may have led to the hydrolysis of imines side products formed from further reactions of the primary amines, and thus enhanced the selectivity toward primary amines. Benzyl alcohol and phenethyl alcohol, which are insoluble in water at room temperature, formed a homogeneous solution on heating, and thus the reaction might be considered "in water". Aliphatic alcohols such as 1-hexanol were not miscible with water even on heating and the reaction took place "on water" [77, 78]. The selectivity for the linear primary amines is improved by the use of co-solvents such as toluene or dioxane in water.

While insufficient mechanistic data exist at present for the direct amination reaction, ongoing studies in our lab suggests the involvement of acridine middle aryl ring (C9) as a cooperative site in synergy with the metal center (reaction of complex **16** with ammonia, either in the presence or absence of alcohol, led to a new complex resulting from hydride transfer to C9; its structure (C9 being CH_2) was solved by an X-ray diffraction study [79]) (Fig. 12). We believe that upon reaction of alkoxide **19** with ammonia, the hydride on the ruthenium is transferred to C9 center of acridine to form the intermediate **20**, as we have observed with the analogous chloride (instead of alkoxide) complex [79]. Presumably, de-coordination of the acridine nitrogen followed by NH_3 coordination positions the acridine ring favorably for hydride transfer. Subsequent ring opening by one of the phosphine arms allows the β-hydride elimination to generate an aldehyde–ammonia intermediate **21**. Subsequently, an imine hydride intermediate **22** is formed, followed by imine insertion into Ru–H, hydride transfer from C(9), and reaction with alcohol, to

Table 5 Selective syntheses of primary amines from alcohol and ammonia catalyzed by PNP pincer complex **16**

$$RCH_2OH + NH_3 \xrightarrow[\text{toluene, reflux}]{\mathbf{16}\text{ (0.1 mol \%)}, \Delta} RCH_2NH_2 + H_2O$$

Entry	RCH_2OH	Time (h)	Conv. of alcohol	RCH_2NH_2	Yield (%)[a]
1	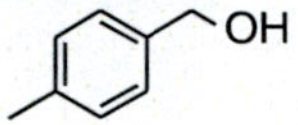	12	100	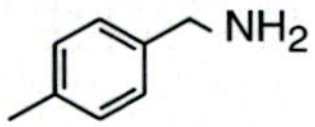	83 (70)[b]
2	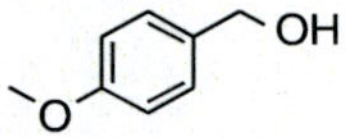	14	100	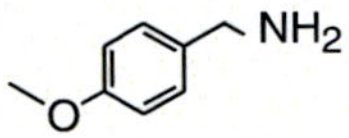	78
3	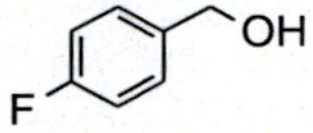	24	100	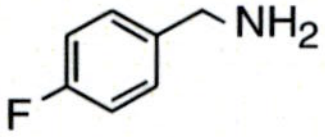	91
4	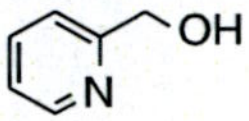	30	100	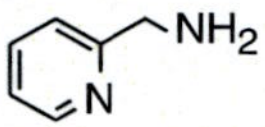	96
5	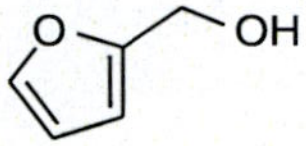	12	100	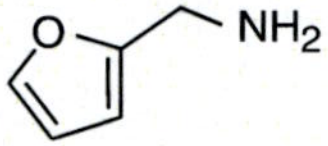	95
6[c]		20	97		61 [35][d]
7	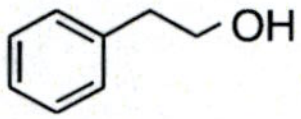	32	100	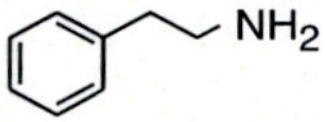	69
8		12	100	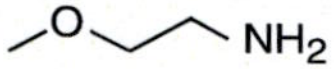	95

(*continued*)

Table 5 (*continued*)

Entry	RCH_2OH	Time (h)	Conv. of alcohol	RCH_2NH_2	Yield (%)[a]
9[c]	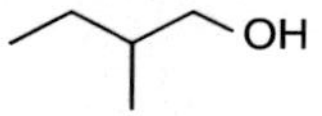	18	93	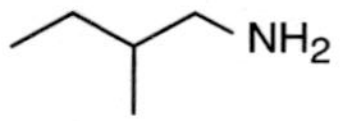	68 (61)
10	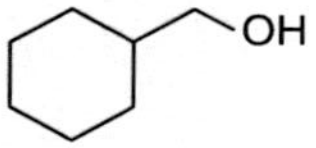	25	95.5	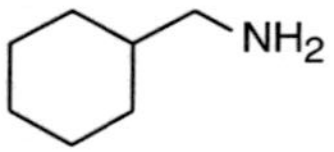	82 (73)
11	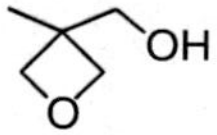	25	96.4	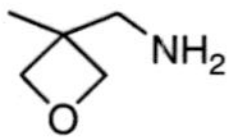	90 (84)

Conditions: Complex **16** (0.01 mmol), alcohol (10 mmol), ammonia (7.5 atm), and toluene (3 ml) were heated in a Fischer–Porter glass pressure tube

[a]Conversion of alcohols and yield of products were determined by GC

[b]Yield in parenthesis represents isolated yields

[c]Neat reaction. Isolated yield is an average of two experiments

[d]Yield of dipentylamine

Table 6 Direct synthesis of amines from alcohols and ammonia catalyzed by the ruthenium complex **16** in and on water

$$RCH_2OH + NH_3 \xrightarrow[\text{Water, 135 °C}]{\mathbf{16}\text{ (0.1 mol\%)}, \Delta} RCH_2NH_2 + H_2O$$

Entry	RCH_2OH	Time (h)	Convn.	RCH_2NH_2	Yield (%)[a]
1	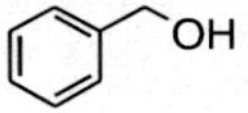	18	100	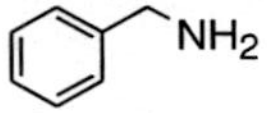	95 (86)
2	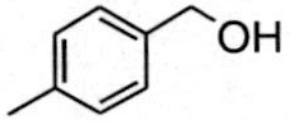	18	100	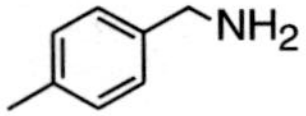	92
3	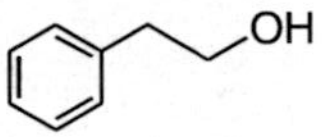	36	100	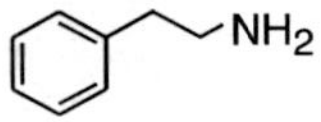	80[b]
4	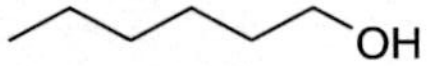	24	92	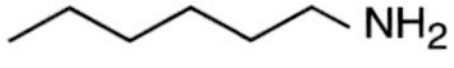	55[c]
5[d]	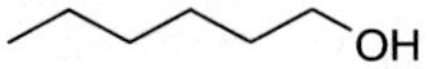	28	89		74
6[e]		30	99	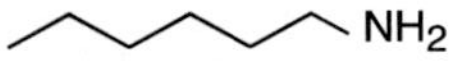	80
7[e]		30	99		70

Conditions: Complex **16 (0.01 mmol), alcohol (10 mmol), ammonia (7.5 atm), and water (3 ml) were heated at reflux in a Fischer–Porter glass pressure tube.** Conversion of alcohols and yield of products were analyzed by GC; Yield in parenthesis is isolated yield

[a]Corresponding imine was the major byproduct in entries 1–3; corresponding acid was the byproduct in entries 6–7

[b]Corresponding acids were found in aqueous layer

[c]Hexamide was found in aqueous layer

[d]Mixture of 2 ml water and 2 ml toluene was used as solvent

[e]Mixture of 1 ml water and 2 ml dioxane was used as solvent

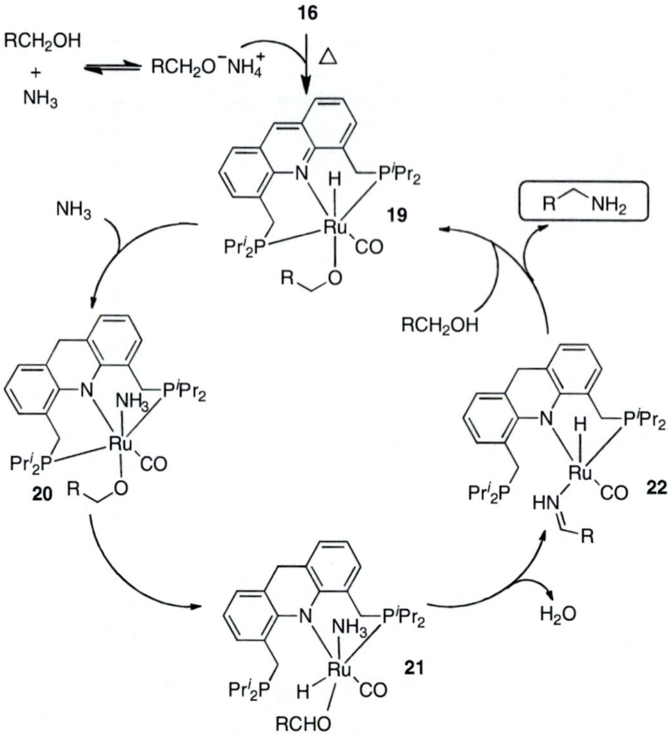

Fig. 12 Proposed catalytic cycle for the synthesis of primary amines from alcohols and ammonia

yield the product amine and regenerating **19** (Fig. 12), representing overall "borrowing hydrogen" from the alcohol to intermediate imine [63]. The selectivity to primary amine is likely a result of the steric preference for coordination of ammonia in intermediate **21** over that of primary amine.

4 Hydrogenation Based on Ligand Dearomatization–Aromatization

4.1 Catalytic Hydrogenation of Esters to Alcohols

The dehydrogenative coupling reaction of alcohols to form esters is driven by the liberated hydrogen gas, and we thought that it might be possible to reverse it by the application of mild hydrogen pressure. Indeed, complex **5** catalyzes the hydrogenation of esters to alcohols under 5 atm of H_2 [51]. The reaction is general and not limited to activated esters (Table 7). It proceeds under relatively mild, neutral conditions, with no additives being required, and provides a "green," mild pathway for the synthesis of alcohols from esters, without the need for the traditionally

Table 7 Mild hydrogenation of esters to alcohols catalyzed by PNN pincer complex **5**

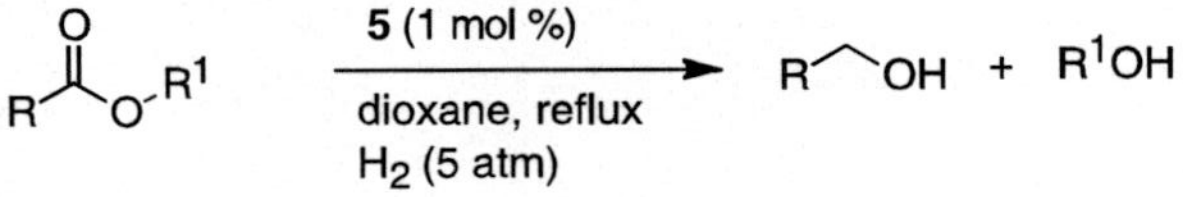

Entry	Ester	Time (h)	Conv. (%)	Alcohols (yields, %)		
				RCH_2OH		R^1OH
1	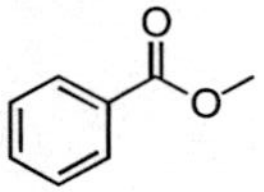	4	100	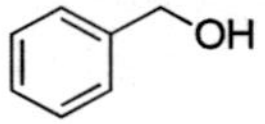	(97)	MeOH (100)
2	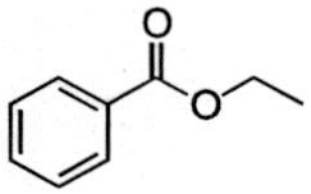	4	99.2	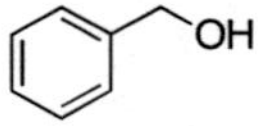	(96)	EtOH (99)
3	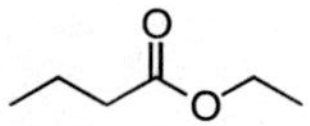	4	100	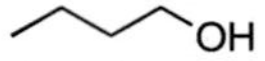	(98)	EtOH (98.6)
4	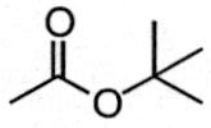	24	10.5	EtOH	(10.5)	^{t}BuOH (10.5)
5	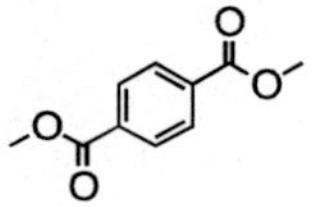	5	100	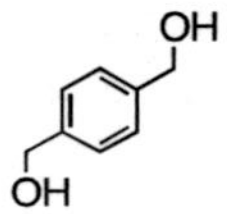	(97)	MeOH (100)
6	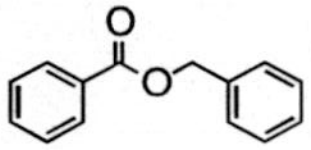	7	98.5	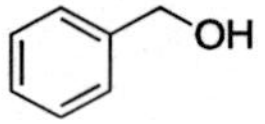	(98)	

(*continued*)

Table 7 *(continued)*

Entry	Ester	Time (h)	Conv. (%)	Alcohols (yields, %)		
				RCH_2OH		R^1OH
7	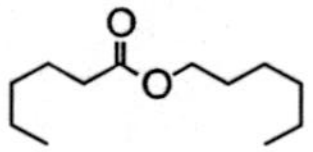	5	82.2	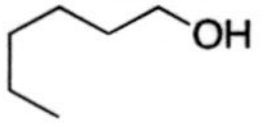	(82.2)	
8	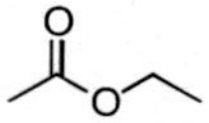	12	86	EtOH	(85.6)	

Conditions: catalyst **5 (0.02 mmol), ester (2 mmol), and dioxane (2 ml) were charged in a 90 ml Fischer–Porter glass pressure tube, and the tube was filled with** H_2 (5 atm). The solution was heated at 115°C (actual solution temperature) with stirring. After cooling to room temperature, conversion of ester and the product alcohols were determined by GC

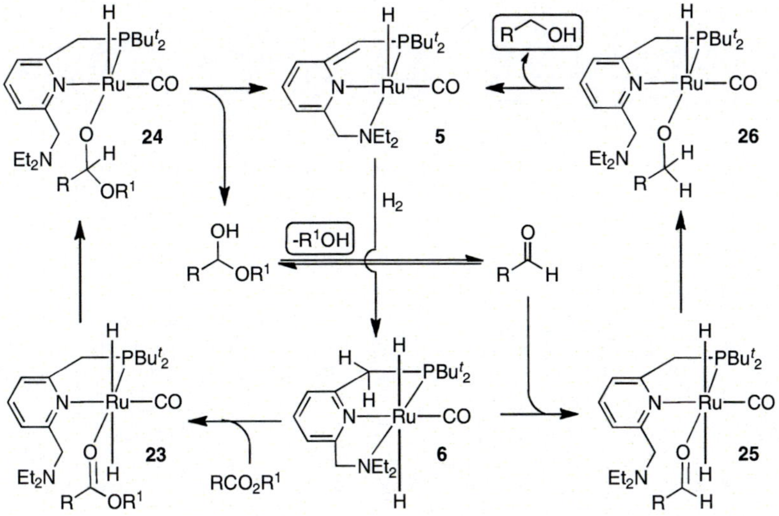

Fig. 13 Mechanism of ester reduction involving metal–ligand cooperation catalyzed by the PNN ruthenium pincer complex **5**

used stoichiometric amounts of hydride reagents, which generate stoichiometric amounts of waste. Activated esters reacted faster than the inactivated esters (Table 7). The reaction is sensitive to steric hindrance of the esters. Hydrogenation of esters, in particular unactivated esters, is still very rare (for a recent review on hydrogenation of polar bonds, including esters, see [80]) [81, 82].

A possible mechanism of the reaction is outlined in Fig. 13. Reaction of 5 with H_2 leads to the *trans*-dihydride complex **6**, which undergoes dissociation of the amine "arm" to provide a site for ester coordination, forming intermediate **23**. Hydride transfer to the ester carbonyl gives intermediate **24**, and subsequent amine arm coordination and dearomatization of the pyridine core regenerate complex **5** and eliminate a hemiacetal, which is in equilibrium with the aldehyde in solution. The aldehyde is then hydrogenated to the corresponding alcohol via a similar cycle involving the intermediates **25** and **26**. Facile ring opening by the hemilabile amine "arm" of complex **5** facilitated the reactions (Fig. 13).

4.2 Catalytic Hydrogenation of Carbon Dioxide to Formate Salts

While we concentrated mainly on the catalytic applications of ruthenium PNP and PNN pincer complexes described above, we have also established the metal–ligand cooperation in C–H and H–H activation reactions by Ir-PNP pincer complexes. The Ir-PNP pincer complex **27**, which was formed by C–H activation of benzene by a dearomatized Ir(I) complex, reacts with H_2 to provide a *trans*-dihydride complex exclusively. Use of D_2 revealed, surprisingly, formation of D–Ir–H in **28**, while one D atom is attached to pyridinylmethylenic carbon (Scheme 9). Further studies

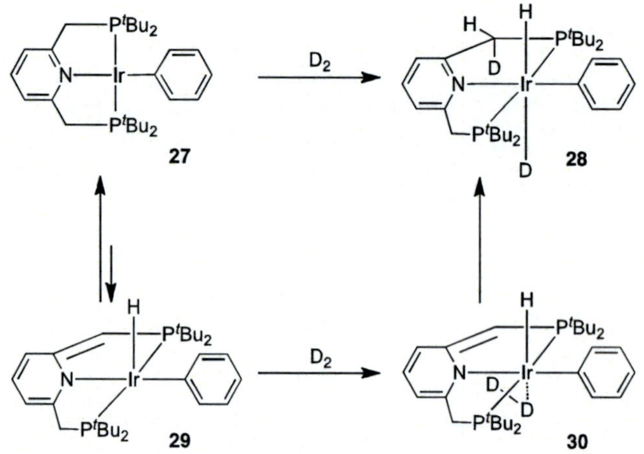

Scheme 9 Reaction of D_2 with the Ir(I) complex **27** actually involves activation by the Ir(III) complex **29**

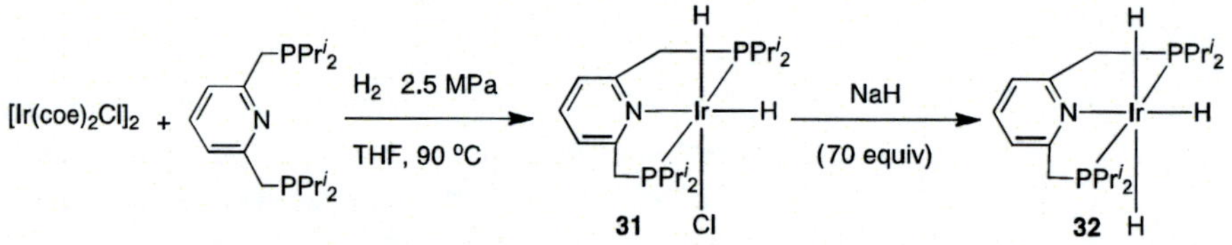

Scheme 10 Synthesis of Nozaki's iridium(III) trihydride PNP pincer complex **32**

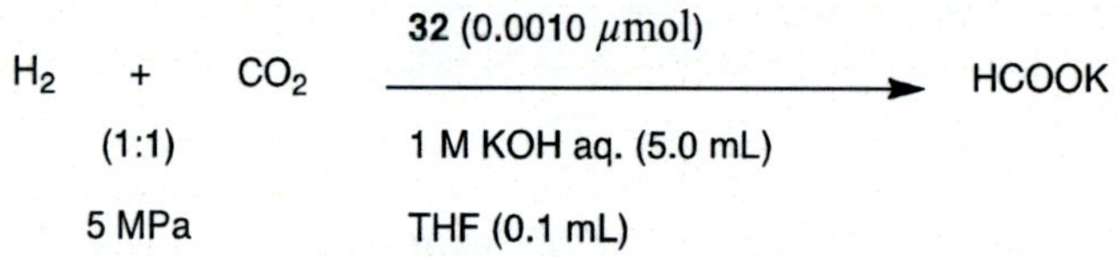

Scheme 11 Hydrogenation of carbon dioxide to formats by iridium(III) trihydride PNP pincer complex **32**

indicated that H_2 is actually activated by the dearomatized Ir(III) complex **29** formed by proton migration from the side arm [56].

Nozaki et al. applied the metal–ligand cooperation of Ir(III) PNP pincer complexes for the highly efficient catalytic hydrogenation of carbon dioxide to formats [83]. The chloroiridium(III) dihydride PNP complex **31** was synthesized by the reaction of $[Ir(coe)_2Cl]_2$ (coe = cyclooctene) with PNP ligand under hydrogen pressure. Reaction of complex **31** with excess of NaH resulted in the formation of stable Ir(III) trihydride PNP pincer complex **32** (Scheme 10).

Complex **32** is a highly active catalyst for the reduction of carbon dioxide to a formate salt by hydrogen and a base (Scheme 11). Heating an aqueous KOH solution of complex **32**, carbon dioxide, and hydrogen resulted in the formation

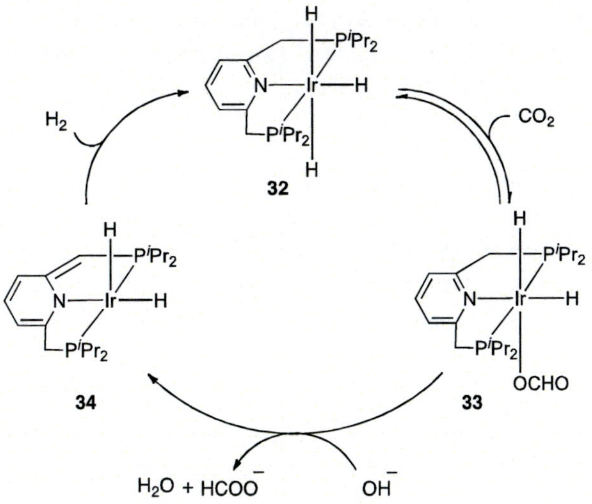

Fig. 14 Proposed catalytic cycle for the hydrogenation of carbon dioxide, facilitated by metal–ligand cooperation of Ir(III) PNP pincer complex **32**

of potassium formate (HCOOK) with excellent TON of 3,500,000 and TOF of 150,000 h^{-1}.

The postulated mechanism for the hydrogenation of CO_2 using the iridium-trihydride complex **32** is shown in Fig. 14. Upon reaction with CO_2, complex **32** generates the intermediate formate complex **33,** which reacts with hydroxide to give the dearomatized amidoiridium dihydride complex **34**. Reaction of **34** with H_2 regenerates complex **32**, closing the catalytic cycle. The dearomatized complex **34** was prepared by the reaction of **32** with $CsOH.H_2O$, which upon reaction with H_2 led to the formation of the iridium-trihydride complex **32**, supporting the proposed catalytic cycle.

The unprecedented high catalytic activity associated with complex **32** in comparison to the other reported catalysts [84–92] for the same transformation is attributed to the stability (as a result of strong coordination between the metal and the PNP pincer ligand) of complex **32** and to metal–ligand cooperation.

5 Conclusion

It is evident that metal–ligand cooperation by pyridine- and acridine-based pincer complexes, involving aromatization–dearomatization of the heteroaromatic core, is a powerful new approach for bond activation and catalyst design. The pincer ligand acts in concert with metal center in bond making and breaking, forming the basis for the design of new chemical reactions. Efficient atom economical and environmentally benign methods were discovered for a range of important products directly from alcohols with the liberation of H_2. While we focused in this chapter on catalytic transformations of alcohols, the new mode of metal–ligand cooperation by aromatization–dearomatization of pincer complexes composed of heteroaromatic ligands has also led to N–H activation of amines and ammonia by the **2**-^{t}Bu-PNP pincer complex [59], and to an unprecedented approach to water splitting involving consecutive thermal liberation of dihydrogen and light-induced liberation of dioxygen promoted by the PNN pincer complex **5** [53]. A related metal–ligand cooperation originating from deprotonation of a nitrogen-containing aliphatic pincer ligand was reported to catalyze ammonia borane decomposition, relevant to hydrogen storage [93]. We believe that future research on metal–ligand cooperation based on pincer complexes of a variety of metals would enhance our fundamental understanding of such fundamental processes, and will provide new opportunities for the design of sustainable catalytic reactions aimed at solving valuable synthetic problems.

Acknowledgments Our research described in this review was supported by the Israel Science Foundation, by the MINERVA Foundation, by the DIP program for German-Israeli Cooperation, by the European Research Council under the FP7 framework (ERC No 246837), and by the Kimmel Center for Molecular design. D.M. is the holder of the Israel Matz Professorial Chair of Organic Chemistry.

References

1. Beller M, Bolm C (eds) (2004) Transition metals for organic synthesis. Wiley-VCH, Weinheim
2. Whittaker JW (2005) Arch Biochem Biophys 433:227
3. Peters JW, Lanzilotta WN, Lemon BJ, Seefeldt LC (1998) Science 282:1853
4. Liu Z-P, Hu P (2002) J Am Chem Soc 124:5175
5. Fan H-J, Hall MBA (2001) J Am Chem Soc 123:3828
6. Greco C, Bruschi M, De Gioia L, Ryde UA (2007) Inorg Chem 46:5911
7. Noyori R, Ohkuma T (2001) Angew Chem Int Ed 40:40
8. Clapham SE, Hadzovic A, Morris RH (2004) Coord Chem Rev 248:2201
9. Ikariya T, Blacker J (2007) Acc Chem Res 40:1300
10. Grutzmacher H (2008) Angew Chem Int Ed 47:1814
11. Ikariya T, Gridnev ID (2010) Top Catal 53:894
12. Yamakawa M, Ito H, Noyori R (2000) J Am Chem Soc 122:1466
13. Noyori R, Koizumi M, Ishii D, Ohkuma T (2001) Pure Appl Chem 73:227
14. Abbel R, Abdur-Rashid K, Faatz M, Hadzovic A, Lough AJ, Morris RHA (2005) J Am Chem Soc 127:1870
15. Casey CP, Bikzhanova GA, Cui Q, Guzei IL (2005) J Am Chem Soc 127:14062
16. Rybtchinski B, Milstein D (1999) Angew Chem Int Ed 38:870
17. Jensen CM (1999) Chem Commun:2443
18. Albrecht M, van Koten G (2001) Angew Chem Int Ed 40:3750
19. Vigalok A, Milstein D (2001) Acc Chem Res 34:798
20. van der Boom ME, Milstein D (2003) Chem Rev 103:1759
21. Singleton JT (2003) Tetrahedron 259:1837
22. Milstein D (2003) Pure Appl Chem 75:445
23. Rybtchinski B, Milstein D (2004) ACS Symp Ser 885:70
24. Szabo KJ (2006) Synlett:811
25. Rybtchiski B, Milstein D (2007) In: Morales-Morales D, Jensen CM (eds) The chemistry of pincer compounds. Elsevier BV, Oxford, pp 87–105, Chapter 5
26. Milstein D (2010) Top Catal 53:915
27. Kawatsura M, Hartwig JF (2001) Organometallics 20:1960

28. Stambuli JP, Stauffer SR, Shaughnessy KH, Hartwig JF (2001) J Am Chem Soc 123:2677
29. Gibson DH, Pariya C, Mashuta MS (2004) Organometallics 223:2510
30. Kloek SM, Heinekey DM, Goldberg KI (2006) Organometallics 25:3007
31. Kloek SM, Heinekey DM, Goldberg KI (2007) Angew Chem Int Ed 46:4736
32. Hanson SK, Heinekey DM, Goldberg KI (2008) Organometallics 27:1454
33. Pelczar EM, Emge TJ, Krogh-Jespersen K, Goldman AS (2008) Organometallics 27:5759
34. van der Vlugt JI, Pidko EA, Vogt D, Lutz M, Spek AL, Meetsma A (2008) Inorg Chem 47: 4442
35. Tanaka R, Yamashita M, Nazaki K (2009) J Am Chem Soc 131:14168
36. van der Vlugt JI, Siegler MA, Janssen M, Spek AL (2009) Organometallics 28:7025
37. van der Vlugt JI, Reek JNH (2009) Angew Chem Int Ed 48:8832
38. Hermann D, Gandelman M, Rozenberg H, Shimon LJW, Milstein D (2002) Organometallics 21:812
39. Ben-Ari E, Gandelman M, Rozenberg H, Shimon LJW, Milstein D (2003) J Am Chem Soc 125:4714
40. Zhang J, Gandelman M, Shimon LJW, Rozenberg H, Milstein D (2004) Organometallics 23: 4026
41. Ben-Ari E, Cohen R, Gandelman M, Shimon LJW, Martin JML, Milstein D (2006) Organometallics 25:3190
42. Zhang J, Gandelman M, Herrman D, Leitus G, Shimon LJW, Ben-David Y, Milstein D (2006) Inorg Chim Acta 359:1955
43. Feller M, Karton A, Leitus G, Martin JML, Milstein D (2006) J Am Chem Soc 128:12400
44. Feller M, Ben-Ari E, Gupta T, Shimon LJW, Leitus G, Diskin-Posner Y, Weiner L, Milstein D (2007) Inorg Chem 46:10479
45. Zhang J, Gandelman M, Shimon LJW, Milstein D (2007) Dalton Trans:107
46. Zhang J, Gandelman M, Shimon LJW, Milstein D (2008) Inorg Chem 27:3526
47. Feller M, Iron MA, Shimon LJW, Diskin-Posner Y, Leitus G, Milstein D (2008) J Am Chem Soc 130:14374
48. Zang J, Gandelman M, Shimon LJW, Milstein D (2008) Organometallics 27:3526
49. Feller M, Ben-Ari E, Iron MA, Diskin-Posner Y, Leitus G, Shimon LJW, Milstein D (2010) Inorg Chem 49:1615
50. Zhang J, Leitus G, Ben-David Y, Milstein D (2005) J Am Chem Soc 127:12429
51. Zhang J, Leitus G, Ben-David Y, Milstein D (2006) Angew Chem Int Ed 45:1113
52. Gunanathan C, Ben-David Y, Milstein D (2007) Science 317:790
53. Kohl SW, Weiner L, Schwartsburd L, Konstantinovski L, Shimon LJW, Ben-David Y, Iron MA, Milstein D (2009) Science 324:74
54. Li J, Shiota Y, Yoshizawa K (2009) J Am Chem Soc 131:13584
55. Yang X, Hall MB (2010) J Am Chem Soc 132:120
56. Ben-Ari E, Leitus G, Shimon LJW, Milstein D (2006) J Am Chem Soc 128:15390
57. Iron MA, Ben-Ari E, Cohen R, Milstein D (2009) Dalton Trans:9433
58. Zeng G, Guo Y, Li S (2009) Inorg Chem 48:10257
59. Khaskin E, Iron MA, Shimon LJW, Zhang J, Milstein D (2010) J Am Chem Soc 132:8542
60. Vuzman D, Povernov E, Shimon LJW, Diskin-Posner Y, Milstein D (2008) Organometallics 27:2627
61. Blum Y, Shvo Y (1985) J Organomet Chem 282:C7
62. Murahashi S-I, Naota T, Ito K, Maeda Y, Taki H (1987) J Org Chem 52:4319
63. Watson AJA, Williams JMJ (2010) Science 329:635 (and references cited therein)
64. Zweifel T, Naubron J-V, Grutzmacher H (2009) Angew Chem Int Ed 48:559
65. Watson AJA, Maxwell AC, Williams JMJ (2009) Org Lett 11:2667
66. Nordstrom LU, Vogt H, Madsen R (2008) J Am Chem Soc 130:17672
67. Ghosh SC, Senthilkumar M, Zhang Y, Xu XY, Hong SH (2009) Adv Synth Catal 351:2643
68. Shimizu K, Ohshima K, Satsuma A (2009) Chem Eur J 15:9977
69. Zhang J, Senthilkumar M, Ghosh SC, Hong SH (2010) Angew Chem Int Ed 49:6391

70. Dam JH, Osztrovszky G, Nordstrøm LU, Madsen R (2010) Chem Eur J 16:6280
71. Zhang Y, Chen C, Ghosh SC, Li Y, Hong SH (2010) Organometallics 29:1374
72. Gnanaprakasam B, Zhang J, Milstein D (2010) Angew Chem Int Ed 48:1468
73. Gunanathan C, Milstein D (2008) Angew Chem Int Ed 47:8661
74. Gunanathan C, Shimon LJW, Milstein D (2009) J Am Chem Soc 131:3146
75. Hillebrand S, Barkowska B, Bruckmann J, Kruger C, Haenel MW (1998) Tetrahedron Lett 39:813
76. Ito T, Horino H, Koshiro Y, Yamamoto A (1982) Bull Chem Soc Jpn 55:504
77. Narayan S, Muldoon J, Finn MG, Fokin VV, Kolb HC, Sharpless KB (2005) Angew Chem Int Ed 44:3275
78. Shapiro N, Vigalok A (2008) Angew Chem Int Ed 47:2849
79. Gunanathan C, Gnanaprakasam B, Iron MA, Shimon LJW, Milstein D (2010) J Am Chem Soc 132:14763
80. Ito M, Ikariya T (2007) Chem Commun:5134
81. Saudan LA, Saudan CM, Debieux C, Wyss P (2007) Angew Chem Int Ed 46:7473
82. Kuriyama W, Ino Y, Ogata O, Sayo N, Saito T (2010) Adv Synth Catal 352:92
83. Tanaka R, Yamashita M, Nozaki K (2009) J Am Chem Soc 131:14168
84. Ezhova NN, Kolesnichenko NV, Bulygin AV, Slivin-skii EV, Han S (2002) Russ Chem Bull Int Ed 51:2165
85. Graf E, Leitner W (1992) J Chem Soc Chem Commun:623
86. Gassner F, Leitner W (1993) J Chem Soc Chem Commun:1465
87. Angermund K, Baumann W, Dinjus E, Fornika R, Görls H, Kessler M, Krüger C, Leitner W, Lutz F (1997) Chem Eur J 3:755
88. Jessop PG, Ikariya T, Noyori R (1994) Nature 368:231
89. Jessop PG, Hsiao Y, Ikariya T, Noyori R (1996) J Am Chem Soc 118:344
90. Urakawa A, Jutz F, Laurenczy G, Baiker A (2007) Chem Eur J 13:3886
91. Munshi P, Main AD, Linehan JC, Tai CC, Jessop PG (2002) J Am Chem Soc 124:7963
92. Himeda Y, Onozawa-Komatsuzaki N, Sugihara H, Kasuga K (2007) Organometallics 26:702
93. Käß M, Friedrich A, Drees M, Schneider S (2009) Angew Chem Int Ed 48:905

Top Organomet Chem (2011) 37: 85–125
DOI: 10.1007/3418_2011_7

Published online: 21 April 2011

Shvo's Catalyst in Hydrogen Transfer Reactions

Madeleine C. Warner, Charles P. Casey, and Jan-E. Bäckvall

Abstract This chapter reviews the use of Shvo's catalyst in various hydrogen transfer reactions and also discusses the mechanism of the hydrogen transfer. The Shvo catalyst is very mild to use since no activation by base is required in the transfer hydrogenation of ketones or imines or in the transfer dehydrogenation of alcohols and amines. The Shvo catalyst has also been used as an efficient racemization catalyst for alcohols and amines. Many applications of the racemization reaction are found in the combination with enzymatic resolution leading to a dynamic kinetic resolution (DKR). In these dynamic resolutions, the yield based on the starting material can theoretically reach 100%. The mechanism of the hydrogen transfer from the Shvo catalyst to ketones (aldehydes) and imines as well as the dehydrogenation of alcohols and amines has been studied in detail over the past decade. It has been found that for ketones (aldehydes) and alcohols, there is a concerted transfer of the two hydrogens involved, whereas for typical amines and imines, there is a stepwise transfer of the two hydrogens. One important question is whether the substrate is coordinated to the metal or not in the hydrogen transfer step(s). The pathway involving coordination to activate the substrate is called the inner-sphere mechanism, whereas transfer of hydrogen without coordination is called the outer-sphere mechanism. These mechanistic proposals together with experimental and theoretical studies are discussed.

M.C. Warner and J.-E. Bäckvall (✉)
Department of Organic Chemistry, Arrhenius Laboratory, Stockholm University, SE-100 44 Stockholm, Sweden
e-mail: jeb@organ.su.se

C.P. Casey (✉)
Department of Chemistry, University of Wisconsin–Madison, 1101 University Avenue, Madison, WI, USA
e-mail: casey@chem.wisc.edu

Keywords Dynamic kinetic asymmetric transformation (DYKAT) · Dynamic kinetic resolution (DKR) · Hydrogenation · Imine reduction · Ketone reduction · Mechanism of carbonyl reduction · Mechanism of imine reduction · Mechanism of dihydrogen activation · Ruthenium catalysis · Shvo's catalyst · Transfer hydrogenation

Contents

1 Introduction

Shvo's catalyst **1** is a cyclopentadienone-ligated diruthenium complex, $[Ru_2(CO)_4(\mu\text{-}H)(C_4Ph_4COHOCC_4Ph_4)]$. It was first synthesized in 1984 by Shvo et al. [1, 2]. Since then it has been widely applied in various hydrogen transfer reactions, including hydrogenation of carbonyl compounds [2, 3], transfer hydrogenation of ketones and imines [4, 5], disproportion of aldehydes to esters [6], and Oppenauer-type oxidations of alcohols [7–9] and amines [10–12]. Shvo's complex **1** has also been found to be effective as a racemization catalyst for secondary alcohols and amines, and complex **1** has therefore been used together with enzymes in several dynamic kinetic resolution (DKR) protocols [13–18].

Shvo's diruthenium complex **1** is activated by heat and dissociates into the two catalytically active monoruthenium complexes: an isolable 18 electron complex **2** and a proposed highly reactive 16 electron complex (**A**) that has never been observed directly (Scheme 1). The 18 electron species acts as a hydrogenation catalyst, whereas the 16 electron species acts as a dehydrogenation catalyst [19, 20]. The two catalytically active complexes interconvert during racemization and hydrogen transfer (Scheme 1).

Shvo's hydride catalyst **2** was the first reported example of a ligand-metal bifunctional catalyst. Noyori introduced this nomenclature for catalysts that transfer a hydride from a metal center and a proton from a ligand [21, 22]. In the case of catalyst **2**, simultaneous transfer of hydride from the metal center

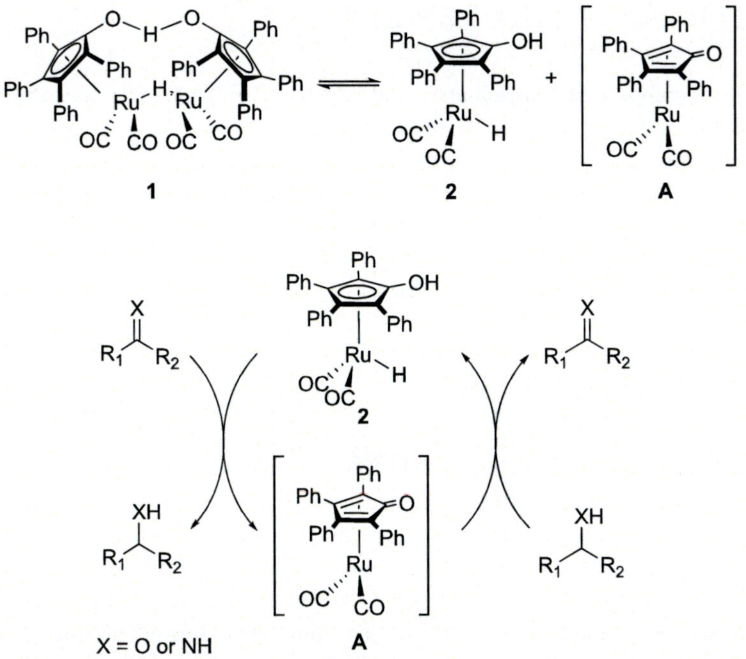

Scheme 1 Thermal activation of Shvo's catalyst **1** into monomers **2** and **A**, and their mechanism of racemization via hydrogenation/dehydrogenation

and a proton from the hydroxycyclopentadienyl ligand is proposed to occur in the hydrogenation of ketones; this process is reversible and accounts for transfer hydrogenation.

The mechanism for hydrogen transfer by complex **1** has been extensively studied, and it has been found that the mechanism for imines/amines differs from that for alcohols/ketones. The hydrogenation of ketones and the dehydrogenation of alcohols have been proposed to operate through a concerted outer-sphere mechanism, i.e., the proton and the hydride are transferred to the substrate simultaneously [19, 23]. In contrast, it is thought that the hydrogenation of imines and the dehydrogenation of amines proceed in a stepwise manner, i.e., protonation of the substrate precedes the hydride transfer [24]. These mechanistic insights are supported by studies on the kinetic isotope effects of each reaction [20, 23, 24]. The hydrogenation of imines and the dehydrogenation of amines have been studied in more detail and some data support an inner-sphere mechanism [25–27], whereas other data support an outer-sphere mechanism (see Sect. 4.3) [28, 29].

The aim of this chapter is to provide an overview of the mechanistic aspects and the applications of the Shvo catalyst **1** in hydrogen transfer reactions. Several reviews have recently appeared on the applications of the Shvo catalyst, and this topic was reviewed in 2005 [30], 2009 [31], and 2010 [32]. Mechanistic studies on the Shvo catalyst in hydrogen transfer reactions were reviewed in 2006 [33], 2009 [34] and 2010 [32].

2 Hydrogen Transfer Reactions

The Shvo catalyst **1** can participate in the transfer of hydrogen from one molecule to another. Such hydrogen transfer reactions are useful in synthetic organic chemistry for the reduction of ketones (aldehydes) and imines, and for the oxidation of alcohols and amines. In the former case (transfer hydrogenation), a hydrogen donor such as isopropanol or formic acid is used, which reduces the carbonyl compound or imine to alcohol or amine, respectively. In the oxidation of alcohols and amines (transfer dehydrogenation), a hydrogen acceptor such as acetone or a quinone is used.

2.1 Transfer Hydrogenation

In transfer hydrogenation, the Shvo hydride **2** delivers hydrogen to the substrate, which can be a carbonyl compound, an imine, or a C–C multiple bond. After transfer of hydrogen from bifunctional catalyst **2** (as H^- from Ru and H^+ from oxygen) to the substrate to give the reduced product, the unstable 16-electron complex **A** is formed. The latter complex now reacts with the hydrogen donor, e.g., isopropanol or formic acid and abstracts two hydrogen atoms. This regenerates the Shvo hydride **2**. An attractive feature with the Shvo catalyst **1** is that no base is required for its activation.

2.1.1 Transfer Hydrogenation of Carbonyl Compounds

In 1996, Shvo and coworkers reported a catalytic transfer hydrogenation of ketones and aldehydes with formic acid and catalyst **1** (1) [35]. The reaction was carried out at 100°C and gave turnover numbers of up to 8,000. Low catalytic loading (0.013–0.03 mol%) was used and high selectivity was obtained in the transfer hydrogenation, which was found to be more efficient than the corresponding hydrogenation (H_2) using catalyst **1**, reported previously by the same group [6].

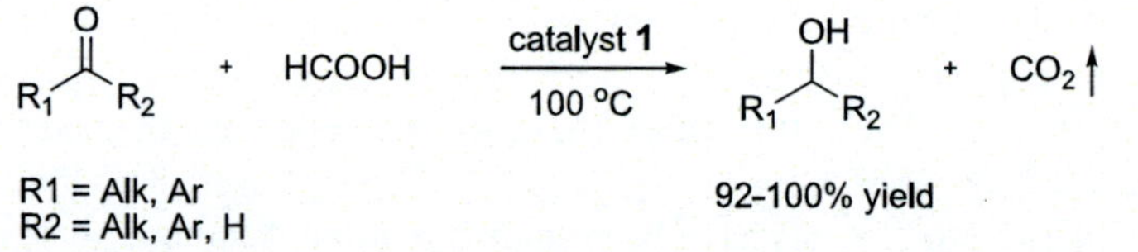

(1)

The Shvo catalyst **1** was successfully used in the transfer hydrogenation of 1,3-diones to the corresponding 1,3-diols with isopropanol as the hydrogen donor (2) [36, 37]. This reaction is synthetically useful for the reduction of cyclic diones since reduction of these diketones by $LiAlH_4$ preferentially gives the allylic alcohol [36]. Also piperidine-3,5-diones were efficiently reduced to the corresponding diols by isopropanol using **1** as catalyst [37], and these diols were subsequently used in dynamic kinetic asymmetric transformations (DYKATs) to provide stereodefined 3,5-disubstituted piperidines [36, 37].

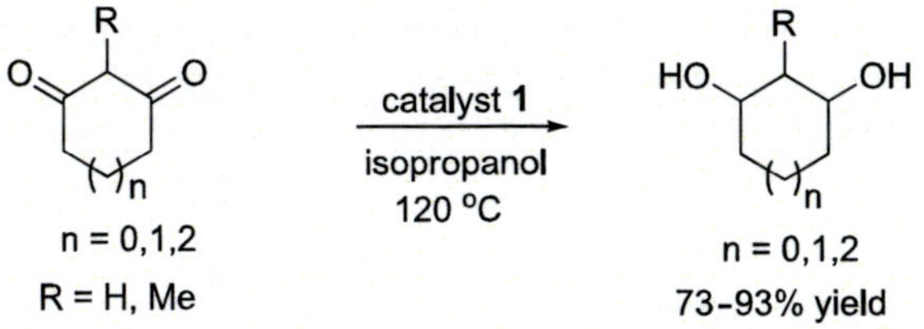

(2)

The hydrogen transfer from the tolyl variant of **2** to benzaldehyde was studied by Casey and coworkers [19]. Selective deuterium labeling of this hydride showed that the transfer of the two hydrogens (Ru–H, O–H) to the aldehyde is concerted (see Sect. 4.3.1).

2.1.2 Transfer Hydrogenation of Imines

Transfer hydrogenation of imines to amines by isopropanol using $RuCl_2(PPh_3)_3$ in the presence of base (K_2CO_3) has been reported to give amines in high yields and selectivity [38]. In this reaction, aldimines reacted faster than ketimines. In 2002, a transfer hydrogenation of imines by isopropanol with 0.3–1.0 mol% of catalyst **1** was reported (3) [4]. Interestingly, in contrast to the previous [38] ruthenium-catalyzed transfer hydrogenation of imines, ketimines reacted faster than aldimines with **1** as the catalyst [4].

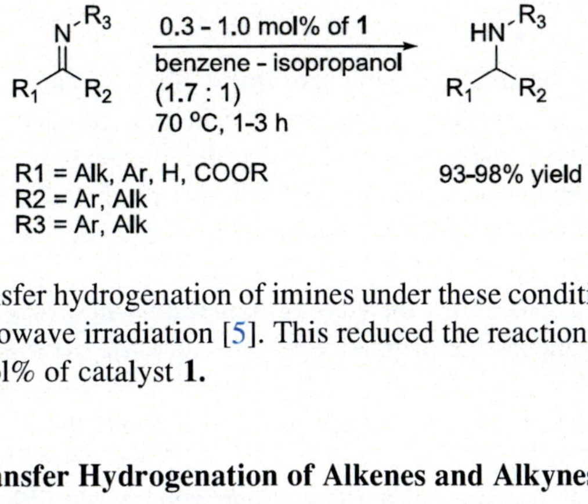

(3)

The transfer hydrogenation of imines under these conditions was also carried out under microwave irradiation [5]. This reduced the reaction time to 10–20 min with 0.1–0.5 mol% of catalyst **1.**

2.1.3 Transfer Hydrogenation of Alkenes and Alkynes

Transfer hydrogenation of carbon–carbon multiple bonds such as alkenes is rare with the Shvo catalyst **1**. Only in cases where the carbon–carbon multiple bond is polarized by an electron-withdrawing group has transfer hydrogenation catalyzed by **1** been observed. Thus, methyl vinyl ketone was selectively transfer hydrogenated into 2-butanone in high yield using formic acid as hydrogen donor and **1** as catalyst (4) [35].

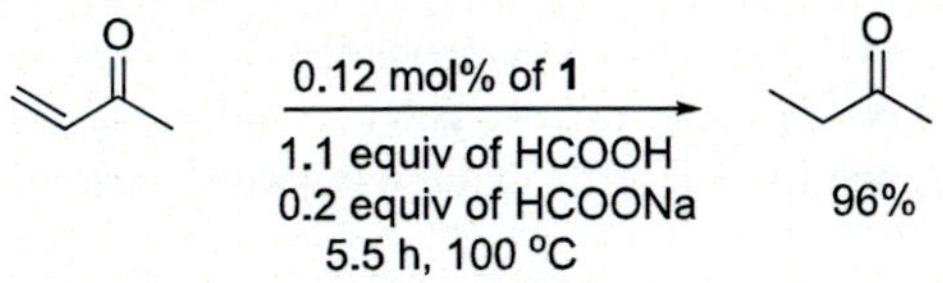

(4)

However, there are reports on the hydrogenation of alkenes and alkynes using the Shvo catalyst **1** [6, 30, 31, 39]. The reactions were carried out at a hydrogen pressure of 500 psi at 145°C.

2.2 *Transfer Dehydrogenation*

The Shvo catalyst has successfully been used in transfer dehydrogenation of alcohols and amines. In these reactions, the crucial step is dehydrogenation of an alcohol to a ketone or dehydrogenation of an amine to an imine by the 16-electron complex **A**. The dehydrogenation step is thought to be preceded by coordination of the substrate to the ruthenium, which brings the substrate in proximity to the catalyst. In this process, there is also a hydrogen bonding between the OH (NH) and the keto group of the catalyst. The alcohol complex is not stable, whereas many amine complexes have been isolated. There are two principal pathways proposed for dehydrogenation of these substrates, which will be discussed in Sect. 4.3. In the inner-sphere pathway a β-elimination would occur, and in the outer-sphere pathway decoordination of the substrate followed by transfer of the two hydrogens from the substrate to the catalyst would occur while the substrate is held in proximity to the catalyst by a hydrogen bond.

2.2.1 Oxidation of Alcohols

Secondary alcohols can be oxidized with the aid of Shvo's catalyst **1**. In these reactions, the 16 electron complex **A** is dehydrogenating the alcohol and **A** is converted to the hydride complex **2**. In 1994, it was demonstrated that a benzoquinone can be used as hydrogen acceptor for the hydride **2** produced in the dehydrogenation of the alcohol [40, 41]. The hydroquinone produced was in turn oxidized by MnO_2 [40] or by air (with the aid of a metal macrocycle) [41, 42].

An efficient and mild procedure for alcohol oxidation with Shvo's catalyst **1**, and with the use of acetone as hydrogen acceptor and solvent, was reported to give good to high yields of ketones [43]. This method was subsequently applied to the oxidation of 3β steroidal alcohols (5-en-3β-ols) to the corresponding rearranged enones (4-en-3-ones) [44].

The mechanism of the catalytic transfer dehydrogenation of alcohols by the Shvo catalyst **1** was studied by Johnson and Bäckvall [23] using tetrafluorobenzoquinone as the oxidant (Scheme 2). It was found that dehydrogenation by **A** is the slow step in the catalytic oxidation of *p*-$FC_6H_4CH(OH)Me$, and there was a zero-order dependence on the quinone. The use of specifically deuterated derivatives *p*-$FC_6H_4CD(OH)Me$ and $FC_6H_4CH(OD)Me$ gave individual kinetic isotope effects on both C–H ($k_H/k_D = 2.57$) and O–H ($k_H/k_D = 1.87$). Furthermore, dideuterated derivative $FC_6H_4CD(OD)Me$ gave a combined kinetic isotope effect k_H/k_D of 4.61. It was concluded that these isotope effects are consistent with a concerted transfer of hydrogens [23]. It was shown by Williams that even with the use of a mild oxidant

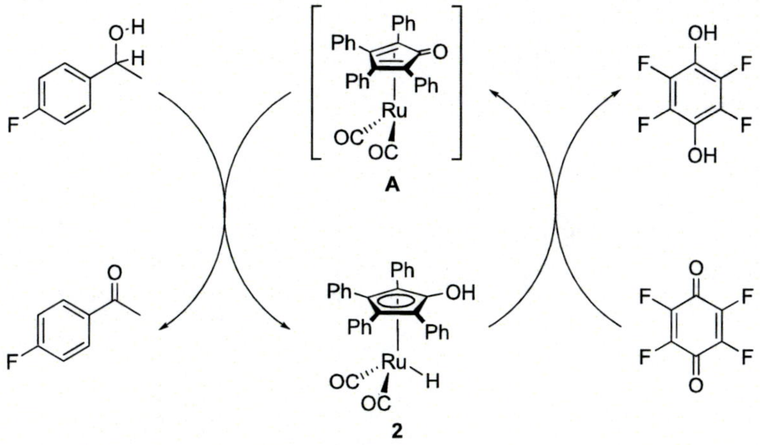

Scheme 2

like acetone in place of tetrafluorobenzoquinone in this catalytic alcohol oxidation, the reoxidation of **2** to **A** is still not rate determining [45].

In a recent application, the Shvo catalyst **1** was used to oxidize enantiomerically pure bicyclic diols (*R*,*R*)-**3** to the hydroxyketones (*R*)-**4** in acetone [46]. The acetone acts as the hydrogen acceptor and, interestingly, under these conditions, it was possible to stop the reaction after oxidation of one of the hydroxyl groups to give the hydroxyketones **4** in 82–85% yield and 97–98% *ee* (5). The hydroxyketones obtained are useful starting materials for the synthesis of natural products and biologically active compounds, and (*R*)-**4b** was used for the synthesis of Sertraline [46].

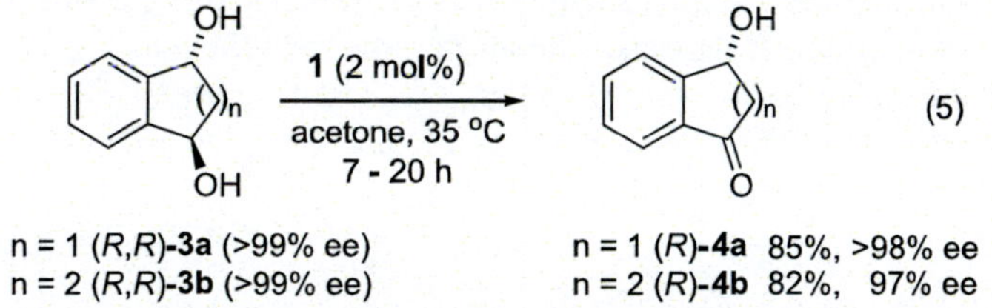

The temperature used in the oxidation of the diols **3**, 35°C, is probably the lowest temperature reported for a hydrogen transfer reaction with the Shvo catalyst.

The aerobic oxidation of secondary alcohols using Shvo's catalyst **1** was recently combined with an efficient hybrid electron transfer mediator **5** (6) [47, 48]. This leads to a facile aerobic oxidation via transfer dehydrogenation, whereas the previous system (cf. electron transfer system of Scheme 3) [41, 42] with separate quinone and metal macrocycle now has been modified by tethering the quinone to the metal macrocycle (Scheme 3).

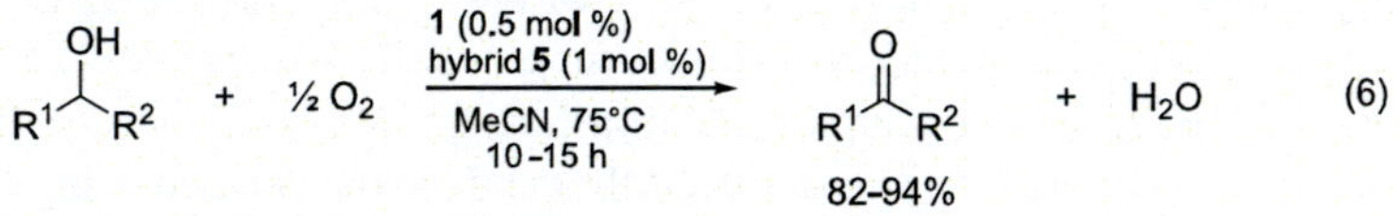

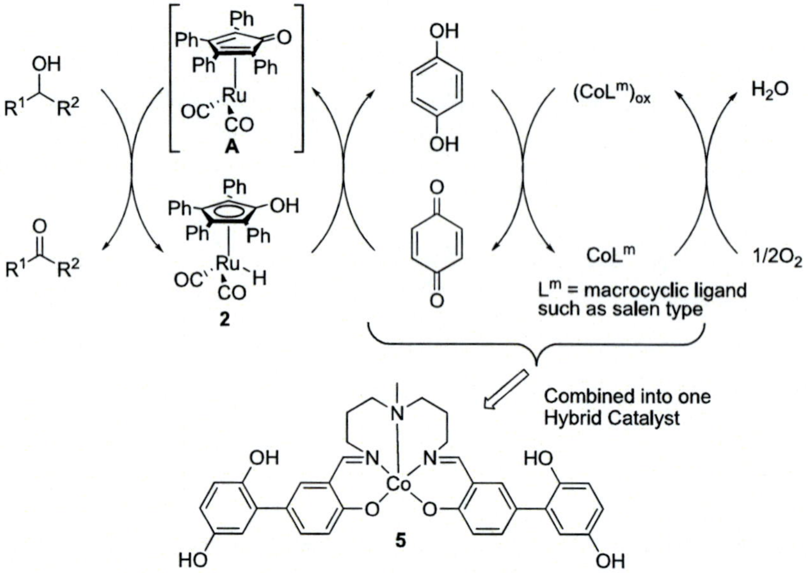

Scheme 3

The isotope effect of the oxidation in (6) was studied using 1-deutereio-1-phenylethanol. The isotope effect determined from separate reactions of deuterated and non-deuterated substrate gave an isotope effect of 2.83 ± 0.30. The deuterium isotope effect determined from a competitive experiment with a 1:1 mixture of deuterated and non-deuterated substrate gave the same isotope effect within experimental error, 2.89 ± 0.40. This shows that there is no strong coordination of the alcohol to ruthenium. It was also shown that the dehydrogenation of the alcohol by **A** is irreversible by the use of an enantiomerically pure alcohol; oxidation of (*S*)-1-phenylethanol showed that the alcohol stays enantiomerically pure throughout the reaction. All data are consistent with a rate-determining dehydrogenation.

2.2.2 Oxidation of Amines

Amines can be oxidized with the aid of the Shvo catalyst **1**, in an analogous way as secondary alcohols are oxidized to ketones. The amine is coordinated to the 16-electron complex **A**, and a subsequent dehydrogenation via either an outer-sphere pathway or an inner-sphere pathway gives the imine and hydride complex **2**. A range of secondary benzylic amines were oxidized to imines via transfer dehydrogenation with Shvo's catalyst **1** (2 mol%) using 2,6-dimethoxybenzoquinone (DMBQ) or catalytic DMBQ/MnO_2 as oxidant [10]. The reaction was run in refluxing toluene for 2–6 h and gave good to high yields of imines (70–95%). The dehydrogenation catalyzed by **1** was subsequently extended to an aerobic oxidation of imines using a coupled electron transfer system similar to that used in Scheme 3 [12]. The quinone used was DMBQ and the oxygen-activating complex was a cobalt(salen)-type complex. The aerobic oxidation was run in toluene at 110°C (7) and afforded imines in high yields (84–99%).

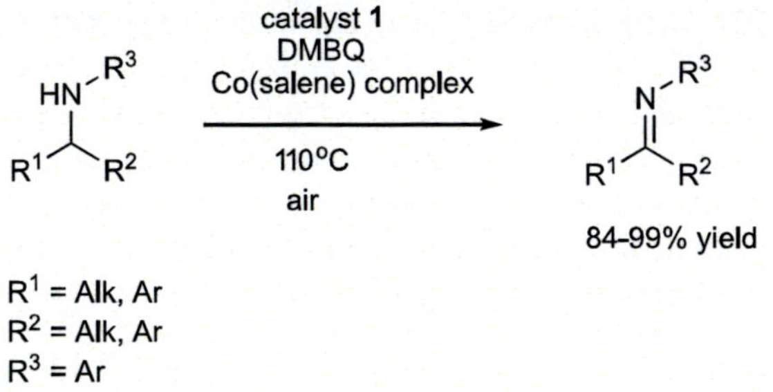

(7)

A hydrogen transfer-based oxidative approach to benzoxazoles starting from a primary amine and an *o*-aminophenol was developed with Shvo's complex **1** as catalyst (8) [49]. The oxidant and hydrogen acceptor used was DMBQ.

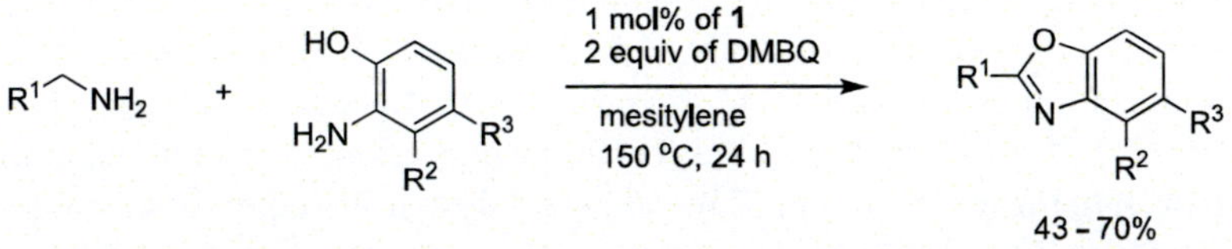

(8)

The reaction proceeds via a ruthenium-catalyzed transfer dehydrogenation of the amine to imine followed by condensation of the imine with the aminophenol to give a 1,2-dihydrobenzoxazole. Subsequent dehydrogenation of the latter compound with Shvo's catalyst **1** and DMBQ gives the benzoxazole (Scheme 4).

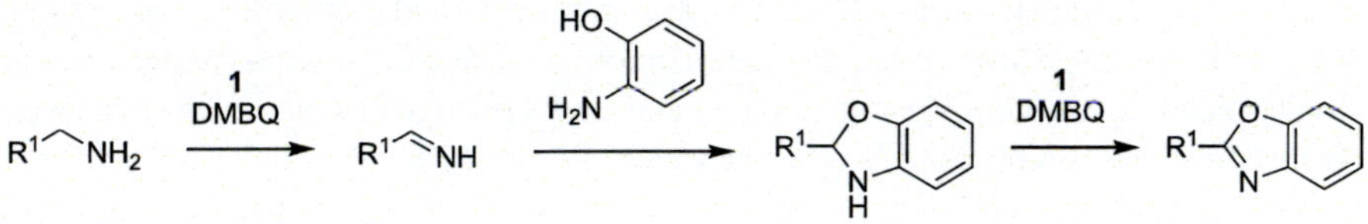

Scheme 4

A similar dehydrogenation of an amine (primary or secondary amine) to imine with catalyst **1** was used by Beller to alkylate indoles (9) [50]. The imine reacts with the indole to give an aminoalkylated intermediate, which undergoes elimination of ammonia or amine to give an oxidized indole intermediate. The hydrogens removed from the amine are now added back to the latter intermediate to give the indole product; so in this case the overall reaction is not an oxidation.

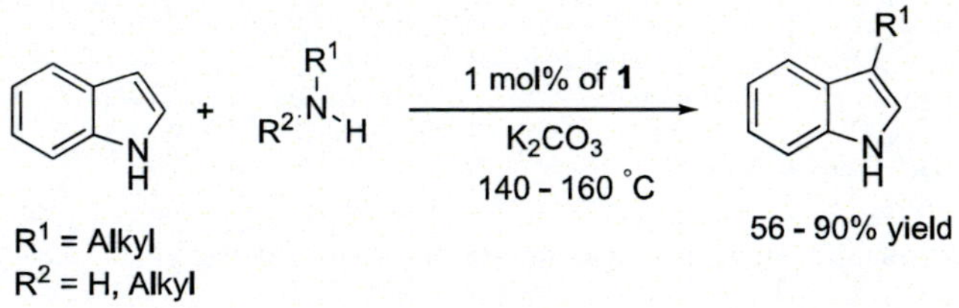

(9)

2.3 *Racemization Reactions*

Racemization is defined as the interconversion of enantiomers to give a 1:1 mixture of enantiomers (called a racemate). Complete racemization always leads to total loss of optical activity. Often when racemization is mentioned in the literature, it is with the idea of avoiding it. However, racemization can also be essential for the preparation of enantiomerically enriched compounds [51]. Hydrogen transfer can be used as an efficient method for racemization. Shvo's complex **1** is widely used as a racemization catalyst for both alcohols and amines (Fig. 1).

2.3.1 Racemization of Alcohols

Shvo's complex **1** catalyzes the racemization of stereocenters α to the OH group. Racemization proceeds via hydrogen transfer reactions. Dehydrogenation of the enantiomerically pure alcohol by complex (**A**) produces a ketone (Fig. 1). Subsequent re-addition of the hydride by **2** to either side of the prochiral ketone leads to racemization of the alcohol substrate (Fig. 1, cf. Scheme 1, X=O).

Shvo's catalyst **1** was found effective for the racemization of enantiomerically pure (*R*)-1-phenylethanol [13]. With 2 mol% of catalyst and 1 equiv of acetophenone in *t*-BuOH at 70°C, the substrate was completely racemized within 45 h.

2.3.2 Racemization of Amines

Shvo's complex **1** also catalyzes the racemization of stereocenters α to NHR″ groups (Fig. 1, cf. Scheme 1, X=NH). In 2002, Bäckvall and coworkers reported on an efficient ruthenium-catalyzed racemization of aromatic amines under transfer hydrogenation conditions using Shvo's catalyst **1** [18]. At first the method suffered from considerable amounts of by-product formation, but addition of a hydrogen donor, 2,4-dimethyl-3-pentanol, was found to suppress the formation of these side products. The presence of electron-donating substituents on the aromatic ring of benzylic amines was found to increase the rate of racemization, whereas electron-withdrawing substituents led to a lower racemization rate. The racemization process was also found to be more efficient at higher temperature, 110°C being the best choice. Interestingly, both primary and secondary amines were efficiently racemized under these conditions.

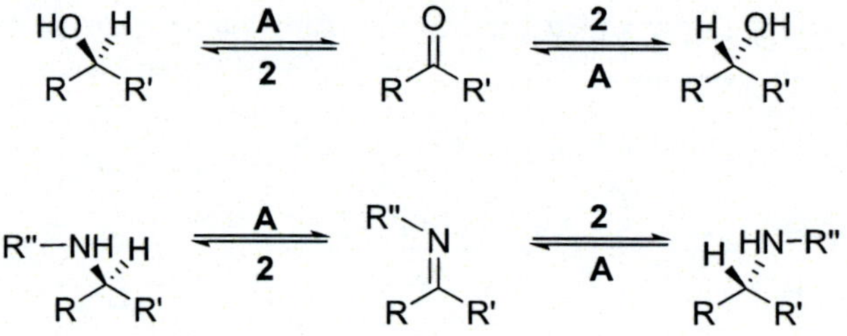

Fig. 1 Principle of racemization via hydrogen transfer by Shvo's catalyst **1**

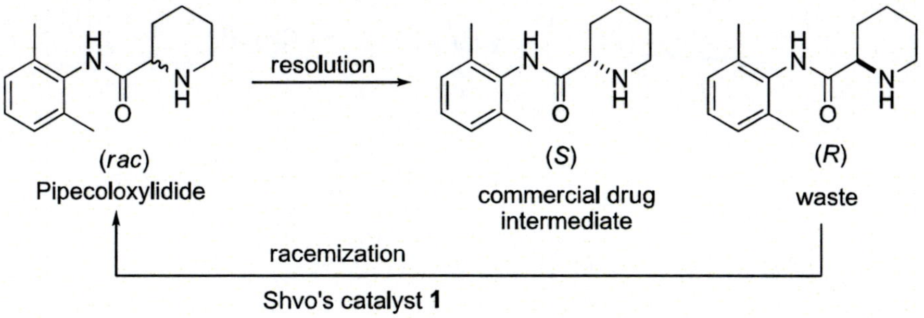

Scheme 5

More recently, a procedure for the racemization of pipecoloxylidide was developed by Bäckvall and coworkers [52]. Pipecoloxylidide is a pharmaceutical precursor of several local anesthetics of the amino amide group type. The (*S*)-enantiomer of these drugs is the desired enantiomer since it shows high activity and is less toxic. It is therefore highly desirable to recycle the undesired (*R*)-enantiomer to increase the overall efficiency of the process (Scheme 5). It was found that with 2 mol% of catalyst **1** in dibutyl ether at 140°C, enantiomerically pure pipecoloxylidide was almost completely racemized after 1 h (2% *ee*).

3 Mechanism of Hydrogen Transfer

There are two ways to think of mechanisms of catalytic reactions: a molecular view that involves determining the detailed mechanism of individual steps in a process, and a more empirical view that defines the kinetic mechanism of the catalytic process. Both approaches have been applied to the Shvo hydrogenation catalyst system. Shvo's initially proposed overall mechanism (Scheme 1) involves three processes: (*i*) reversible dissociation of the bridging hydride **1** to give the active reducing agent monoruthenium hydride **2** and an unobserved reactive intermediate **A**, (*ii*) ketone reduction by **2** which generates alcohol and reactive intermediate **A**, and (*iii*) H_2 reaction with **A** to regenerate **2**. Many subsequent studies have supported the basic outlines of Shvo's original proposal. We will first discuss efforts to determine the detailed molecular mechanism of individual steps in the Shvo mechanism and then return to discussion of the overall kinetics of the catalytic process.

3.1 Mechanism of Carbonyl Reduction by 2 (and Alcohol Oxidation by A)

The diruthenium bridging hydride **1** is unreactive toward aldehydes and ketones at room temperature and above and is not the active reducing agent in the Shvo system

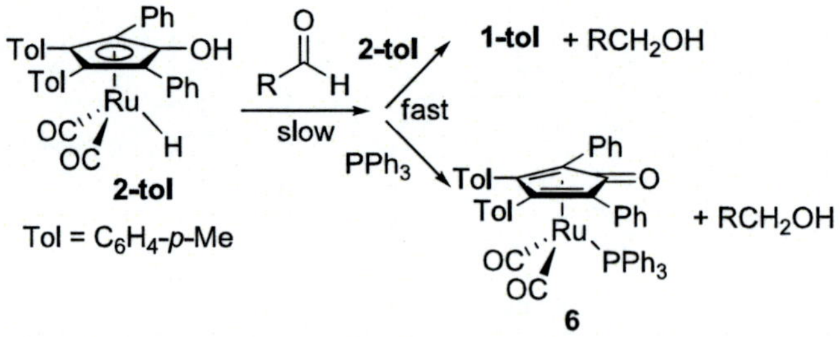

Scheme 6

[2, 3, 19]. However, it thermally dissociates to monoruthenium hydride **2**, which is the active reducing agent in the Shvo system. Monohydride **2** can be quantitatively generated by heating solutions of **1** under 4 atm of H_2; this allows independent study of the reduction of carbonyl compounds by **2**.

Hydride **2-tol** was used in mechanistic studies by the Casey group because its simplified NMR spectrum aided kinetic studies without influencing the mechanism. Hydride **2-tol** reduces benzaldehyde very rapidly at −40°C in toluene and at 0°C in THF (Scheme 6) [19]. In the absence of trapping agents, **2-tol** reacts with aldehydes to produce alcohols and diruthenium complex **1-tol**, thought to arise by reaction of dienone intermediate **A′** with a second equivalent of **2-tol**. When the reduction of aldehydes by **2-tol** was carried out in the presence of PPh_3 as a trapping agent, phosphine complex **6** was the isolated product. The rate of reduction was first order in both aldehyde and **2-tol** and was independent of PPh_3. The rate of reduction was faster than the rate of ^{13}CO exchange with **2-tol** in the dark [19, 53], and the rate of benzaldehyde reduction by **2-tol** was not inhibited by CO pressure; this provides evidence that CO dissociation is not involved in the reaction. In THF, $\Delta H^{\ddagger} = 12$ kcal mol^{-1} and $\Delta S^{\ddagger} = -28$ eu; this large negative entropy of activation is consistent with an associative reaction [19]. Primary kinetic isotope effects were found for both RuD ($k_{\mathbf{RuH}}/k_{\mathbf{RuD}} = 1.5$ in THF) and CpOD ($k_{\mathbf{OH}}/k_{\mathbf{OD}} = 2.2$ in THF) [19]. The combined isotope effect ($k_{\mathbf{RuHOH}}/k_{\mathbf{OHRuD}} = 3.6$) was approximately the product of individual isotope effects, consistent with transfer of both hydrogens in the same step [54].

Based on this data, Casey and Singer proposed an unusual mechanism that did not involve prior coordination of the substrate to the metal hydride (Scheme 7) [19]. They proposed simultaneous transfer of proton and hydride to a carbonyl compound outside the coordination sphere of the metal; this is now called the outer-sphere mechanism. Their initial suggestion that this directly produced intermediate **A′** and alcohol was later elaborated to include an initially formed hydrogen-bonded intermediate **B′** in which the alcohol is hydrogen bonded to the dienone carbonyl. Intermediate **B′** is subsequently converted to a solvent cage intermediate **C′** in which the hydrogen bond has been broken but **A′** and alcohol remain within a solvent cage.

Bäckvall suggested an alternative mechanism that avoided the unusual feature of the outer-sphere mechanism [20]. He suggested what is now termed the inner-sphere mechanism (Scheme 8) which begins with reversible Cp-ring slippage and

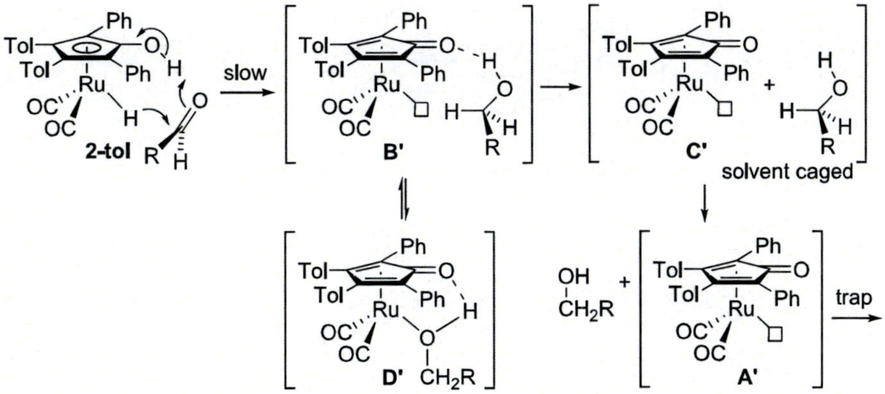

Scheme 7

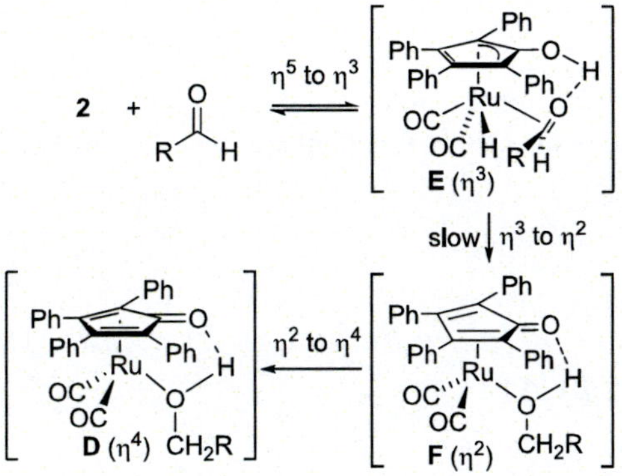

Scheme 8

coordination of aldehyde to ruthenium to give **E**, followed by rate-limiting simultaneous proton and hydride transfer to give **F** (a coordinatively unsaturated alcohol complex with an η^2-dienone ligand), and finally ring slippage to generate **D**, an alcohol complex.

In principle, the outer- and inner-sphere mechanism might be distinguished since alcohol complex **D** is the initial product only of the inner-sphere mechanism. If an aldehyde reduction was carried out in the presence of a trapping alcohol, the inner-sphere mechanism would require the exclusive formation of a complex with the alcohol formed in the reduction, whereas the outer-sphere mechanism suggests that intermediate **A′** could also be trapped by an external alcohol. Unfortunately, alcohol complexes have never been seen in the Shvo system; they would be expected to dissociate very rapidly since the cationic complex **7** which is expected to have an even more tightly bound alcohol than **D** undergoes rapid alcohol exchange at low temperature (Scheme 9) [55]. It should be mentioned that related iron–alcohol complex **8** has recently been fully characterized [56]. One impetus to study the mechanism of imine reduction was to take advantage of the kinetic stability of the

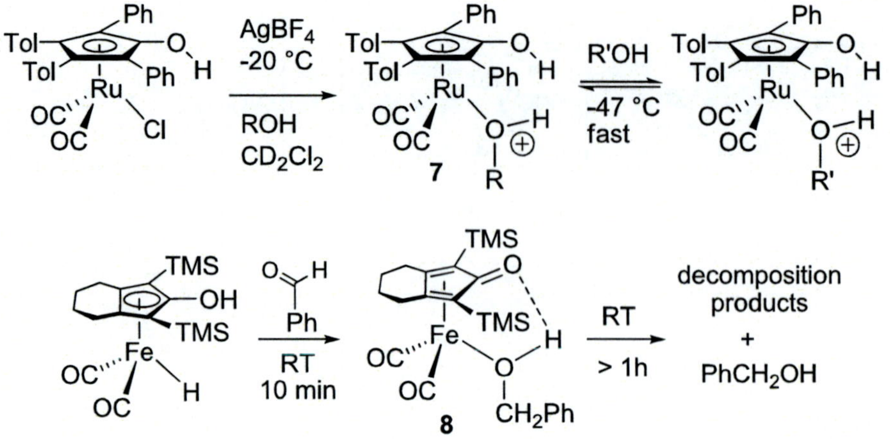

Scheme 9

resulting amine complexes to carry out related trapping experiments to distinguish between inner- and outer-sphere mechanisms. These experiments will be discussed in a later section.

3.2 Mechanism of Imine Reduction by 2 (and Amine Oxidation by A)

There are two major differences between the reduction of imines and aldehydes by **2-tol**: imines are reduced more rapidly than aldehydes and give amine complexes as stable products (Scheme 10) [57]. Substituents on the imine affect the rate of reduction and the stability of the amine complex product. PhCH=NPh reacts with **2-tol** in THF at −40°C at a rate about 150 times faster than benzaldehyde. The more basic imine PhCH=NCH$_2$Ph reacts with **2-tol** in THF at −57°C at a rate about 15 times faster than PhCH=NPh. The stability of the amine complexes depends on the basicity and steric requirements of the complexed amine. Dialkylamine complex **10** is stable to 80°C, while the less basic aryl(alkyl)amine complex **9** is stable to only 50°C. The complex of the very bulky *t*-Bu(PhCH$_2$)NH complex **11** decomposed above 0°C. In general, amine complexes with greater number of substituents are less stable. Beller has isolated the NH$_3$ complex **12** from catalytic systems; it is unusually stable and acts as a viable catalyst precursor [58].

Kinetic deuterium isotope effects on the rate of reduction of imines by **2-tol** also vary dramatically as a function of the substituent on nitrogen, indicating changes in the rate-limiting step. For PhCH=NC$_6$F$_5$ with an electron-withdrawing group on nitrogen, primary kinetic isotope effects were found for both RuD (k_{RuH}/k_{RuD} = 1.5 in THF) and CpOD (k_{OH}/k_{OD} = 2.2 in THF); these isotope effects are similar to those seen for aldehydes and ketones and are consistent with rate-limiting simultaneous

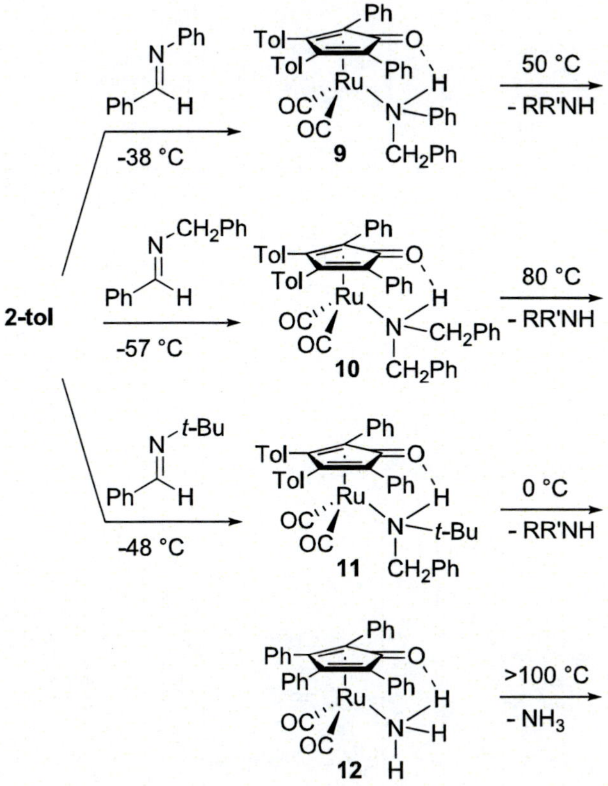

Scheme 10

transfer of both hydrogens [58]. Bäckvall reported negligible isotope effects ($k_{\mathrm{RuHOH}}/k_{\mathrm{RuDOD}} = 1.05$ in CD_2Cl_2) for the reduction of the ketimine *p*-$CH_3OC_6H_4C$-Me=NPh by **2** [25, 26]; this clearly indicates a change in mechanism to one in which hydrogen transfer is not rate limiting.

For the oxidation of *p*-$CH_3OC_6H_4$CHMeNHPh to *p*-$CH_3OC_6H_4$CMe=NPh by the coordinatively unsaturated species **A**, Bäckvall found a primary isotope effect for CD bond cleavage of 3.2 and a negligible isotope effect (1.0 ± 0.1) for ND bond cleavage [11]; these results can be explained in terms of rate-limiting CH bond breaking followed by proton transfer from nitrogen to oxygen (either inner or outer sphere). Similar transition states involving only the α-CH bond of tertiary amines are required to explain Shvo's observation of the transalkylation of tertiary amines catalyzed by **1** (Scheme 11) [59] and Beller's observation that tertiary amines can undergo a net transfer of an alkyl group to indoles [51].

With electron donor alkyl groups on the imine nitrogen, imine reduction by **2-tol** has been shown to be reversible, and a new rate-limiting step, amine coordination, has been proposed [58]. During the reduction of *p*-$CH_3C_6H_4CH$=$NCHMe_2$ by **2-tol** up to ~15% of ketimine, *p*-$CH_3C_6H_4CH_2N$=CMe_2 was observed in addition to amine complex **13**. When the reduction of the ketimine by **2-tol** was studied independently,

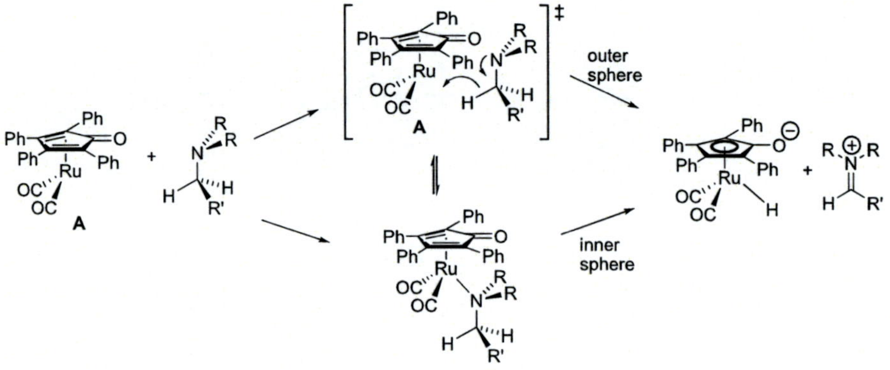

Scheme 11

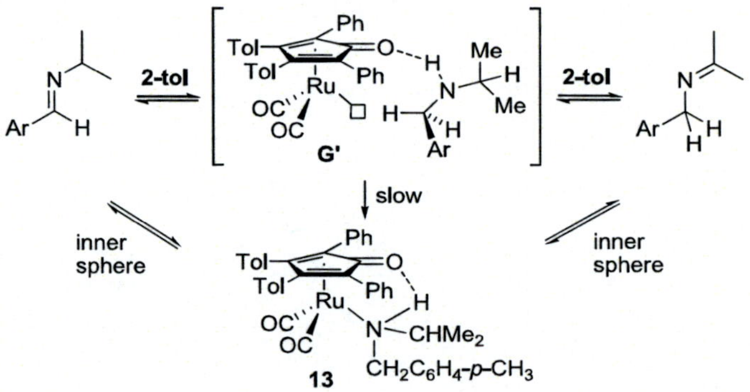

Scheme 12

rapid isomerization to imine preceded the formation of **13**. Similar isomerizations were observed in the transfer hydrogenation of imines with the Shvo catalyst **1** [4]. These observations require rapid and reversible imine reduction and rate-limiting amine complex formation (Scheme 12). Inverse kinetic deuterium isotope effects ($k_{RuH}/k_{RuD} = 0.64$ and $k_{OH}/k_{OD} = 0.9$) were observed in the reduction of the related PhCH=NCMe$_3$ by **2-tol**; the inverse RuD isotope effect is attributed to an equilibrium isotope effect favoring deuterium on carbon because of the greater strength of the CH bond compared to the RuH bond.

In an effort to distinguish between outer-sphere and inner-sphere mechanisms for reductions by **2-tol**, the reduction of imines was carried out in the presence of amines as potential trapping agents. The outer-sphere mechanism suggested that unsaturated intermediate **A′** might be competitively trapped by either the newly reduced amine or by the added amine; in contrast, the inner-sphere mechanism predicts exclusive formation of the complex of the newly reduced amine (Scheme 13).

When the reduction of PhCH=NPh was carried out at −60°C in the presence of aniline, only **9**, the complex of the newly reduced amine, was observed (Scheme 14) [28].

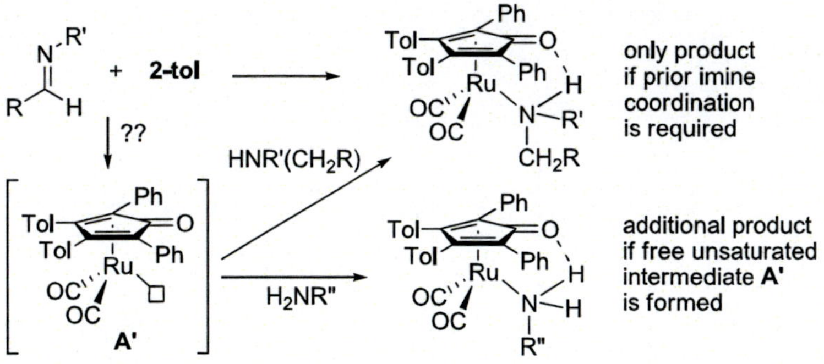

Scheme 13

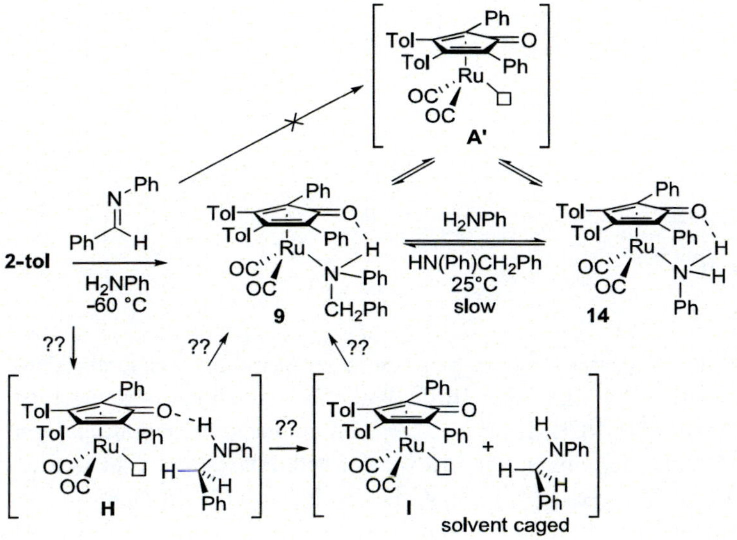

Scheme 14

Upon warming to room temperature, **9** was slowly converted to aniline complex **14**. In independent work, it was similarly found that hydrogenation of PhCMe=NMe by **2** in the presence of the external amine trap *p*-$MeOC_6H_4CHMeNHMe$ afforded 90–100% of the complex with the newly reduced amine [25]. These intermolecular trapping experiments are consistent with an inner-sphere mechanism and rules out the direct formation of the free coordinatively unsaturated intermediate **A′**. However, they leave open the possibility that the initial product is a hydrogen-bonded intermediate **H** or a solvent-caged intermediate **I,** which collapses to amine complex **9** more rapidly than the amine leaves the vicinity of the ruthenium.

To test for a solvent-caged intermediate, an intramolecular trapping experiment was devised. Reduction of *p*-$NH_2C_6H_4N$=CHPh (15) with **2-tol** at −60°C gave a 1:1 mixture of the amine complex of the newly formed amine **16** and of the intramolecular trapping product **17** [28]. The ratio of products was shown to be

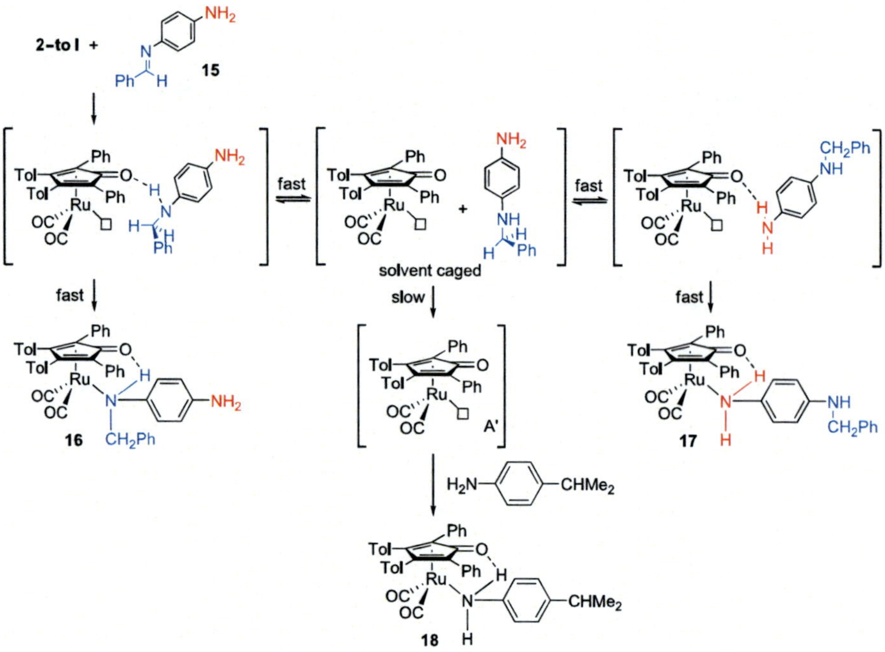

Scheme 15

kinetically controlled; warming to room temperature was required to give an equilibrium 6 : 94 mixture of **16**:**17**. When the reduction of **15** was carried out in the presence of H_2NCHMe_2 as a potential intermolecular trapping agent, none of the intermolecular trapping product **18** was formed. Interestingly, the isomerization of **16** to **17** in the presence of *p*-$NH_2C_6H_4CHMe_2$ occurred intramolecularly without the formation of **18**; **16** to **17** equilibrated within 15 min at room temperature but formation of **18** was seen only after 24 h. Scheme 15 provides a rationale in terms of an outer-sphere mechanism.

Bäckvall suggested the possibility that the intramolecular trapping product **17** might have been formed by an inner-sphere mechanism in which the initially formed coordinatively unsaturated η^2-dienone amine complex **J** rearranged via migration across the aromatic π-system. The proposed inner-sphere mechanism is shown in Scheme 16.

To test this possibility, Bäckvall studied the reduction of the *N*-aryl ketimine **19** in which the imine and intramolecular amine trap are linked by a saturated cyclohexane ring. Reduction of **19** by **2** occurred at −80°C to give exclusively **20**, in which the newly reduced amine was coordinated to ruthenium (Scheme 17) [25]. Upon warming to −8°C, **20** rearranged to **21**. ^{15}N NMR experiments were also carried out where the imine nitrogen part of **19** was marked with ^{15}N. At −80°C, only **20** could be observed in the ^{15}N NMR. Bäckvall concluded that these experiments support the inner-sphere mechanism.

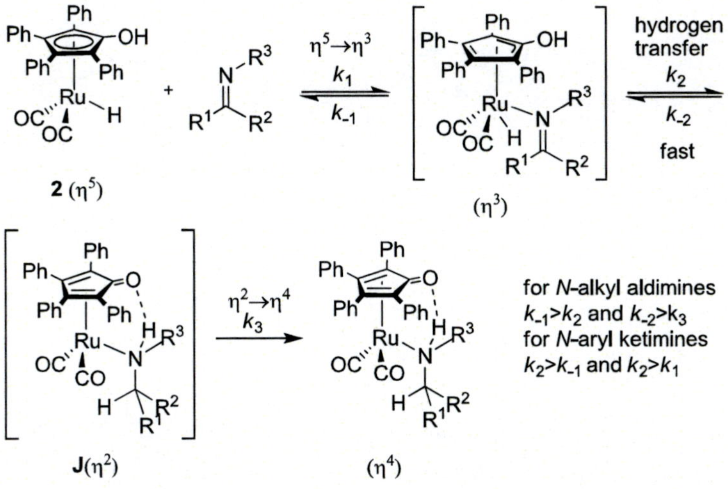

Scheme 16 Proposed inner-sphere mechanism for hydrogenation of imines by **2**

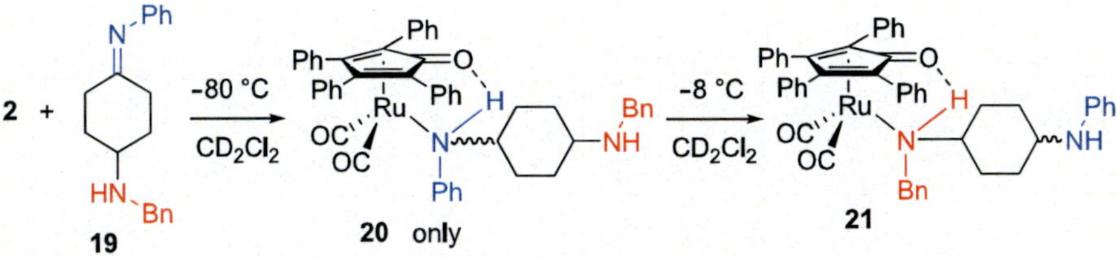

Scheme 17

Casey and Clark considered the possibility that the above reaction occurred by an outer-sphere mechanism but that the initially formed hydrogen-bonded intermediate collapsed to amine complex **20** more rapidly than the hydrogen bond broke to give a solvent-caged intermediate that could produce the intramolecular trapping product **21**. They considered the possibility that the substituents on nitrogen might influence the relative rates of amine coordination and hydrogen bond cleavage of the initially formed hydrogen-bonded intermediate. They studied the *N*-alkyl ketimimine **22** in which the imine and intramolecular amine trap are also linked by a saturated cyclohexane ring. Reduction of **22** by **2-tol** in toluene-d_8 gave an 87:13 mixture of **23** (the ruthenium complex of the newly formed amine) and **24** [the intramolecular trapping product (Scheme 18)] [29]. The same reduction of **22** in CD_2Cl_2 at $-20°C$ gave a 91:9 ratio of **23** and **24**. This provides support for an outer-sphere reduction mechanism.

The high proportion of amine complex formed from the newly formed amine (87–100%) in the intramolecular trapping experiments where the amine trap is linked by a saturated cyclohexane ring would seem to support the inner-sphere mechanism. However, the intramolecular trapping experiments can all be explained in terms of an outer-sphere mechanism if one assumes that a coordinatively unsaturated dienone intermediate with the newly formed amine weakly hydrogen

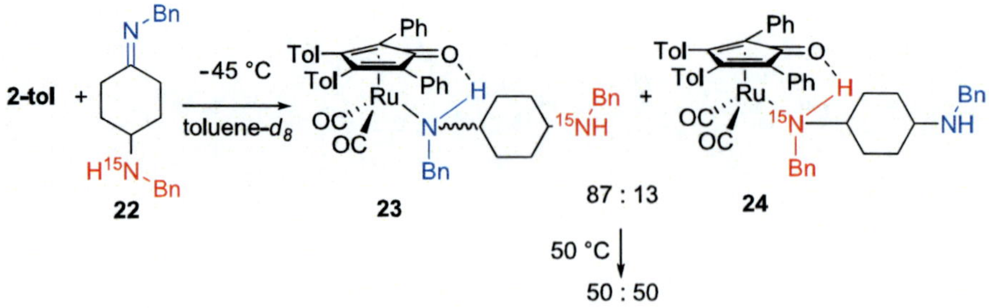

Scheme 18

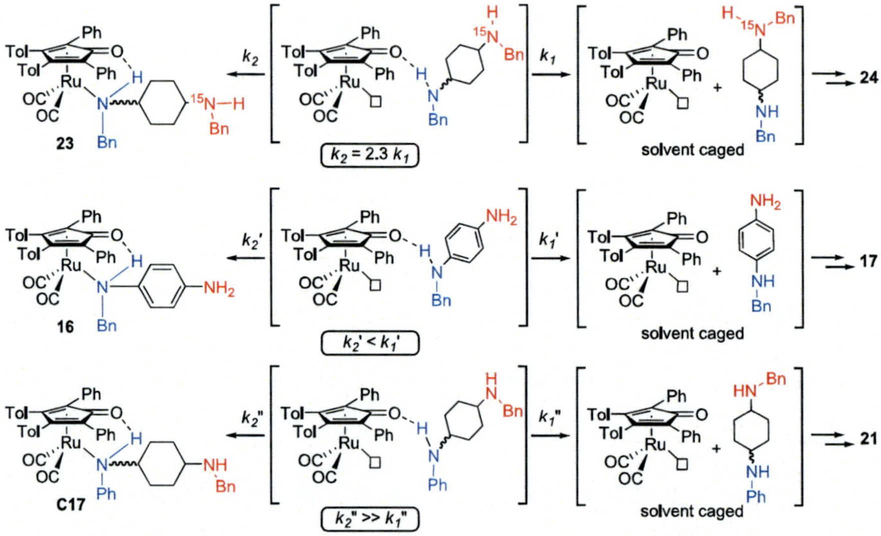

Scheme 19

bonded to the dienone carbonyl is formed (Scheme 19) [29]. The product ratios depend on the relative rates of coordination of the newly formed amine (k_2) and breaking of the hydrogen bond to the newly formed amine (k_1). Since both of these rates should be faster for a more electron-rich amine, which would be both more nucleophilic and also a weaker hydrogen bond donor, it is impossible to predict the dependence of this rate ratio (and the amount of trapping product) on the structure of the amine.

In the reduction of imines by **2-tol**, the amine is generated within a solvent cage and complexation to ruthenium occurs before diffusion bring them apart. The possibility that amine coordination to ruthenium might be even faster than nitrogen inversion, together with the configurational stability of the amine complex product, provided an opportunity to determine the stereochemistry of imine reduction by **2-tol**. Amine nitrogen inversion is extremely rapid ($\Delta G^{\ddagger}$ ~7.5 kcal mol^{-1}) and had previously precluded determination of the stereochemistry of

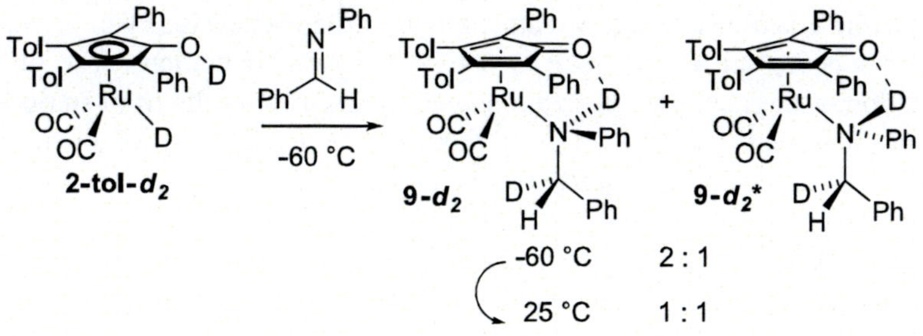

Scheme 20

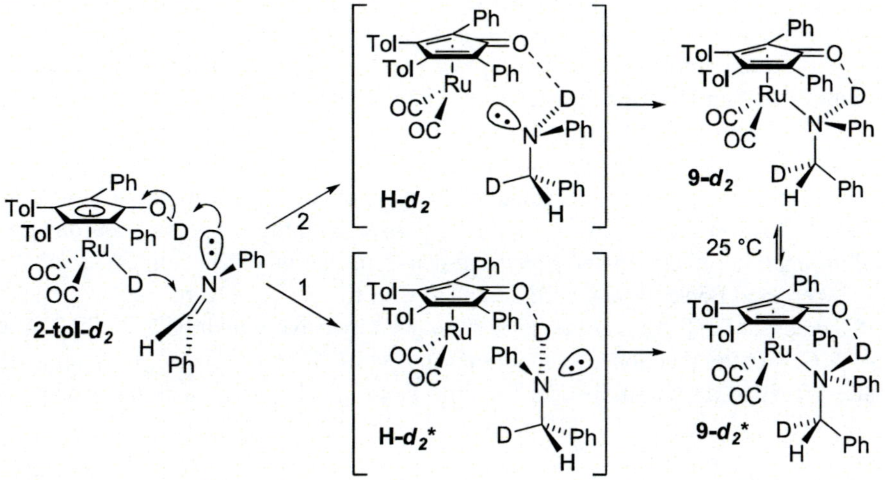

Scheme 21

reduction of imines by any reagent or catalytic system [60, 61]. Deuterium labeling experiments established that the reduction of *N*-aryl imines by **2-tol-d_2** is largely stereospecific [62]. Reduction of PhCH=NPh by deuterated **2-tol-d_2** at −60°C produced a 2:1 mixture of stereoisomers of deuterated **9-d_2:9-d_2*** (Scheme 20); the major stereoisomer **9-d_2** has D and Ru added to the same face of the imine C=N bond. Upon warming to room temperature, isomerization occurred to give a 1:1 equilibrium mixture of the diastereomers. The observation of two diastereomers with predominant *trans* addition is consistent with an outer-sphere mechanism. One would expect that the inner-sphere mechanism via coordination of the imine would give *trans* addition exclusively, since ruthenium and hydride would add *cis* to the imine double bond, and the proton would protonate the lone pair of the nitrogen. For the inner-sphere mechanism, the formation of the significant amounts of *cis* products might be attributed to isomerization of the product upon inadvertent warming of the samples.

The outer-sphere mechanism can lead to either diastereomer, depending on the orientation assumed by the newly generated nitrogen lone pair (Scheme 21).

When hydride is transferred from ruthenium to carbon, a new lone pair is generated perpendicular to the C=N plane in one of the two orientations. The preferred orientation of the newly generated lone pair apparently results from steric effects pushing the amine substituents away from the sterically crowded dienone. The observed stereospecificity requires that amine inversion is slow compared to amine coordination.

The coordinatively unsaturated intermediates with an amine hydrogen bonded to the dienone carbonyl display unusually high reactivity. The inter- and intramolecular trapping experiments require that the collapse of this intermediate by coordination of nitrogen to ruthenium occur faster than breaking the weak hydrogen bond between the amine and a dienone carbonyl and much faster than hydrogen bond breaking followed by diffusion apart. The stereospecificity of the reduction of *N*-aryl imines by **2-tol-d_2** requires that the coordination of the newly formed amine to ruthenium occurs more rapidly than lone pair inversion at the nitrogen center. The isomerization and deuterium scrambling that accompany the reduction of *N*-alkyl imines by **2-tol** require that the unsaturated intermediate transfer hydride back from carbon to ruthenium faster than amine coordination to nitrogen. The "slow" reactions of the intermediate including amine escape from the solvent cage and amine inversion are only slow in comparison with the even faster coordination of nitrogen to ruthenium and reversible dehydrogenation of the amine.

Åberg and Bäckvall designed an experiment to test the solvent cage hypothesis (Scheme 22) [27]. Ketoamine **25** was reduced by Shvo's hydride **2** in CD_2Cl_2 at −20°C, and the reaction was monitored by NMR. In this reaction, ketoamine **25** also serves as an intermolecular trapping molecule. It was argued that if there is

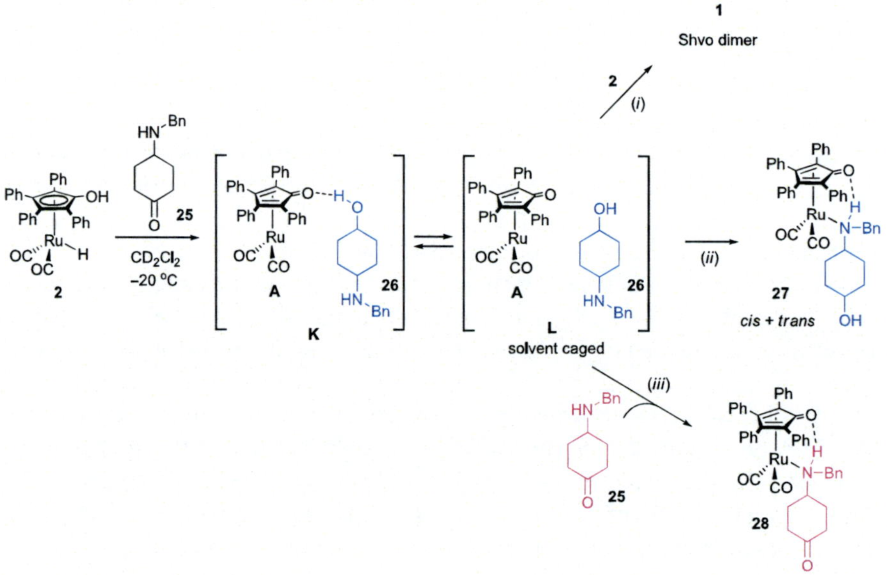

Scheme 22

a solvent cage effect, the newly reduced molecule **26** inside a solvent cage with **A** (intermediate **L**) would coordinate to the ruthenium via its amine function, before coordination of **25**. The results show that there is no cage effect, and the two amine complexes **27** and **28** were formed in a statistic ratio with respect to the concentration of **26** and **25**, respectively, in solution. In terms of an outer-sphere mechanism, this result can be explained if the stronger hydrogen bond between the newly formed alcohol and **A** (intermediate **K**) gives a longer lived intermediate that is trapped intermolecularly by either **25** or **26**.

Density functional theory (DFT) calculations are often used in conjunction with mechanistic investigations in organometallic chemistry. They are least useful when they simply are found to be consistent with a mechanistic hypothesis. They are much more useful either when they show that a mechanistic hypothesis is energetically unreasonable or when several plausible mechanistic pathways are investigated and one has a significantly lower energy barrier.

Casey and Cui carried out DFT computations at the B3LYP/LANL2DZ level of theory on the simplified model reaction of $(C_5H_4OH)Ru(CO)_2H$ (**2-H**) with formaldehyde [28]. Their calculations support a concerted outer-sphere mechanism with an early transition state in which the CpO–H and the Ru–H distances are lengthened by only 0.1 Å; the calculated activation barrier of 13.8 kcal mol^{-1} is similar to the experimental $\Delta H^{\ddagger}$ of 12 kcal mol^{-1} found for the reaction of **2-tol** with benzaldehyde. Casey and Cui also carried out computations on the reduction of $H_2C{=}NCH_3$ by **2-H** [28]; the transition state showed much more advanced transfer of the CpO–H hydrogen to N (1.364 Å O—H, 1.134 Å H—N), while the Ru–H distance lengthened by only 0.1 Å. The activation enthalpy for reduction of $H_2C{=}NCH_3$ (4.8 kcal mol^{-1}) was substantially lower than that calculated for reduction of formaldehyde, in agreement with the more rapid reduction of imines compared to aldehydes by **2-tol**.

Privalov and Bäckvall [25, 63] used DFT computations to study the feasibility of an inner-sphere mechanism for reduction of $Me_2C{=}NMe$ by **2**. They included solvent in their computations and found two possible rate-limiting steps of similar energy for the inner-sphere mechanism: imine coordination to ruthenium via ring slippage (15 kcal $mole^{-1}$) and hydride transfer to give an (η^2-cyclopentadienone) ruthenium amine intermediate (12 kcal $mole^{-1}$).

Ujaque and Lledos carried out a more comprehensive theoretical analysis on various mechanisms for reductions by **2** [64, 65]. An inner-sphere mechanism involving initial CO dissociation was readily eliminated based on a computed CO dissociation energy of 46 kcal mol^{-1}. The energy barrier for the outer-sphere reduction of formaldehyde by **2** was computed as 8 kcal mol^{-1}, while the lowest energy inner-sphere mechanism had a barrier of 35 kcal mol^{-1}. The outer-sphere mechanism for imine reduction by **2** was shown to have a lower barrier (9 kcal mol^{-1}) than the best inner-sphere ring-slip mechanism (~26 kcal mol^{-1}).

For the inner-sphere mechanism of imines, DFT computations show a higher barrier than the experimentally determined barrier, although not unreasonably high (15 kcal/mol). On the other hand, the corresponding DFT computations for the outer-sphere mechanism show a barrier similar to that found experimentally.

3.3 The Dihydrogen Activation Step

The loss of H_2 from **2-tol** was investigated to obtain information about the microscopic reverse of the reaction, the H_2 activation step in the Shvo catalytic hydrogenation system [66]. The loss of dihydrogen from **2-tol** was studied in the presence of PPh_3 as a trapping agent for the unsaturated intermediate **A′** generated by H_2 loss (Scheme 23). The reaction required heating at 90°C in toluene, and the rate of hydrogen loss was shown to be first order in [**2-tol**] and zero order in [PPh_3] ($\Delta H^{\ddagger} = 26$ kcal mol^{-1}, $\Delta S^{\ddagger} = -4$ eu). Utilizing labeled starting material **2-tol-OHRuD**, it was established that less than 10% RuD/OH exchange occurred prior to loss of dihydrogen. These data are consistent with a mechanism involving irreversible formation of a transient dihydrogen ruthenium complex **J**, loss of H_2 to give unsaturated ruthenium complex **A′**, and trapping by PPh_3.

The initial hypothesis that dihydrogen complex **M** was formed by direct transfer of hydrogen from the CpOH to the RuH was investigated using DFT computations [66]. The calculated barrier of 43 kcal mol^{-1} was so much greater than the experimental barrier of 26 kcal mol^{-1} that this hypothesis was rejected. An alternative pathway in which a water molecule mediates the proton transfer via transition state **N** had a calculated barrier of ~20 kcal mol^{-1} (Scheme 24).

These computations prompted investigation of the loss of H_2 from **2-tol** in the presence of EtOH. Added EtOH resulted in **2-tol-OHRuD/2-tol-ODRuH** interchange 250 times faster than dihydrogen elimination [66]. The rate of dihydrogen generation increased as the concentration of [EtOH] increased and until it reached a saturation limit at about four to five times the rate in the absence of added alcohol. This rate dependence is consistent with a shift from one kinetic mechanism in the absence of added EtOH to another mechanism in the presence of added EtOH. At high [EtOH], a rapid equilibrium is set up with undetectable amounts of a reversibly formed dihydrogen complex **M–HD** and loss of dihydrogen is rate limiting (Scheme 25).

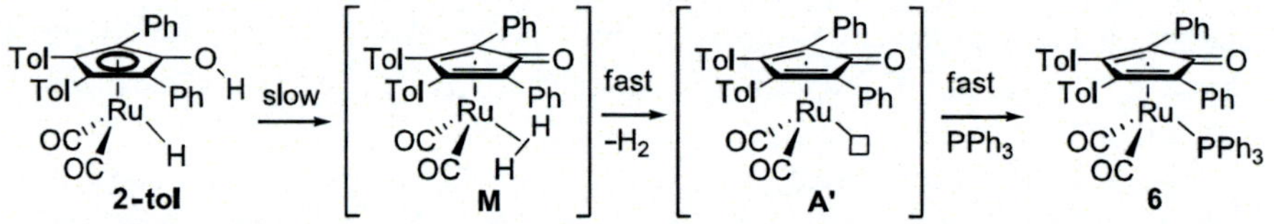

Scheme 23

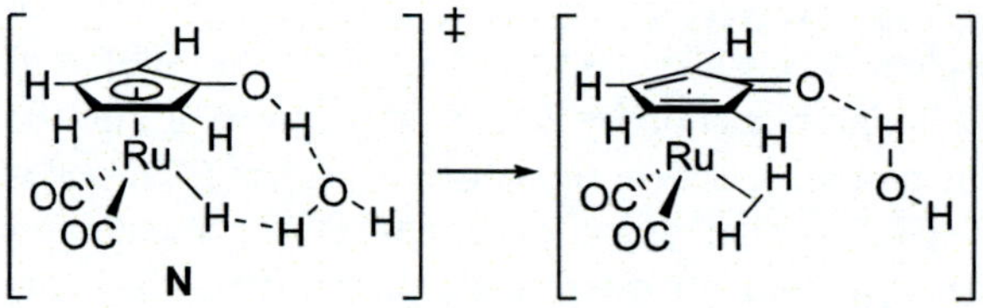

Scheme 24

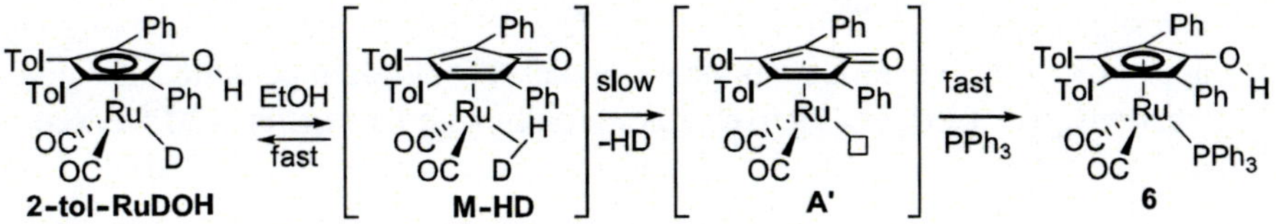

Scheme 25

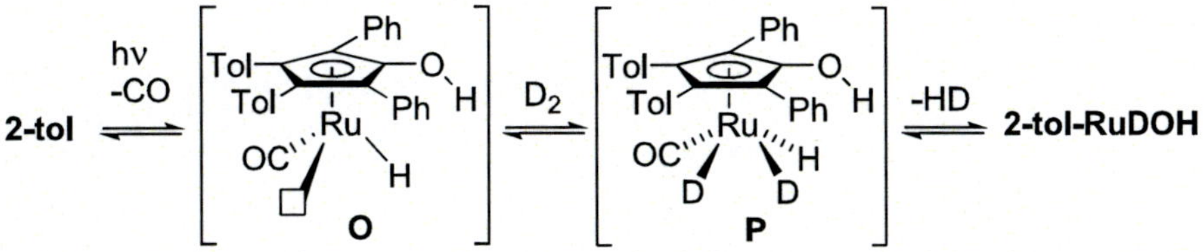

Scheme 26

In the absence of added EtOH, the CpOH acts as the H relay agent to slowly form dihydrogen complex **M**, which eliminates dihydrogen faster than it reverts to **2-tol**. This requires a transition state that involves two molecules of ruthenium hydride **2-tol** – an apparent conflict with the observation of a first order rate dependence on [**2-tol**]. This contradiction was resolved by pulse gradient spin echo NMR experiments that established that **2-tol** is a hydrogen-bonded dimer in toluene [66]; the observed first order kinetics result from the ground state hydrogen-bonded dimer of **2-tol** reacting via a transition state which also involves two molecules of **2-tol** to make dihydrogen complex **M** which rapidly decomposes.

In the course of studying the reactions of **2-tol**, the exchange of D_2 with the RuH of **2-tol** (but not with the OH of **2-tol**) was discovered [53]. Under conditions similar to those used for H_2 elimination, less than 10% H/D exchange between **2-tol** and D_2 was seen in the presence of PPh_3 at 90°C. However, in the absence of PPh_3, H/D exchange between the RuH functionality of **2-tol** and D_2 occurred rapidly at room temperature (Scheme 26). The rate of H/D exchange was more rapid under fluorescent laboratory lighting. H/D exchange between **2-tol** and D_2 was also strongly inhibited under a CO atmosphere. These inhibition results implicate coordinatively unsaturated intermediate **O** in the H/D exchange process. Reversible oxidative addition of D_2 to **O** to give trihydride **P** would explain the exchange process.

If the generation of intermediate **O** were rate limiting for RuH/D_2 exchange, then the rate of ^{13}CO exchange with **2-tol** should have been similar to the rate of RuH/D_2 exchange, but the rate of ^{13}CO exchange was at least 300 times slower than RuH/D_2 exchange. Consequently, there must be a chain process that results in multiple RuH/D_2 exchanges every time CO dissociates to give unsaturated intermediate **O**. A chain mechanism involving reversible addition of RuH of **2-tol** to **O** to form H_2RuRu intermediate **Q** was suggested (Scheme 27) [67].

Formic acid can be used as the ultimate reducing agent in transfer hydrogenations catalyzed by the Shvo complex **1** [3, 35]. Shvo proposed that a formate complex served as a precursor to ruthenium hydride **2**, which was the active reducing agent.

To investigate the possibility that a formate complex might directly reduce carbonyl compounds, Casey synthesized formate complex **29** at low temperature and studied its reactions [67]. Complex **29** exchanges rapidly with DCO_2H at $-20°C$, decomposes to **2-tol** slowly at 0°C, and does not react directly with benzaldehyde (Scheme 28). These results demonstrate that formate complex **29** is not itself a reducing agent but simply a precursor of ruthenium hydride **2-tol**. Proposed mechanisms for these reactions are shown in Scheme 29.

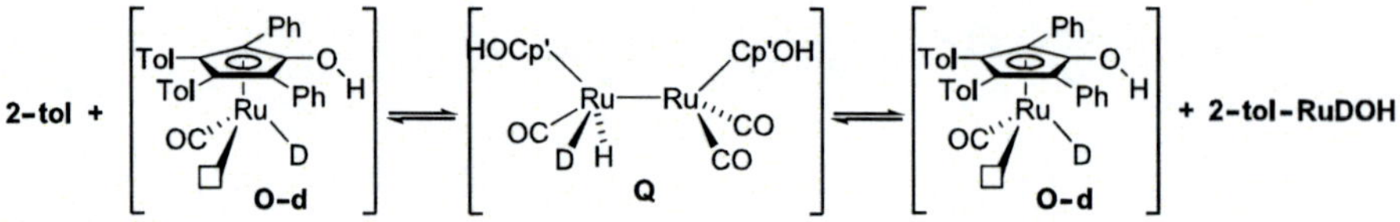

Scheme 27

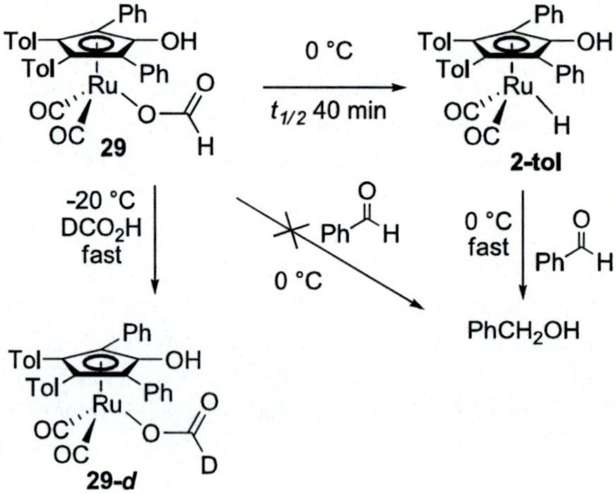

Scheme 28

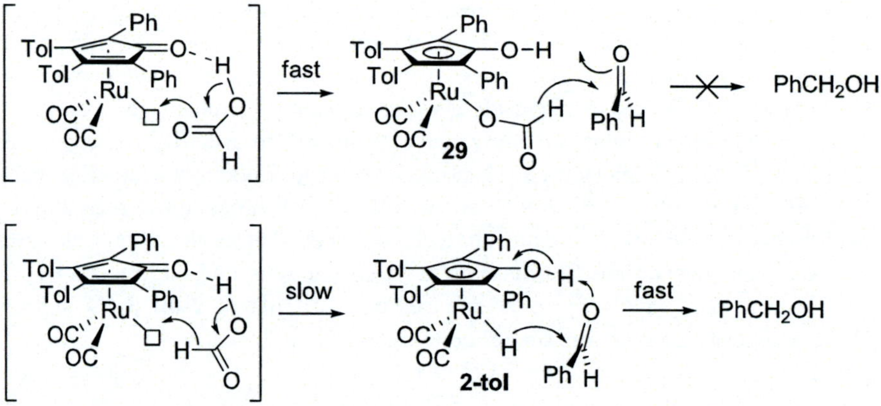

Scheme 29

3.4 *Kinetic Mechanism for Carbonyl Reduction*

Casey and Beetner used in situ IR spectroscopy to monitor the hydrogenation of benzaldehyde catalyzed by the Shvo hydrogenation catalyst **1-tol** [68]. The disappearance of benzaldehyde and the concentrations of the ruthenium species present throughout the hydrogenation reaction were observed (Fig. 2). In toluene, bridging diruthenium hydride **1-tol** was the only observable ruthenium species until nearly all of the substrate was consumed. The dependence of the hydrogenation rate on substrate, H_2 pressure, and total ruthenium concentration were all found to be less than first order.

A full kinetic model of the hydrogenation based on rate constants for individual steps in the catalysis was developed (Scheme 30) [69]; this kinetic model simulates the rate of carbonyl compound hydrogenation and of the amounts of ruthenium species **1-tol** and **2-tol** present during hydrogenation [69].

Since the rates of the individual steps were measured at very different temperatures, measured activation parameters were used to estimate rate constants at the temperature of a given catalytic reaction [69]. The rate of dissociation of bridging hydride **1-tol** at 60°C (k_1) was obtained by measuring the rate of reaction with PPh_3 to give **2-tol** and phosphine complex **C3**; the rate was first order in [**1-tol**] and independent of [PPh_3]. The rate of benzaldehyde reduction by **2-tol** (k_3) had been measured below 0°C, and the rate of loss of H_2 from **2-tol** had been measured at 90°C. The rates of reaction of reactive intermediate **A′** with H_2 (k_{-2}) and with **2-tol**

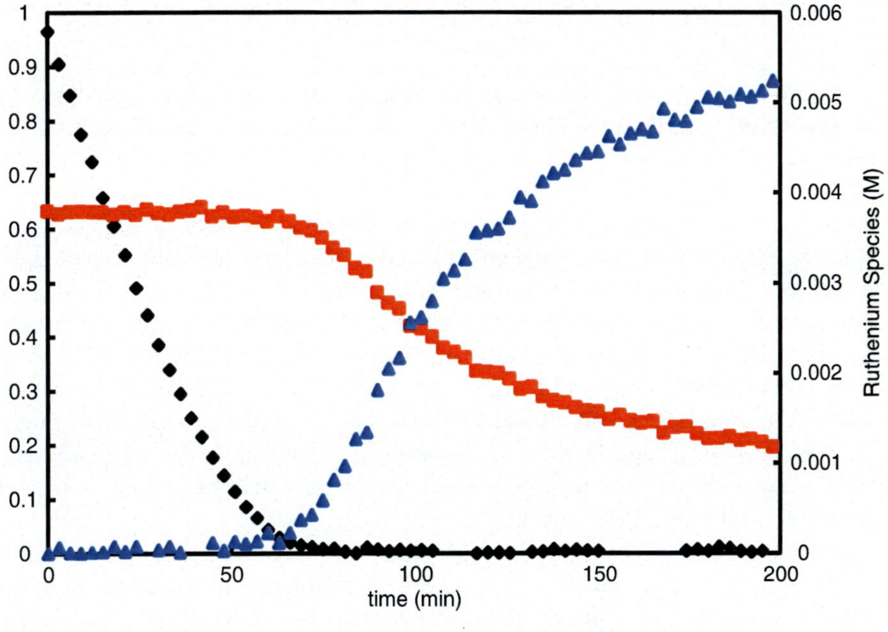

Fig. 2 Concentrations of benzaldehyde (*black diamond*), **1-tol** (*black square*), and **2-tol** (*black triangle*) during hydrogenation of benzaldehyde with [**1-tol**] under 35 atm H_2 at 60°C in toluene

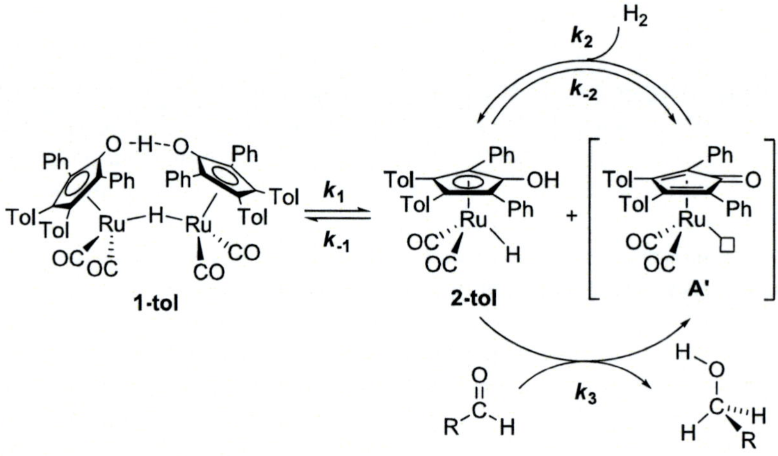

Scheme 30

$$\textbf{1-tol} + H_2 \underset{k_{-1}}{\overset{k_1}{\rightleftharpoons}} \textbf{2-tol} + \textbf{A'} + H_2 \underset{k_{-2}}{\overset{k_2}{\rightleftharpoons}} 2\ \textbf{2-tol}$$

$$K_{eq} = \frac{k_1 k_2}{k_{-1} k_{-2}} = \frac{[\textbf{2-tol}]^2}{[\textbf{1-tol}][H_2]} \qquad \frac{k_2}{k_{-1}} = \frac{K_{eq} k_{-2}}{k_1}$$

Scheme 31

to give **1-tol** (k_{-1}) cannot be measured directly, but their ratio was obtained by measuring K_{eq} for the reaction of **1-tol** with H_2 to give 2 equivalents of **2-tol** (Scheme 31).

With all these rate constants, it was possible to predict the time course of the reaction (Fig. 3). Remarkably good agreement (within a factor of 3) was found between experimental hydrogenation rates and simulated rates calculated using the reactions in Scheme 30 and rate constants extrapolated to 60°C estimated from kinetic and equilibrium measurements made at much lower or much higher temperature. This close agreement between experimental and simulated hydrogenations provides increased confidence in the fundamental soundness of the kinetic model. The simulations also mirrored the less than first order dependences of rates on [PhCOH], [H_2], and [**1-tol**]$_0$. In addition, the simulations showed dominant presence of **1-tol** during catalysis (even though **2-tol** would have been favored if **1-tol** and **2-tol** were in equilibrium).

Scheme 30 provides an excellent quantitative picture of the kinetics of hydrogenation. The hydrogenation of benzaldehyde in toluene proceeds by turnover limiting dissociation of diruthenium bridging hydride **1-tol** to give the active reducing monoruthenium hydride reducing agent **2-tol**, followed by hundreds of cycles in which **2-tol** reduces benzaldehyde and generates reactive intermediate **A′**

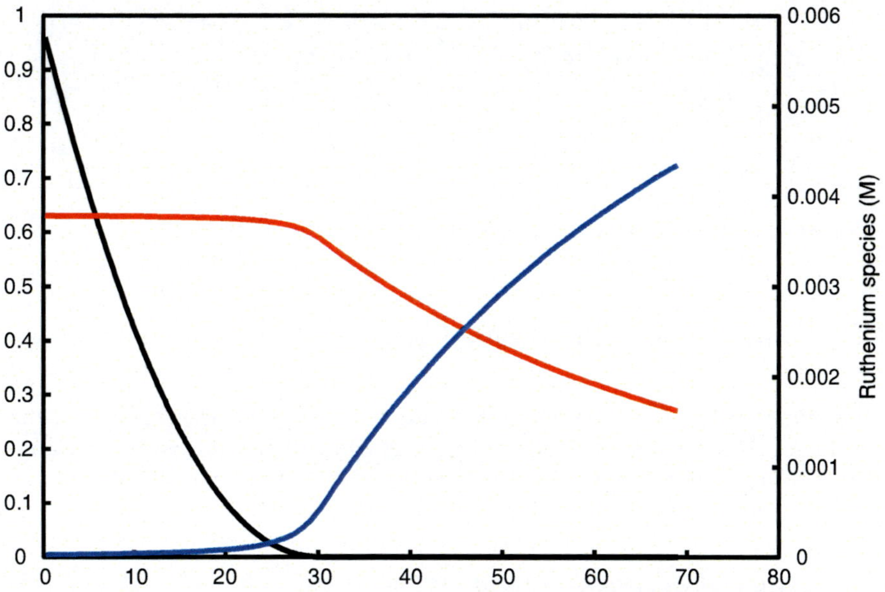

Fig. 3 Kinetic modeling simulation of concentrations of benzaldehyde (*black*), **1-tol** (*red*), and **2-tol** (*blue*) during hydrogenation of benzaldehyde with [**1**] under 35 atm H_2 at 60°C in toluene (see Fig. 2 for comparison with experimental line shapes, note the difference in time scales)

which reacts with H_2 to regenerate **2-tol**. The reaction of **A′** with H_2 occurs several hundred times faster than reaction with **2-tol** to generate dormant diruthenium complex **1-tol**. On each hydrogenation cycle, **A′** has the opportunity to be converted to **1-tol**, and eventually this occurs. The very rapid reaction of **2-tol** with benzaldehyde suppresses its concentration far below its equilibrium value.

The Shvo catalytic system **1** ⇔ **2** often makes inefficient use of ruthenium since so much of the ruthenium is present as the dormant species **1**, and so little is present as the active reducing agent **2**. To develop new more active catalysts, systems with structures that interfere with the formation of unreactive M–H–M systems but maintain high reactive of the M–H species are needed (Scheme 32). Destabilization of M–H–M formation was achieved by introducing a bulky NHPh group on the Cp ring in place of -OH in **30**, but the low NH acidity resulted in slow stoichiometric reduction of carbonyl groups by **30** [69]. Protonation of **30** to give **31**, which possesses a much more acidic $-NH_2^+$ group, gave a very active catalyst, but acid-catalyzed side reactions led to some ether formation. Complex **32**, in which PPh_3 is substituted for one CO, also prevented formation of unreactive M–H–M complexes [70, 71]. **32** is a very selective catalyst for hydrogenation of aldehydes over ketones; it is also a faster catalyst than **1** ⇔ **2** for aldehyde hydrogenation below 60°C but is slower at higher temperatures. The economical iron catalyst **33**, which does not form M–H–M complexes, is an active catalyst for ketone hydrogenation at room temperature and 3 atm [72].

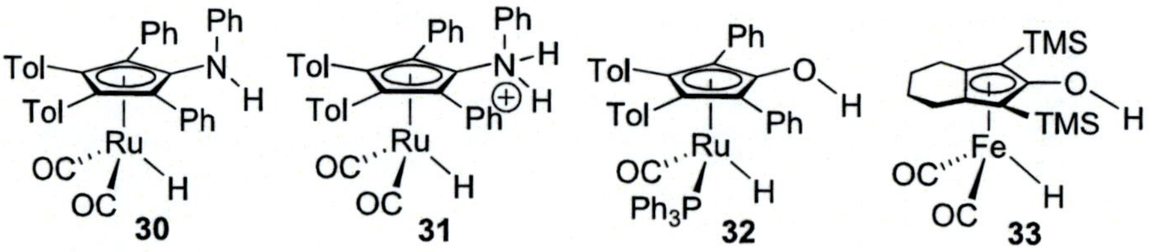

Scheme 32 Catalysts designed to avoid formation of unreactive H–M–H complexes

4 Applications in Enzymatic DKR

The Shvo catalyst has found many applications as a racemization catalyst in enzymatic resolutions of alcohols and amines leading to a DKR. Also diols and aminoalcohols have been used in these applications in DYKATs.

4.1 DKR of Secondary Alcohols

In 1997, Bäckvall and coworkers presented an efficient DKR protocol for both aromatic and aliphatic *sec*-alcohols using Shvo's catalyst **1** for the racemization and *Candida antarctica* lipase B (CALB) for the enzymatic resolution [13, 14]. With 2 mol% of **1,** the corresponding acetates were obtained in high yields and with *ee* values exceeding 99% in most cases (10).

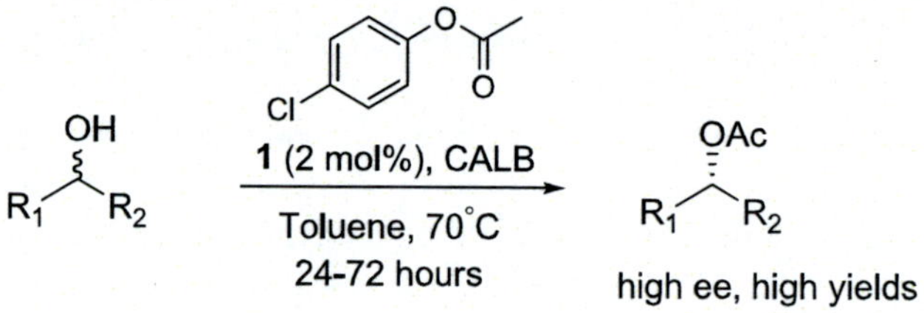

(10)

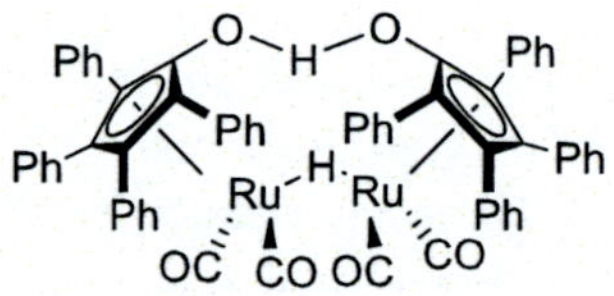

1

In 2000, an efficient DKR protocol for α-hydroxy esters was reported by Bäckvall et al. using Shvo's catalyst **1** for the racemization [73]. *Pseudomonas cepacia* lipase (PS-C) was used for the enzymatic resolution. The protocol was applicable to aromatic hydroxy acids with both electron-withdrawing and σ-donating groups present on the aromatic ring. The corresponding acetates were obtained in high yields (69–80%) and with very good *ee* values (94–98% *ee*) (11). Much higher % *ee* was obtained when the β-carbon of the α-hydroxy ester was more substituted. Two substrates with a less-substituted β-carbon gave only 80% and 30% *ee*, respectively (12).

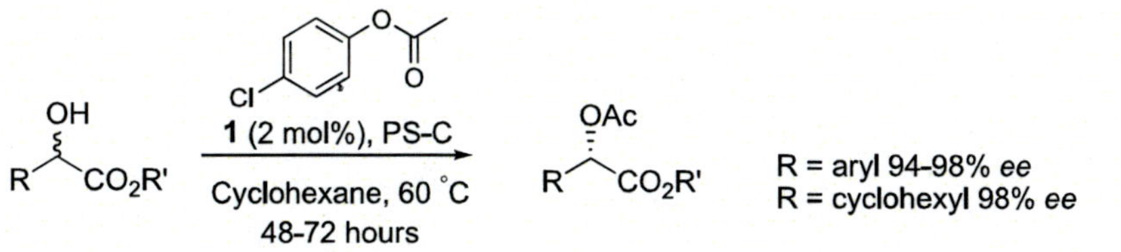

(11)

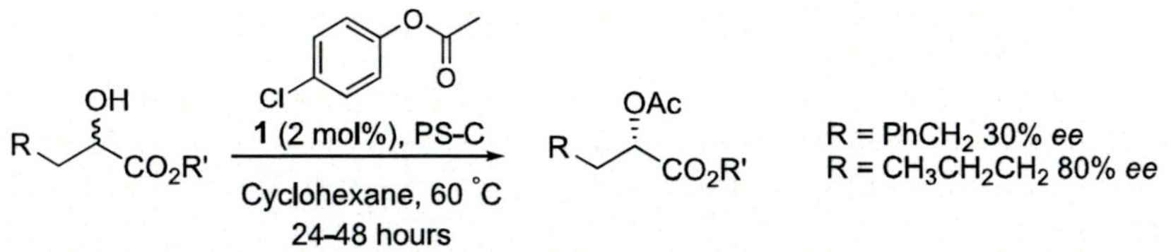

(12)

An efficient DKR of β-azido alcohols was reported in 2001 by Bäckvall and coworkers. The chiral β-azido alcohol products are useful as direct precursors of aziridines and β-amino alcohols. Shvo's catalyst **1** (4–6 mol%) was used for the racemization and CALB for the kinetic resolution [74]. Good isolated yields (70–87%) of the products were obtained with enantiomeric excesses exceeding 99% in many cases (13). In all cases, small amounts of the corresponding ketones (2–8%) were observed as by-products during the hydrogen transfer process, probably due to the high temperature used (80°C).

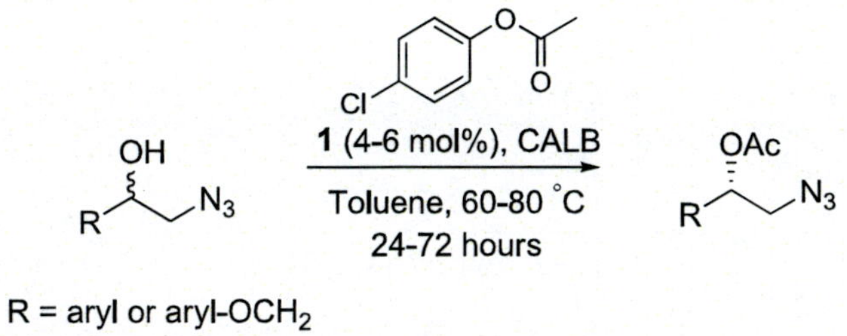

(13)

In 2002, an efficient route to chiral δ-lactones was reported using DKR in the enantio-determining step [75]. DKR of racemic δ-hydroxy esters was performed with Shvo's catalyst **1** for racemization. *Pseudomonas cepacia* lipase was selected as the resolving agent as it exhibited the best efficacy in the kinetic resolution. At first, the procedure suffered from large amounts of ketone formation, but it was found that the addition of a hydrogen source reduced the amount of ketone formed. Molecular hydrogen (1 bar) was chosen as the hydrogen donor as it inhibited the by-product formation completely. For two substrates, the procedure afforded the corresponding enantiomerically pure δ-acetoxy esters in 98% *ee* and with 80% and 81% yield, respectively (14). The synthetic utility of this protocol was demonstrated by a one-pot two-step procedure transforming the chiral δ-acetoxy esters into the corresponding δ-lactones.

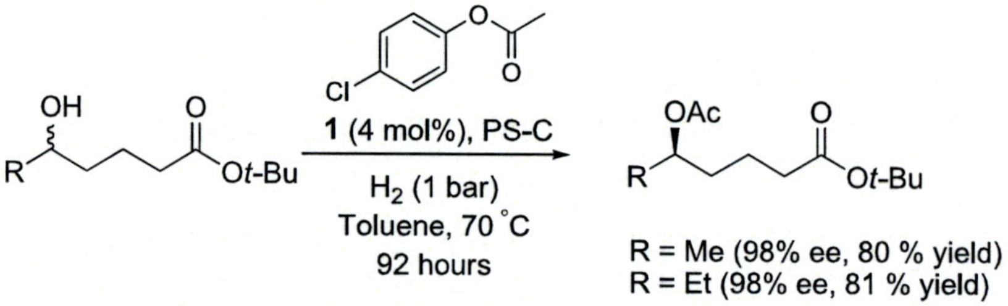

(14)

In a similar application, DKR of a racemic γ-hydroxy acid derivative was performed to access the corresponding chiral γ-lactones. DKR of *N,N*-diisopropyl-4-hydroxypentanamide afforded the corresponding acetate in 98% *ee* and 86% yield [76].

Chiral β-halo alcohols are synthetically interesting intermediates for asymmetric catalysis. Their versatility is attributed to the halogen that can readily act as a good leaving group. Chiral epoxides and chiral β-amino alcohols are two examples of accessible compounds. Bäckvall and coworkers reported an efficient DKR protocol for a wide range of racemic β-chloro alcohols in 2002, where Shvo's catalyst **1** was used for the racemization (15) [77]. The corresponding enantiomerically pure acetates were obtained in good to excellent enantiomeric excess (85–96% *ee*) and in high yields (74–99% by NMR). Subsequent base-induced ring closure of a few β-chloro acetates afforded the chiral epoxides with retained enantiomeric excesses.

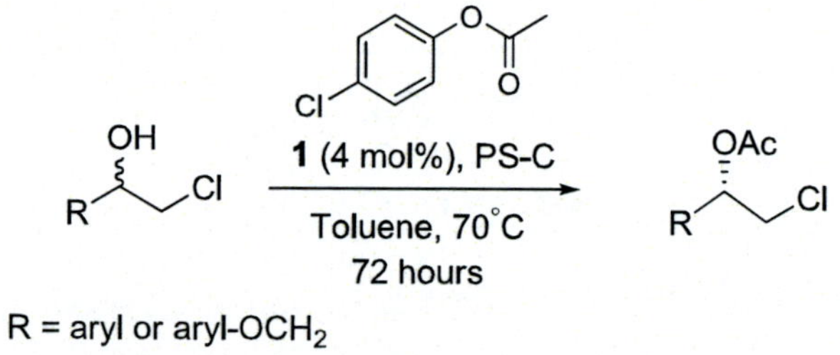

(15)

An efficient DKR protocol for benzoins (1,2-diaryl-2-hydroxyethanones) was presented by Alcántara and coworkers in 2006 using Shvo's catalyst **1** for the racemization [78]. For the enzymatic resolution, *Pseudomonas stutzeri* lipase (Lipase TL) was used, giving the (*S*)-acylated products in excellent enantiomeric excesses (>99% *ee*) and with yields up to 90% (16). The chiral benzoins have synthetic importance as building blocks for different heterocycles.

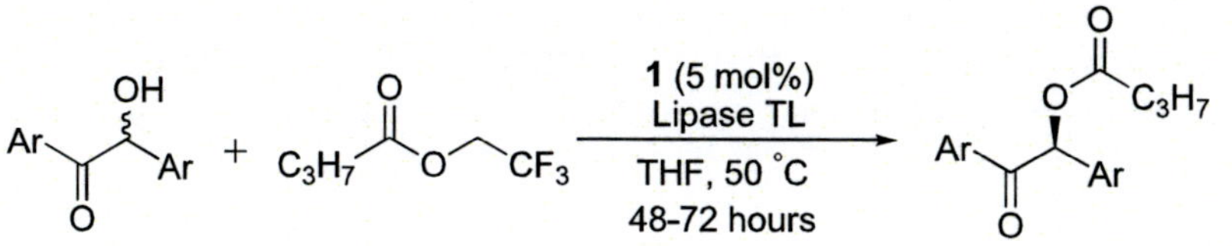

(16)

More recently, Shvo's catalyst **1** was used for a DKR protocol of 1-heteroaryl ethanols [79]. The protocol was found applicable to a wide range of this type of substrate. The desired products were obtained in high to excellent enantiomeric excesses (88 to >99% *ee*) and with moderate to excellent conversions (57–100%).

4.2 DKR of Primary Alcohols

While there are many examples of secondary alcohol DKR protocols, there are only a few examples of DKR of primary alcohols, and these all possess a stereocenter β to the OH group. In one example reported by Bäckvall and coworkers in 2007, Shvo's complex **1** was used [80]. The proposed mechanism for racemization begins with formation of **A** and **2**, which create an equilibrium between primary alcohol and aldehyde via dehydrogenation/hydrogenation. The aldehyde can then undergo enolization, either by applying thermal conditions or via acid/base catalysis. Subsequent hydrogenation/dehydrogenation of the aldehyde by complex **1** would lead to racemization of the alcohol. Hence, complex **1** serves as an indirect racemization catalyst for primary alcohols having a stereocenter β to the OH group (Scheme 33).

In combination with a *Burkholderia cepacia* lipase (PS-D I) for the kinetic resolution, various primary alcohols were converted into the corresponding enantiomerically pure esters in high yields (70–87%) and *ee* values ranging from 70% to 99% (17). The protocol was found applicable to both alkyl- and aryl-substituted primary alcohols.

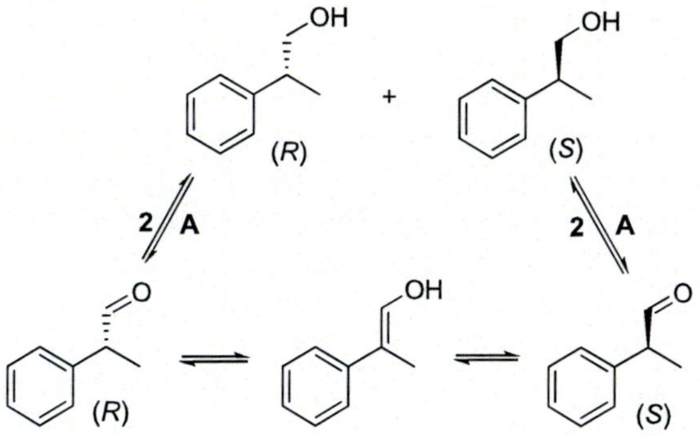

Scheme 33

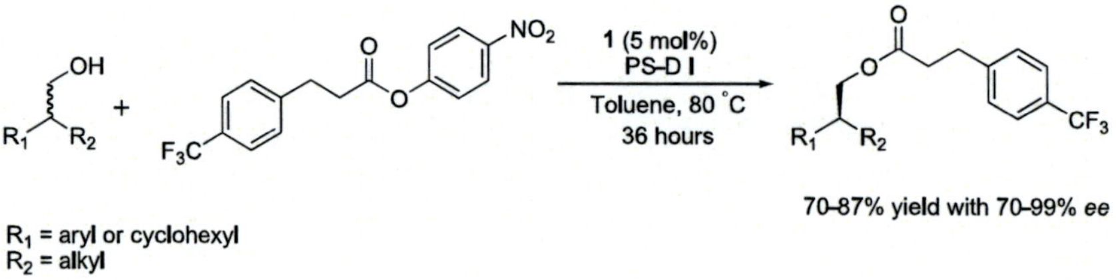

(17)

DKR of primary alcohols has been applied to various derivatives of tropic acid ethyl esters with the use of Shvo's catalyst **1** and *P. cepacia* lipase (PS-C) as the enzyme [81]. The products were obtained with (*S*)-selectivity in high yields (60–88%) and with *ee* values exceeding 90% for several substrates (18). No external hydrogen source was required for the efficiency of the reaction as for many DKR protocols for secondary alcohols. More advantages of the method include low temperature (45°C) and short reaction times (24 h). (*S*)-tropic acid is an important building block for biologically active tropane alkaloids, which are valuable as drugs within the pharmaceutical field. The products are also direct precursors of chiral 2-aryl propionic acids, a structural motif occurring in nonsteroidal anti-inflammatory drugs (NSAIDs) such as ibuprofen and naproxen.

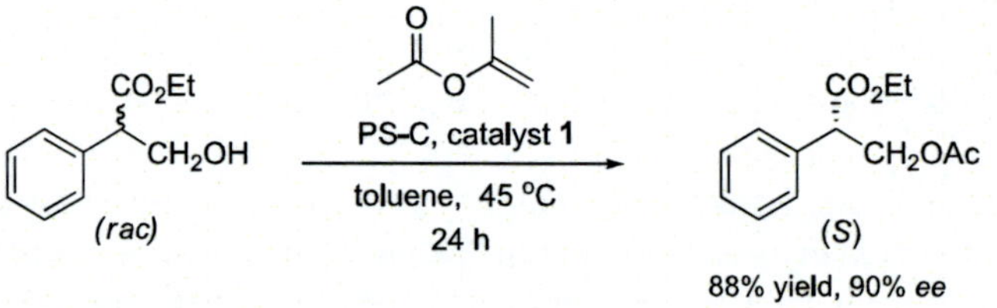

(18)

4.3 DYKAT of Diols

An efficient DKR protocol of symmetrical diols was demonstrated by Bäckvall and coworkers in 1999 [82]. Diols with both aliphatic and aromatic R groups were transformed into the (*R*, *R*)-products using Shvo's catalyst **1** and CALB. In most cases, the *ee* values were >99% and the yields were good to high (47–90%). The selectivity for the (*R*, *R*)-products was consistently high, but for a few of the substrates large amounts of the *meso* products were formed.

Bäckvall and coworkers also reported a DYKAT of unsymmetrical 1,3-diols with one small and one large group [83]. The method makes use of (*i*) selective enzymatic acylation of the least sterically hindered alcohol; (*ii*) epimerization of a secondary alcohol; and (*iii*) intramolecular acyl migration in a *syn*-1,3-diol monoacetate in a one-pot procedure. Shvo's catalyst **1** (4 mol%) was used for the epimerization. CALB was used for the enzymatic resolution along with isopropenyl acetate as the acyl donor. The products were obtained in moderate to high yields (53–73%) and with *ee* values exceeding 99% in all cases. Also, the diastereomeric ratios for the *syn*-diacetates were consistently high, typically >90% *syn* (19).

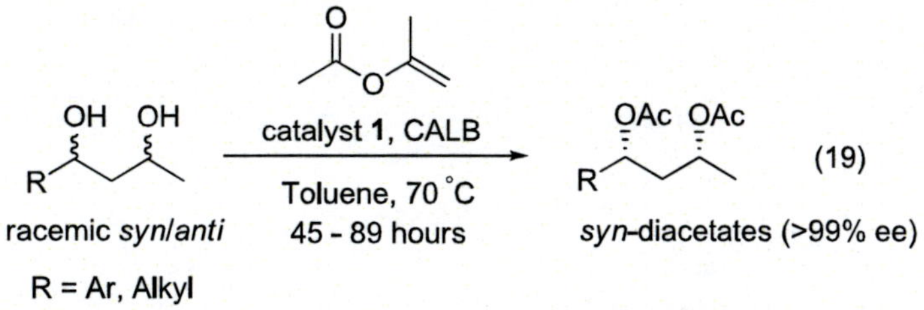

(19)

Enzymatic resolution of 1,4-diols with CALB in combination with a hydrogen transfer process catalyzed by Shvo's complex **1** led to an efficient DYKAT, providing access to chiral γ-acetoxy ketones [84]. A variety of unsymmetrical 1,4-diols were transformed into the products in high to excellent yields (70–96%) and with *ee* values typically exceeding 85% (20). The chiral γ-acetoxy ketones are useful as versatile building blocks for tetrahydrofurans and dihydrofurans.

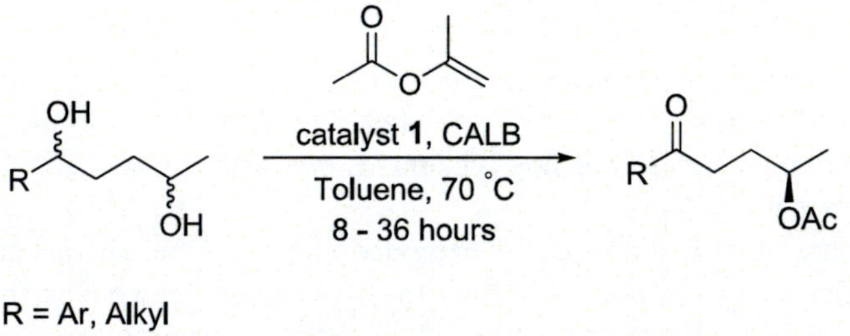

(20)

An efficient DYKAT of a racemic *cis*/*trans* mixture of cyclohexane-1,3-diol using Shvo's complex **1** was reported in 2006 [85]. *Pseudomonas cepacia* lipase (PS-C) was used for the enzymatic resolution along with *para*-chlorophenyl acetate as the acyl donor. The *cis*-diacetate was obtained in 95% yield with high diastereoselectivity (*cis*/*trans* = 97/3).

4.4 DKR of Primary Amines

Enantiomerically pure amines are widely used as building blocks in synthetic organic chemistry, and there is great interest in developing approaches for their synthesis. Enzymatic kinetic resolution of the racemate has had a dominant role in the synthesis of chiral amines, but with the drawback of a maximum theoretical yield of 50%. Therefore, racemization becomes an important tool to provide efficient use of both enantiomers. However, racemization of amines usually requires harsh reaction conditions such as high temperatures or strongly basic media. This limits the substrate scope as these methods generally exhibit low functional group tolerance. Racemization of amines with Shvo's complex **1** involves racemization at the carbon connected to the nitrogen (see Fig. 1).

The racemization of primary amines discussed in Sect. 4.2.3.2 was also combined with an enzymatic kinetic resolution of primary amines in a two-step manner (Scheme 34) [18]. CALB was used for the kinetic resolution. After the first resolution

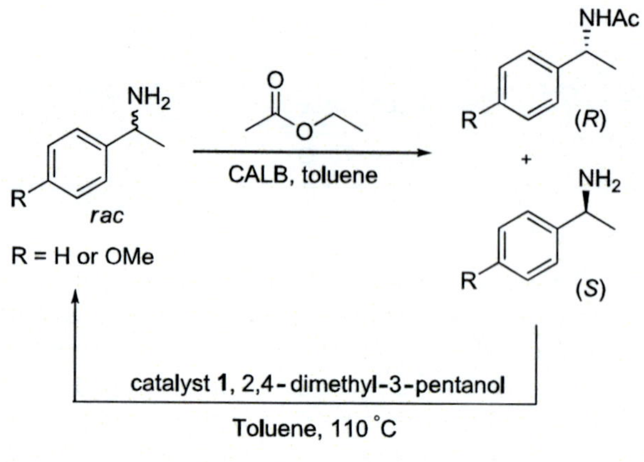

Scheme 34

step, separation of the recovered unreacted amines was performed and these were racemized at 110°C. A second kinetic resolution was then applied to the racemic substrates, affording the corresponding acetamides in excellent enantiomeric excess (>98%) and good yields (66–69%).

The procedure described above was extended into a one-pot chemoenzymatic DKR of primary amines, reported in 2005 by Bäckvall and coworkers [86]. The racemization was investigated first where Shvo's catalyst **1** was found to be reasonably efficient. However, large amounts of by-products were formed (condensation products), which led to a search for a more selective catalyst. An electron-rich variant of the Shvo catalyst was synthesized by replacing all phenyl groups with *para*-methoxyphenyl groups. This catalyst was found to be much more selective and efficient in the racemization, with much less by-product formation. This is most likely due to a fast re-addition of the hydride to the intermediate imine, which depresses formation of condensation products with the imine. After having optimized the racemization protocol, application in DKR was performed using CALB as the enzyme. Isopropyl acetate was chosen as the acyl donor. It was found that addition of Na_2CO_3 was crucial for an efficient DKR, presumably eliminating traces of acid that may interfere with the ruthenium catalyst. The protocol was applicable to a wide range of aliphatic and aromatic substrates with both electron-withdrawing and electron-donating groups present on the aromatic ring. The products were obtained in high yields (69–95%) and with excellent *ee* values (93–99.5%) (21).

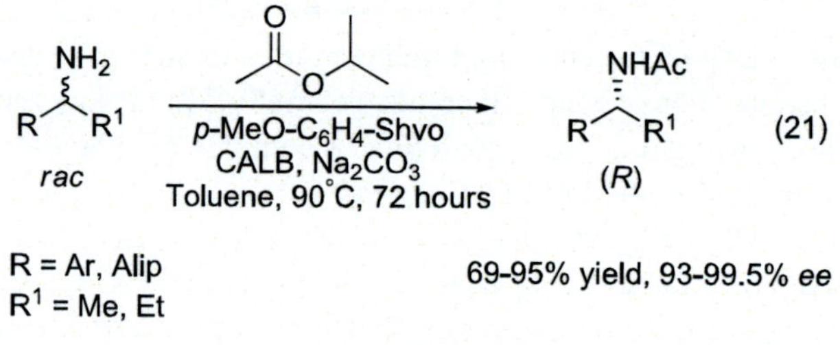

(21)

The DKR procedure described above was improved by Meijer and coworkers in 2007 [87]. The protocol was improved both in terms of reaction time (26 h instead of 72 h) and the required amount of acyl donor (the excess acyl donor could be reduced to 1.1 equiv). This was accomplished using a more effective acyl donor isopropyl 2-methoxyacetate for the enzymatic acylation. CALB was used for the kinetic resolution, and the *para*-methoxyphenyl derivative of the Shvo catalyst was used for racemization (22). All the DKR reactions were performed under reduced pressure (750 mbar) to eliminate the isopropyl alcohol from the reaction mixture. The isopropyl alcohol can be oxidized to acetone, and the latter can in subsequent reaction steps form unwanted condensation products with the amine substrates. The revised protocol afforded the products with excellent selectivity (96–99% *ee*). The yields were slightly lower (56–80%) than those obtained with the Bäckvall protocol [86], mainly due to problems with purification.

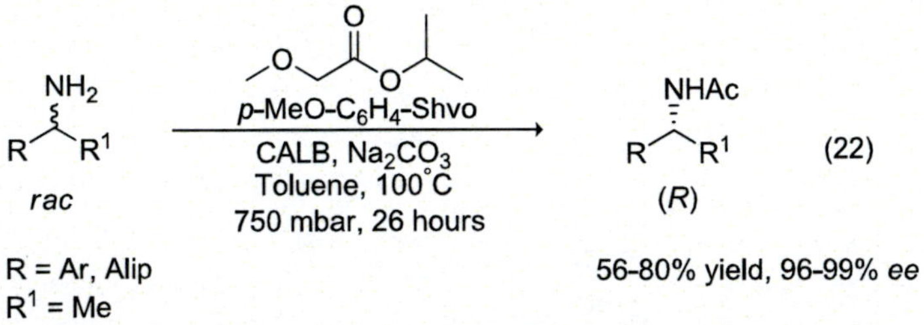

(22)

A drawback of the DKR procedures described above is that the product is a chiral amide, which requires harsh conditions to release the chiral amine. Therefore, development of DKR protocols for racemic amines which transfer a more easily deprotected group is important. A chemoenzymatic DKR of primary amines where the products are benzyl carbamates was reported by Bäckvall and coworkers in 2008 [88]. Using CALB along with dibenzyl carbonate as the acyl donor gave a highly enantioselective reaction ($E = 156$). This enzymatic resolution was also found to be compatible with the *para*-methoxyphenyl derivative of the Shvo catalyst used for racemization. Different aromatic and aliphatic primary amine substrates were efficiently transformed into the corresponding products with high to excellent enantiomeric excess (90–99% *ee*) and with good to excellent yields (60–95%) (23). The benzyloxycarbonyl group can subsequently be removed under very mild conditions via hydrogenolytic cleavage. The deprotection was performed with 1 mol% Pd on charcoal (10%) under a hydrogen atmosphere (1 atm) within 30 min at room temperature. The amines were obtained in quantitative yields and with completely retained *ee* values.

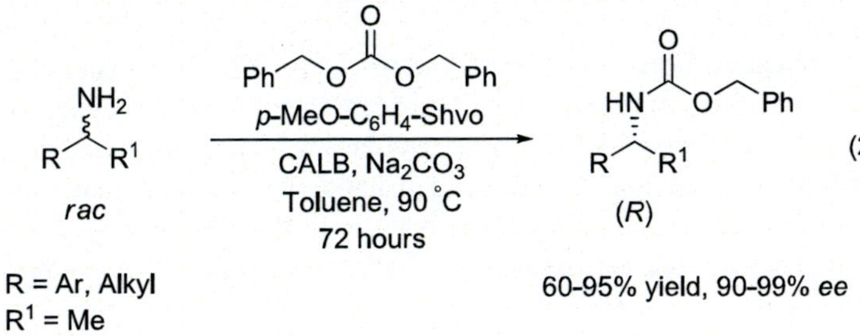

(23)

The substrate scope of the DKR protocol above was extended to include a wide range of primary amines [89]. Functional groups tolerated include fluorine, bromine, nitrile, nitrate, and trifluoromethyl. Aliphatic amines and more sterically hindered substrates were also obtained in high yields and *ee* values. The method was found applicable with both isopropenyl acetate and dibenzyl carbonate as acyl donors. For 18 different substrates, the corresponding enantiomerically pure products were obtained in high to excellent yields (70–95%) and with high to excellent *ee* values (93 to >99%). The chemoenzymatic DKR protocol was also applied to the synthesis of norsertraline, from readily available 1,2,3,4-tetrahydro-1-naphthylamine. Norsertraline is currently in clinical trials for the treatment of central nervous system (CNS) disorders.

More recently, interest in the immobilization of catalysts has increased. *Candida antarctica* lipase A (CALA) immobilized in mesocellular foam (MCF) showed a dramatic increase in the enantioselectivity, as well as an improved thermostability of the enzyme. The immobilized enzyme (CALA/GAmp-MCF) was combined with the *para*-methoxyphenyl derivative of the Shvo catalyst, in an efficient DKR of 3-amino-3-phenylpropanoate at 90°C (24) [90]. The chiral acylated β-amino ester was obtained in 85% yield and with an *ee* value of 89%.

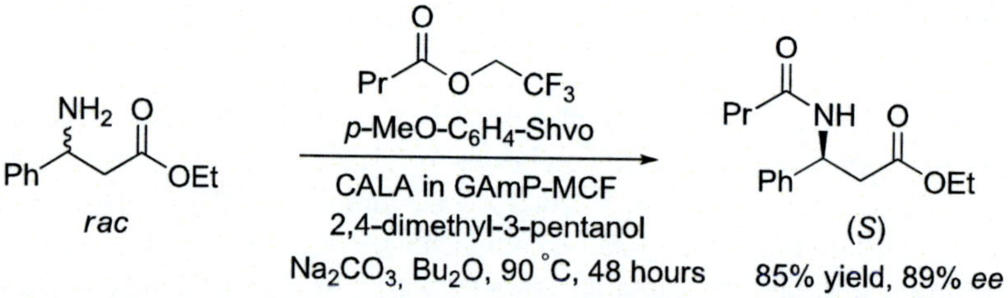

(24)

In a recent publication, DKR of *rac*-1-phenylethylamine was optimized to be applicable on a multigram scale [91]. This is of great importance for industrial applications. The *para*-methoxyphenyl derivative of the Shvo catalyst was used for the racemization. It was found that on a 45 mmol scale, the catalyst loading could be decreased to only 1.25 mol%. For the enzymatic resolution, 10 mg CALB/mmol of substrate was used. With methyl methoxyacetate as the acyl donor, (*R*)-2-methoxy-*N*-(1-phenylethyl)acetamide was obtained with 98% *ee* and in 83% isolated yield.

4.5 DYKAT of 1,3-Aminoalcohols

A recent application deals with the synthesis of chiral 1,3-aminoalcohols through a two-step procedure combining organo-, organometallic, and enzymatic catalysis [92]. In the first step, enantiopure β-aminoketones were prepared through a proline-catalyzed Mannich reaction. A wide range of both electron-deficient and electron-rich β-aminoketones was obtained in >98% *ee*. In the second step, these ketones were subjected to a one-pot tandem hydrogenation/DYKAT procedure to give enantio- and diastereomerically pure 1,3-aminoalcohols. Shvo's complex **1** was used in the second step for both reduction of the ketone and subsequent epimerization of the secondary alcohol formed. CALB along with *p*-chlorophenyl acetate as the acyl donor was found to be selective for the enzymatic resolution. The enantiopure 1,3-aminoacetates were obtained with *ee* values exceeding 99% in all cases and in high yields (73–93%). The diastereomeric ratios exceeded 96:4. The procedure was applicable to both (*S*)- and (*R*)-aminoketones, providing access to both *syn*- and *anti*-1,3-aminoacetates (25).

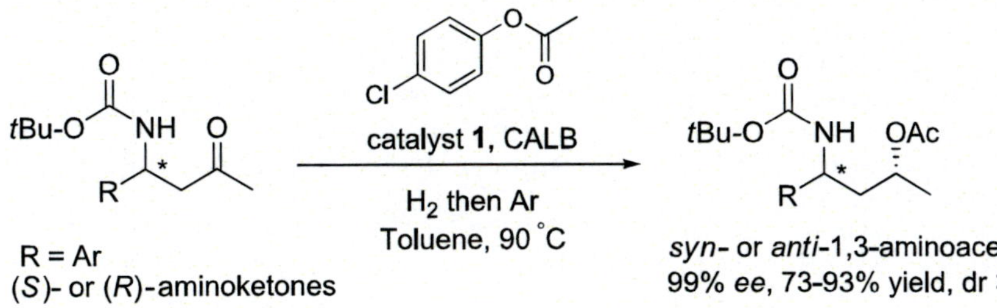

(25)

References

1. Blum Y, Shvo Y (1984) Isr J Chem 24:144
2. Shvo Y, Czarkie D, Rahamim Y, Chodosh DF (1986) J Am Chem Soc 108:7400
3. Menashe N, Shvo Y (1991) Organometallics 10:3885
4. Samec JSM, Bäckvall J-E (2002) Chem Eur J 8:2955
5. Samec JSM, Mony L, Bäckvall J-E (2005) Can J Chem 83:909
6. Blum Y, Czarkie D, Rahamim Y, Shvo Y (1985) Organometallics 4:1459
7. Wang G-Z, Andreasson U, Bäckvall J-E (1994) J Chem Soc Chem Commun:1037
8. Almeida MLS, Beller M, Wang G-Z, Bäckvall J-E (1996) Chem Eur J 2:1533
9. Csjernyik G, Éll AH, Fadini L, Pugin B, Bäckvall J-E (2002) J Org Chem 67:1657
10. Éll AH, Samec JSM, Brasse C, Bäckvall J-E (2002) Chem Commun:1144
11. Éll AH, Johnson JB, Bäckvall J-E (2003) Chem Commun:1652–1653
12. Samec JSM, Éll AH, Bäckvall J-E (2005) Chem Eur J 11:2327
13. Larsson ALE, Persson BA, Bäckvall J-E (1997) Angew Chem Int Ed Engl 36:1211
14. Persson BA, Larsson ALE, Ray ML, Bäckvall J-E (1999) J Am Chem Soc 121:1645
15. Huerta FF, Minidis A, Bäckvall J-E (2001) Chem Soc Rev 30:321
16. Pámies O, Bäckvall J-E (2003) Chem Rev 103:3247
17. Kim MJ, Ahn Y, Park J (2002) Curr Opin Biotechnol 13:578
18. Pamies O, Ell AH, Samec JSM, Hermanns N, Bäckvall J-E (2002) Tetrahedron Lett 43: 4699–4702
19. Casey CP, Singer SW, Powell DR, Hayashi RK, Kavana M (2001) J Am Chem Soc 123: 1090–1100
20. Csjernyik G, Ell AH, Fadini L, Pugin B, Bäckvall J-E (2002) J Org Chem 67:1657–1662
21. Yamakawa M, Ito H, Noyori R (2000) J Am Chem Soc 122:1466–1478
22. Ikariya T, Murata K, Noyori R (2006) Org Biomol Chem 4:393–406
23. Johnson JB, Bäckvall J-E (2003) J Org Chem 68:7681–7684
24. Ell AH, Johnson JB, Bäckvall J-E (2003) Chem Commun:1652–1653
25. Samec JSM, Éll AH, Åberg JB, Privalov T, Eriksson L, Bäckvall J-E (2006) J Am Chem Soc 128:14293–14305

26. Samec JSM, Éll AH, Bäckvall J-E (2004) Chem Commun:2748–2749
27. Åberg JB, Bäckvall J-E (2008) Chem Eur J 14:9169–9174
28. Casey CP, Bikzhanova GA, Cui Q, Guzei IA (2005) J Am Chem Soc 127:14062–14071
29. Casey CP, Clark TB, Guzei IA (2007) J Am Chem Soc 129:11821–11827
30. Karvembu R, Prabhakaran R, Natarajan K (2005) Coord Chem Rev 249:911–918
31. Samec JSM, Bäckvall J-E (2009) Hydroxytetraphenycyclopentadienyl(tetraphenyl-2,4-cyclopentadien-1-one)-μ-hydrotetracarbonyldiruthenium(II). In: Fuchs PL (ed) Encyclopedia of reagents for organic synthesis, vol 7, 2nd edn. Wiley, New York, pp 5557–5564
32. Conley BL, Pennington-Boggio MK, Boz E, Williams TJ (2010) Chem Rev 110:2294–2312
33. Samec J, Bäckvall J-E, Andersson P, Brandt P (2006) Chem Soc Rev 35:237–248
34. Comas-Vives A, Ujaque G, Lledós A (2009) J Mol Struct THEOCHEM 903:123–132
35. Menashe N, Salant E, Shvo Y (1996) J Organomet Chem 514:97
36. Leijondahl K, Fransson AB, Bäckvall J-E (2006) J Org Chem 71:8622
37. Olofsson B, Bógar K, Fransson A-BL, Bäckvall J-E (2006) J Org Chem 71:8256
38. Wang G-Z, Bäckvall J-E (1992) J Chem Soc Chem Commun:980
39. Shvo Y, Goldberg I, Czerkie D, Reshef D, Stein Z (1997) Organometallics 16:133
40. Karlsson U, Wang G-Z, Bäckvall J-E (1994) J Org Chem 59:1196
41. Wang GZ, Andreasson U, Bäckvall J-E (1994) Chem Commun:1037
42. Csernyik G, Éll AH, Fadini L, Pugin B, Bäckvall J-E (2002) J Org Chem 67:1657
43. Almeida MLS, Beller M, Wang G-Z, Bäckvall J-E (1996) Chem Eur J 12:1533
44. Almeida MLS, Kocovský P, Bäckvall JE (1996) J Org Chem 61:6587
45. Thorson MK, Klinkel KL, Wang J, Williams TJ (2009) Eur J Inorg Chem:295–302
46. Krumlinde P, Bogár K, Bäckvall JE (2010) Chem Eur J 16:4031–4036
47. Purse BW, Tran L-H, Piera J, Åkermark B, Bäckvall J-E (2008) Chem Eur J 14:7500–7503
48. Johnston EV, Karlsson EA, Tran L-H, Åkermark B, Bäckvall J-E (2010) Eur J Org Chem:1971–1976.
49. Blacker AJ, Farah MM, Marsden SP, Saidi O, Williams JMJ (2009) Tetrahedron Lett 50:6106–6109
50. Imm S, Bähn S, Tillack A, Mevius K, Neubert L, Beller M (2010) Chem Eur J 16:2705–2709
51. Ebbers EJ, Ariaans GJA, Houbiers JPM, Bruggink A, Zwanenburg B (1997) Tetrahedron 53:9417–9476
52. Thalén LK, Hedberg MH, Bäckvall J-E (2010) Tetrahedron Lett 51:6802–6805
53. Casey CP, Johnson JB, Jiao X, Beetner SE, Singer SW (2010) Chem Commun 46:7915–7917
54. Casey CP, Johnson JB (2005) Can J Chem 83:1339
55. Casey CP, Vos TE, Bikzhanova GA (2003) Organometallics 22:901
56. Casey CP, Guan H (2009) J Am Chem Soc 131:2499
57. Casey CP, Johnson JB (2005) J Am Chem Soc 127:1883–1894
58. Hollmann D, Jiao H, Spannenberg A, Bähn S, Tillack A, Parton R, Altink R, Beller M (2009) Organometallics 28:473–479
59. Shvo Y, Abed M, Blum Y, Laine RM (1986) Isr J Chem 27:267–275
60. Rauk A, Allen LC, Mislow K (1970) Angew Chem Intl Ed Engl 9:400
61. Bushweller CH, O'Neil JW, Bilofsky HS (1972) Tetrahedron 28:2697
62. Casey CP, Bikzhanova GA, Guzei IA (2006) J Am Chem Soc 128:2286–2293
63. Privalov T, Samec JSM, Bäckvall J-E (2007) Organometallics 26:2840–2848
64. Comas-Vives A, Ujaque G, Lledós A (2007) Organometallics 26:4135–4144
65. Comas-Vives A, Ujaque G, Lledós A (2008) Organometallics 27:4854–4863
66. Casey CP, Johnson JB, Singer SW, Cui Q (2005) J Am Chem Soc 127:3100–3109, Correction: J Am Chem Soc (2008) 130:1110
67. Casey CP, Singer SW, Powell DR (2001) Can J Chem 79:1002–1011
68. Casey CP, Beetner SE, Johnson JB (2008) J Am Chem Soc 130:2285–2295
69. Casey CP, Vos TE, Singer SW, Guzei IA (2002) Organometallics 21:5038–5046
70. Casey CP, Strotman NA, Beetner SE, Johnson JB, Priebe DC, Vos TE, Khodavandi B, Guzei IA (2006) Organometallics 25:1230–1235

71. Casey CP, Strotman NA, Beetner SE, Johnson JB, Priebe DC, Guzei IA (2006) Organometallics 25:1236–1244
72. Casey CP, Guan H (2007) J Am Chem Soc 129:5816–5817
73. Huerta FF, Laxmi SYR, Bäckvall J-E (2000) Org Lett 2:1037–1040
74. Pàmies O, Bäckvall J-E (2001) J Org Chem 66:4022–4025, Correction: J Org Chem (2002) 67:1418
75. Pàmies O, Bäckvall J-E (2002) J Org Chem 67:1261–1265
76. Runmo A-BL, Pàmies O, Faber K, Bäckvall J-E (2002) Tetrahedron Lett 43:2983–2986
77. Pàmies O, Bäckvall J-E (2002) J Org Chem 67:9006–9010
78. Hoyos P, Fernández M, Sinisterra JV, Alcántara AR (2006) J Org Chem 71:7632–7637
79. Vallin KSA, Posaric DW, Hamersak Z, Svensson MA, Minidis ABE (2009) J Org Chem 74:9328–9336
80. Strübing D, Krumlinde P, Piera J, Bäckvall J-E (2007) Adv Synth Catal 349:1577–1581
81. Atuu MR, Hossain MM (2007) Tetrahedron Lett 48:3875–3878
82. Persson AB, Huerta FF, Bäckvall J-E (1999) J Org Chem 64:5237–5240
83. Edin M, Steinreiber J, Bäckvall J-E (2004) Proc Natl Acad Sci USA 101:5761–5766
84. Martín-Matute B, Bäckvall J-E (2004) J Org Chem 69:9191–9195
85. Fransson ABL, Xu Y, Leijondahl K, Bäckvall J-E (2006) J Org Chem 71:6309–6316
86. Paetzold J, Bäckvall J-E (2005) J Am Chem Soc 127:17620–17621
87. Veld MAJ, Hult K, Palmans ARA, Meijer EW (2007) Eur J Org Chem:5416–5421
88. Hoben CE, Kanupp L, Bäckvall J-E (2008) Tetrahedron Lett 49:977–979
89. Thalén LK, Zhao D, Sortais J-B, Paetzold J, Hoben C, Bäckvall J-E (2009) Chem Eur J 15:3403–3410
90. Shakeri M, Engström K, Sandström AG, Bäckvall J-E (2010) ChemCatChem 2:534–538
91. Thalén LK, Bäckvall J-E (2010) Beilstein J Org Chem 6:823–829
92. Millet R, Träff AM, Petrus ML, Bäckvall J-E (2010) J Am Chem Soc 132:15182–15184

Top Organomet Chem (2011) 37: 127–160
DOI: 10.1007/3418_2011_3

Published online: 12 March 2011

Liquid-Phase Selective Oxidation by Multimetallic Active Sites of Polyoxometalate-Based Molecular Catalysts

Noritaka Mizuno, Keigo Kamata, and Kazuya Yamaguchi

Abstract The catalytic oxidation is an area of the key technologies for converting petroleum-based feedstocks to useful chemicals such as diols, epoxides, alcohols, and carbonyl compounds. Many efficient homogeneous and heterogeneous oxidation systems based on polyoxometalates (POMs) with green oxidants such as H_2O_2 and O_2 have been developed. This chapter summarizes the remarkable oxidation catalyses by POMs with multimetallic active sites. The multifunctionality of multimetallic active sites in POMs such as cooperative activation of oxidants, simultaneous activation of oxidants and substrates, stabilization of reaction intermediates, and multielectron transfer leads to their remarkable activities and selectivities in comparison with the conventional monometallic complexes. Finally, the future opportunities for the development of shape- and stereoselective oxidation by POM-based catalysts are described.

Keywords Hydrogen peroxide · Molecular oxygen · Multimetallic catalyst · Polyoxometalate · Selective oxidation

Contents

N. Mizuno (✉), K. Kamata, and K. Yamaguchi
Department of Applied Chemistry, School of Engineering, The University of Tokyo, 7-3-1 Hongo, Bunkyo-ku, Tokyo 113-8656, Japan
e-mail: tmizuno@mail.ecc.u-tokyo.ac.jp

1 Introduction

Chemical industries have been contributing to a worldwide economic development over the past century, and yet often cause many serious environmental problems. The demands for environment-friendly chemical processes and products require the development of novel and cost-effective approaches to pollution prevention. For fine chemicals manufacture, antiquated "stoichiometric" technologies (e.g., stoichiometric oxidations with permanganate or dichromate reagents) have still widely been used [1–4]. The replacement of these stoichiometric methodologies with cleaner catalytic ones can reduce the use and generation of toxic and hazardous substances. In these contexts, the catalysis can be expected to remain a cornerstone in building a sustainable chemical community through "green chemistry" [1–4].

The catalytic oxidation is an area of the key technologies for converting petroleum-based feedstocks to useful chemicals such as diols, epoxides, alcohols, and carbonyl compounds [5–11]. Millions of tons of these compounds are annually produced worldwide and applied to all areas of chemical industries ranging from pharmaceutical to large-scale commodities. During recent years, much attention has been paid to the development of catalytic oxidations with environmentally benign oxidants. The choice of the oxidants determines the practicability and efficiency of the oxidation reactions. While a large number of oxidants have been extensively investigated for the catalytic liquid-phase oxidation processes, some oxidants produce toxic and environmentally, politically, and economically unacceptable byproducts. In addition, active oxygen contents of most oxidants are relatively low ($\leq$30%). In these contexts, H_2O_2 and O_2 (or air) are the most attractive oxidants because of their high contents of active oxygen species (O_2:100% for dioxygenase-type, 50% for monooxygenase-type; H_2O_2:47%) and coproduction of only water (no co-products in some cases).

Multimetallic catalysts, alloy catalysts, intermetallic compounds, fuel cell catalysts, colloidal intermediates, metal complexes, and metal clusters have received considerable attention [12–21] because the metal–metal cooperating bifunctional catalysts which can activate reactants simultaneously showed high catalytic activity and stereoselectivity under mild conditions [19]. In fact, there have been many bifunctional multimetallic catalysts in which multimetallic alloy- and electro-catalysts offer a way to fine-tune the catalytic properties of metals, atomic composition, and microstructures [16–18, 20]. Cooperative multimetallic activation of oxidants via the multielectron transfer is also a common feature in biological oxidation catalysis [14]. Artificial multimetallic complexes with two or more metal atoms that contain bridging active oxygen groups also show remarkable oxygenation activities. For example, multimetallic d^0-transition metal complexes containing μ-η^1:η^2-peroxo groups, bio-inspired Cu-based dinuclear μ-η^2:η^2-peroxo complexes, and related bis (μ-oxo) species show high activity for oxidations such as oxygen transfer to C=C double bonds and activation of both aromatic and aliphatic C–H bonds [5–21]. Despite many advances in the field of organometallic complexes, degradation of most organic ligands is almost inevitable under oxidative conditions, and in some case, catalytic lifetimes are quite limited. Therefore, the development of selective oxidation systems with H_2O_2 and O_2 using robust inorganic catalysts containing structurally well-defined multimetallic active sites can contribute to the development of "green" or "environmentally conscious" chemical processes.

"Polyoxometalates" (POMs), which are a large family of anionic metal–oxygen clusters of early transition metals, stimulated many current research activities in broad fields of science such as catalysis, materials, and medicine, because their chemical properties such as redox potentials, acidities, and solubilities in various media can be finely tuned by choosing constituent elements and counter cations [22–37]. Such advantages of inorganic molecular catalysts can introduce bifunctional well-defined single or multiple active sites. In addition, POMs are stable under thermal and oxidative conditions. Therefore, POMs have especially received much attention in the area of the oxidation catalysis. Various types of POMs can act as effective catalysts for the H_2O_2- and O_2-based green oxidations. In this chapter, (1) molecular design of POM catalysts focusing on the structures and the liquid-phase oxidations with H_2O_2 and O_2 (Sect. 2), and (2) oxidation catalysis by multimetallic active sites of POM catalysts (Sect. 3) are described. The multi-functionality of multimetallic active sites in POMs such as *cooperative activation of oxidants, simultaneous activation of oxidants and substrates, stabilization of reaction intermediates, and multielectron transfer* leads to their remarkable activities and selectivities (Fig. 1).

2 Molecular Design of Homogeneous POM Catalysts

POM-based catalysts for the homogeneous oxidations using H_2O_2 and O_2 oxidants can be classified into four groups according to the structures of POMs (Fig. 2): (1) isopoly- and heteropolyoxometalates, (2) peroxometalates, (3) lacunary POMs, and (4) transition metal-substituted POMs. The details on structures, physical and chemical properties, and applications of POMs have been summarized in excellent books and review articles [22–29]. In Sect. 2, the liquid-phase homogeneous oxidations by functionalized POM-based catalysts combined with H_2O_2 and O_2 oxidants are described according to the above classification, and the design of POM-based heterogeneous catalysts is also described.

Typical examples for the homogeneously and heterogeneously catalyzed oxidation reactions are listed in Tables 1 [38–55] and 2 [56–63], respectively.

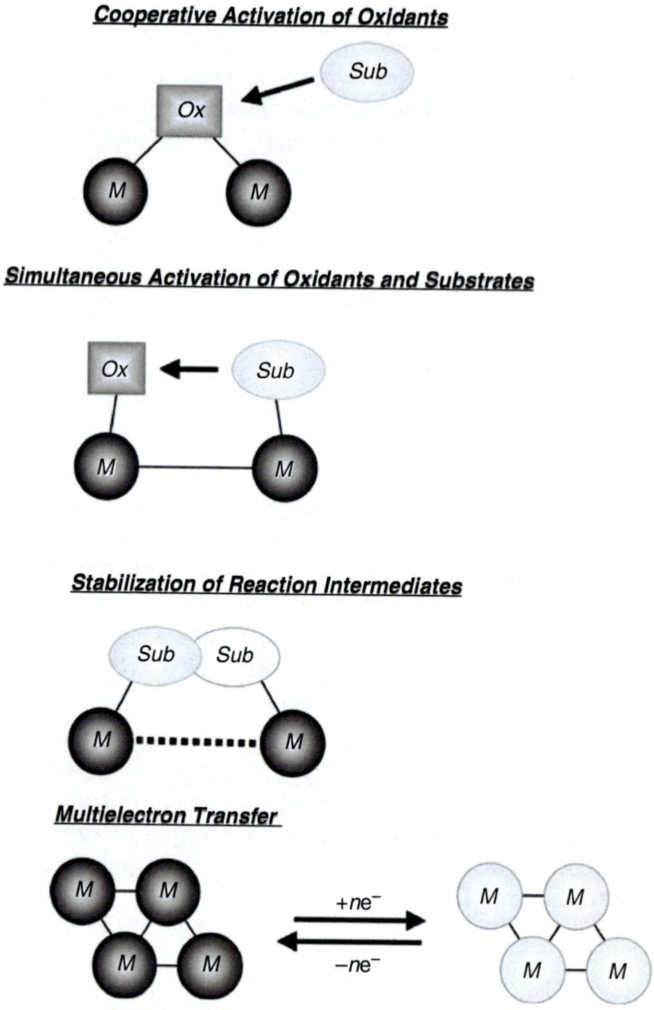

Fig. 1 Bifunctional oxidation catalysis by the multimetallic active sites of POM-based compounds. *Ox* oxidant, *Sub* substrate, *M* metal

2.1 Isopoly- and Heteropolyoxometalates

The tungsten-based isopolyoxometalates and heteropolyoxometalates could act as efficient, stable, selective photooxidation catalysts for the oxidation of alkanes under UV irradiation [38, 64, 65]. The organic products generated in these reactions depended mainly on the ground-state redox potentials of POMs. Under aerobic conditions, autoxidation initiated by radicals generated in the photo-induced redox chemistry was observed. On the other hand, alkanes were converted into the corresponding alkenes under anaerobic conditions. The tungsten-based heteropolyoxometalates also showed high activities for the H_2O_2-based oxidation in

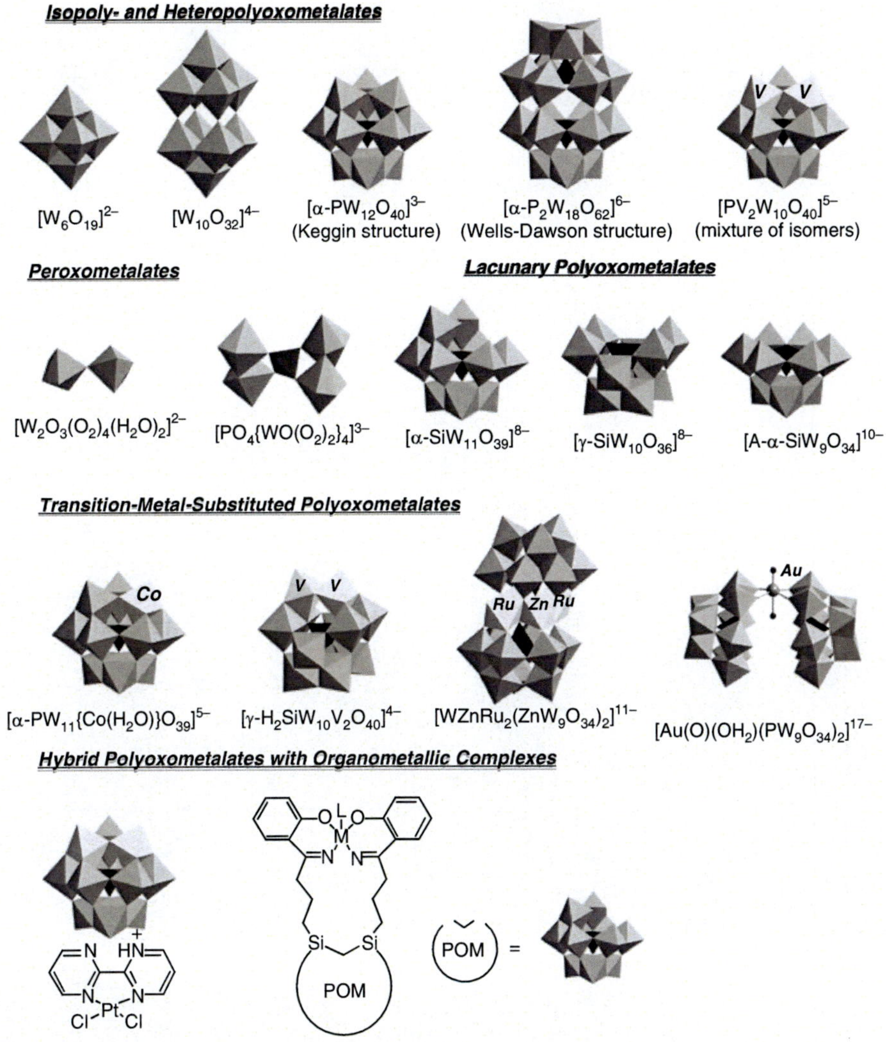

Fig. 2 Classification of homogeneous oxidation POM catalysts

comparison with molybdenum analogs. An effective H_2O_2-based epoxidation of terminal alkenes catalyzed by $H_3[PW_{12}O_{40}]$ in the presence of cetylpyridinium chloride (CPC) as a phase transfer agent has been reported [39]. The system could also be applied to epoxidations of allylic alcohols [39, 66], monoterpenes [44], and α,β-unsaturated carboxylic acids [67], and to oxidations of alcohols [39, 66], amines [68], and alkynes [69], as well as to oxidative transformations of diols [39]. The subsequent spectroscopic and kinetic studies on the epoxidation show that the peroxotungstate $[PO_4\{WO(O_2)_2\}_4]^{3-}$ is a truly catalytically active species for the H_2O_2-based epoxidation with $H_3[PW_{12}O_{40}]$ (Scheme 1) [70–73].

Table 1 Examples of H_2O_2- and O_2-based homogeneous oxidations catalyzed by POMs

Catalyst	Reaction yield (%)	Oxidant	Solvent	Temperature (°C)	References
Isopoy- and heteropolyoxometalates					
$[W_{10}O_{32}]^{4-}$	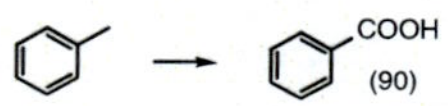	$O_2/h\nu$	CH_3CN	25	[38]
$[PW_{12}O_{40}]^{3-}$/CPC (CPC = cetylpyridinium chloride)		H_2O_2	$CHCl_3$	60	[39]
$[PV_2Mo_{10}O_{40}]^{5-}$	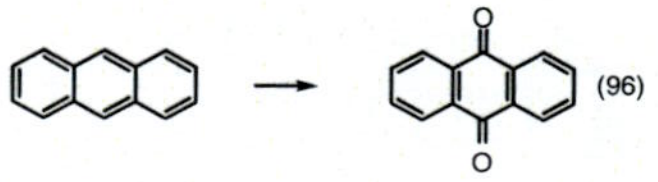	O_2/quinone	CH_3CN	**Reflux**	[40]
$[PV_2Mo_{10}O_{40}]^{5-}$	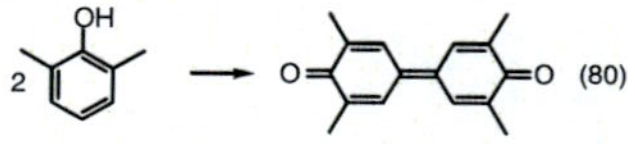	O_2	*n*-Hexanol	60	[41]
Peroxometalates					
$[W_2O_3(O_2)_4(H_2O)_2]^{2-}$		H_2O_2	H_2O	32	[42, 43]
$[PO_4\{WO(O_2)_2\}_4]^{3-}$	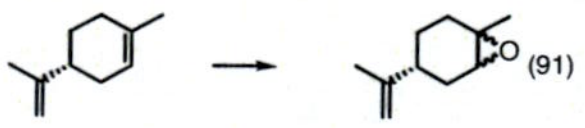	H_2O_2	$CHCl_3$	rt	[44]
$[SeO_4\{WO(O_2)_2\}_2]^{2-}$	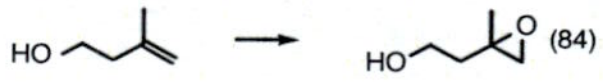	H_2O_2	CH_3CN	32	[45]
$[SeO_4\{WO(O_2)_2\}_2]^{2-}$		H_2O_2	CH_3CN	25	[46]

Lacunary polyoxometalates					
$[\beta_3\text{-}CoW_{11}O_{35}(O_2)_4]^{10-}$	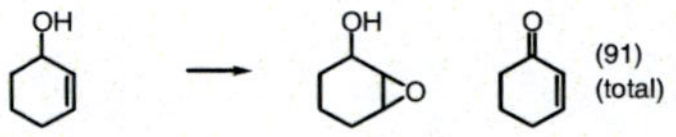	H_2O_2	$CHCl_3$/buffered H_2O	rt	[47]
$[\gamma\text{-}SiW_{10}O_{34}(H_2O)_2]^{4-}$		H_2O_2	CH_3CN	32	[48]
$[\gamma\text{-}SiW_{10}O_{34}(H_2O)_2]^{4-}$		H_2O_2	CH_3CN	32	[49]
Transition-metal-substituted polyoxometalates					
$[WZnRu_2^{III}(ZnW_9O_{34})_2]^{11-}$	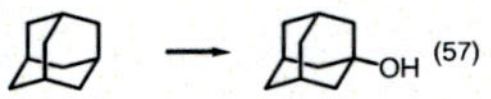	O_2	$ClCH_2CH_2Cl$	80	[50, 51]
$[WZn_3(ZnW_9O_{34})_2]^{12-}$	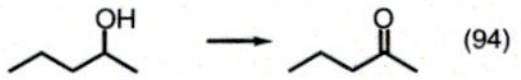	H_2O_2	H_2O	65	[52]
$[Fe_2^{III}(NaOH_2)_2(P_2W_{15}O_{56})_2]^{12-}$	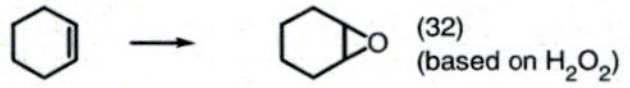	H_2O_2	CH_3CN	25	[53]
$[\gamma\text{-}SiW_{10}\{Fe^{III}(OH_2)\}_2O_{38}]^{6-}$	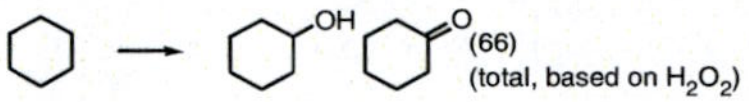	H_2O_2	CH_3CN	32	[54]
$[(CH_3CN)_xFeSiW_9V_3O_{40}]^{6-}$	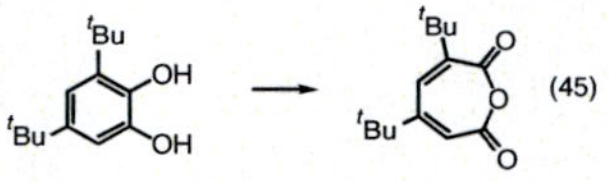	O_2	$ClCH_2CH_2Cl$	65	[55]

Table 2 Examples of H_2O_2- and O_2-based heterogeneous oxidations catalyzed by POMs

Catalyst	Reaction yield (%)	Oxidant	Solvent	Temperature (°C)	References
Immobilization					
$Na_5[PV_2Mo_{10}O_{40}]/C$		O_2	Toluene	100	[56]
$[W_2O_3(O_2)_4(H_2O)_2]^{2-}/IM\text{-}SiO_2$ (IM = *N*-octylimidazolium)	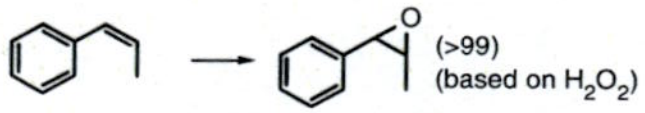	H_2O_2	CH_3CN	60	[57]
$[PW_{12}O_{40}]^{3-}$/MIL-101 (MIL-101 = mesoporous chromium terephthalate coordination polymer)	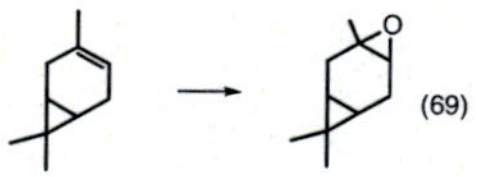	H_2O_2	CH_3CN	50	[58]
NiAl-LDH-WO_4^{2-}	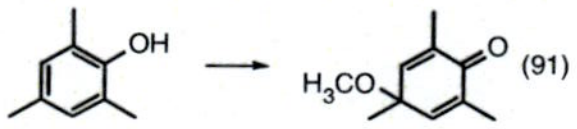	H_2O_2	$H_2O/CH_3OH/THF$	60	[59]
MCM-NHP$[OWO(O_2)_2]_2$	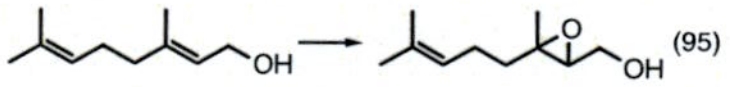	H_2O_2	CH_3CN	50	[60]
Solidification					
$Ag_5[PV_2Mo_{10}O_{40}]$		O_2	CF_3CH_2OH	rt	[61]

NIPAM-$[PW_{12}O_{40}]^{3-}$ (NIPAM = *N*-isopropylacyrlamide-based polymer)	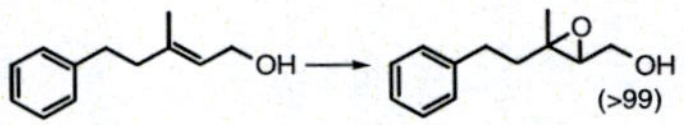	H_2O_2	Solvent-free	rt	[62]
$[Ni(tacn)_2]_2[\gamma\text{-}H_2SiW_{10}V_2O_{40}]$	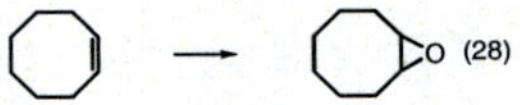	H_2O_2	$CH_3CN/^tBuOH$	10	[63]

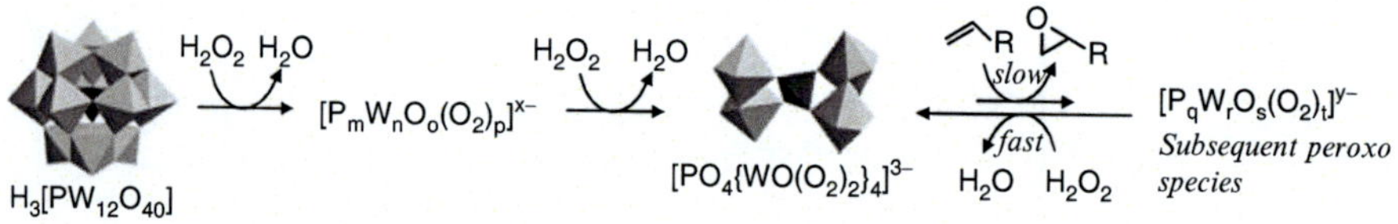

Scheme 1 Proposed mechanism for the epoxidation of alkenes with H_2O_2 catalyzed by $H_3[PW_{12}O_{40}]$ [73]

The V- and Mo-based heteropolyoxometalates are applicable to aerobic oxidation of hydrocarbons because of their high reduction potentials [36, 37]. In fact, the $[PV_nMo_{12-n}O_{40}]^{(3+n)-}/O_2$/hydrocarbon redox systems serves as an excellent catalyst for the aerobic oxidations of a wide variety of hydrocarbons, since the redox potentials of $[PMo_{12-n}V_nO_{40}]^{(3+n)-}$ (ca. 0.7 V vs. SHE) is lower than that of O_2 and is higher than that of common hydrocarbons. A wide variety of functionalized hydrocarbons including alkanes, alcohols, amines, and arenes could be oxidized with O_2 in the presence of mixed-addenda POMs $[PV_nMo_{12-n}O_{40}]^{(3+n)-}$ [74–86]. In these cases, the oxidation includes one (or more) electron transfer from substrates to $[PV_nMo_{12-n}O_{40}]^{(3+n)-}$, affording the corresponding oxidized products with the formation of the reduced POMs. The resulting POMs can smoothly be reoxidized by O_2 although the detailed mechanism for the reoxidation is still unclear. An equilibrated ensemble of POMs with $[AlVW_{11}O_{40}]^{6-}$ which is existing as thermodynamically controlled self-assembly could act as a catalyst for the selective delignification (oxidative degradation of lignin derivatives) of lignocellulose fibers [87]. Equilibration reactions kept the pH of the system near 7 during the catalysis, which avoided the degradation of cellulose by acids or bases (Scheme 2). Similar "self-assembled catalysts" have been reported for the H_2O_2-based alcohol oxidation [52]. It would be very advantageous for practical applications because of operationally simplicity and minimal organic wastes if the reaction could be carried out with such "self-assembled catalysts" by simply mixing the required component of precursors (even without isolation of truly active catalysts).

2.2 *Peroxometalates*

The activation of H_2O_2 by transition metals leads to the generation of a large number of metal peroxo complexes with various coordination modes (η^2-O_2, μ-η^1:η^1-O_2, μ-η^1:η^2-O_2, μ-η^2:η^2-O_2, OOH, etc.) [88–91]. Such structures of metal-coordinated active oxygen species play an important role in various oxidative transformations of organic substrates. Various peroxometalates have been isolated and characterized crystallographically [92, 93]. Among them, the H_2O_2-based oxidation systems by peroxotungstate catalysts have attracted much attention because of the high reactivity and efficiency of H_2O_2 utilization. The tetra-*n*-hexylammonium

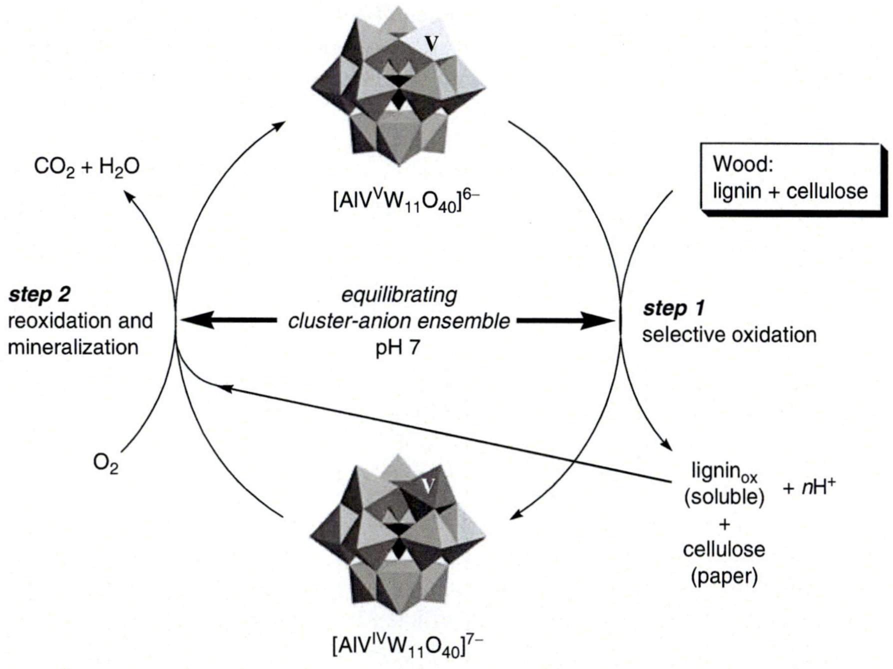

Scheme 2 Proposed mechanism for the selective delignification of wood (lignocellulose) fibers with an equilibrated ensemble of POMs [87]

salt of $[PO_4\{WO(O_2)_2\}_4]^{3-}$ was isolated and characterized crystallographically [94]. The anion consisted of PO_4^{3-} and two $[W_2O_2(O_2)_4]$ units.

As mentioned above, $[PO_4\{WO(O_2)_2\}_4]^{3-}$ has been postulated to be a catalytically active species for the $H_3[PW_{12}O_{40}]/H_2O_2$ system [39]. The catalytic system consisting of $[PO_4\{WO(O_2)_2\}_4]^{3-}$ and $Pd(OAc)_2$ in methanol was also effective for epoxidation of propylene with O_2 instead of H_2O_2 [95]. The nature of the hetero-atoms in the peroxotungstates significantly affects their electrophilic oxygen-transfer reactivities [45, 46, 96, 97].

2.3 Lacunary POMs

Lacunary POMs are of great importance because they have been used as precursors for various compounds such as transition-metal-substituted POMs and inorganic–organic hybrid materials [22–37]. Although the lacunary POMs can activate H_2O_2 to form peroxo species at vacant sites [47], the oxidation catalysis by lacunary POMs has scarcely been reported [73, 98–100]. A tetraprotonated divacant silicodecatungstate $[\gamma\text{-}SiW_{10}O_{34}(H_2O)_2]^{4-}$ catalyzed oxygen-transfer reactions of various substrates such as alkenes, allylic alcohols, sulfides, and silanes with H_2O_2

[48, 49, 101–103] (see Sect. 3). Experimental and theoretical studies on the structure of $[\gamma\text{-}SiW_{10}O_{34}(H_2O)_2]^{4-}$ and the epoxidation mechanism have been reported [104–109]. The inorganic–organic hybrid POMs have attracted much attention because the electron and steric properties of the POMs significantly affect the catalytic properties of the metal centers [110–112]. A hybrid catalyst consisted of a metallosalen complex attached covalently to a lacunary POM $[SiW_{11}O_{39}]^{8-}$ via two propyl spacers. In contrast to a usual Mn^{3+} salen species, the metallosalen complex modified with POM led to formation of a more oxidized Mn^{4+} salen–POM or Mn^{3+} salen cation radical–POM compound [110].

2.4 *Transition-Metal-Substituted POMs*

Transition-metal-substituted POMs are oxidatively and hydrolytically stable compared with organometallic complexes, and their active sites can be controlled. The advantages have been applied to the development of biomimetic catalysis relevant to the enzyme analogs. To date, various kinds of metal-substituted POMs have been synthesized and some of them can efficiently activate O_2 and H_2O_2 [30–37, 113–131]. Although terminal oxo complexes of the late-transition-metal elements have been proposed as possible active species for the catalytic oxidations, the late-transition-metal-oxo complexes have scarcely been known. Hill and coworkers reported the synthesis and characterization of Pt^{4+}-, Pd^{4+}-, and Au^{3+}-oxo complexes, $[M(O)(OH_2)\{WO(OH_2)\}_n(PW_9O_{34})_2]^{m-}$ (M = Pt, Pd, and Au, n = 0–2), stabilized by electron-accepting polyoxotungstate ligands [132–134]. The Au-oxo complex, $[Au(O)(OH_2)\{WO(OH_2)\}_2(PW_9O_{34})_2]^{9-}$, readily reacts with triphenylphosphine to produce triphenylphosphine oxide, $[P_2W_{20}O_{70}(OH_2)_2]^{10-}$, and $Au^+(PPh_3)_n$.

2.5 *Heterogeneous Catalysts*

Until now, many soluble transition-metal-based homogeneous catalysts including POMs have been developed for oxidations using green oxidants H_2O_2 and O_2. They are usually dissolved in reaction solutions making all catalytic sites accessible to substrates and show high catalytic activity and selectivity. Despite these advantages, homogeneous catalysts have a share of only ca. 20% in industrial processes because of complicate procedure for catalyst/product(s) separation, i.e., problem of product contamination and reuse of expensive catalysts are very difficult [135]. Therefore, the developments of easily recoverable and recyclable heterogeneous catalysts have received a particular research interest for fine chemical syntheses.

While many efficient green oxidations based on POMs have been developed as mentioned above, most of them are homogeneous and share common drawbacks of catalyst/product separation and reuse of catalysts. The practical application of POM-based oxidations will need the development of easily recoverable and recyclable catalysts. Recently, the developments of heterogeneous oxidation catalysts based on POMs and the related compounds have been attempted, and the strategies can approximately be classified into the following two categories (Fig. 3): "solidification" of POMs (formation of insoluble solid ionic materials with appropriate counter cations) and "immobilization" of POMs through adsorption, covalent linkage, and ion exchange. The details on the developments of heterogeneous POM catalysts for the H_2O_2- and O_2-based green oxidations have been summarized in the review articles [30–37].

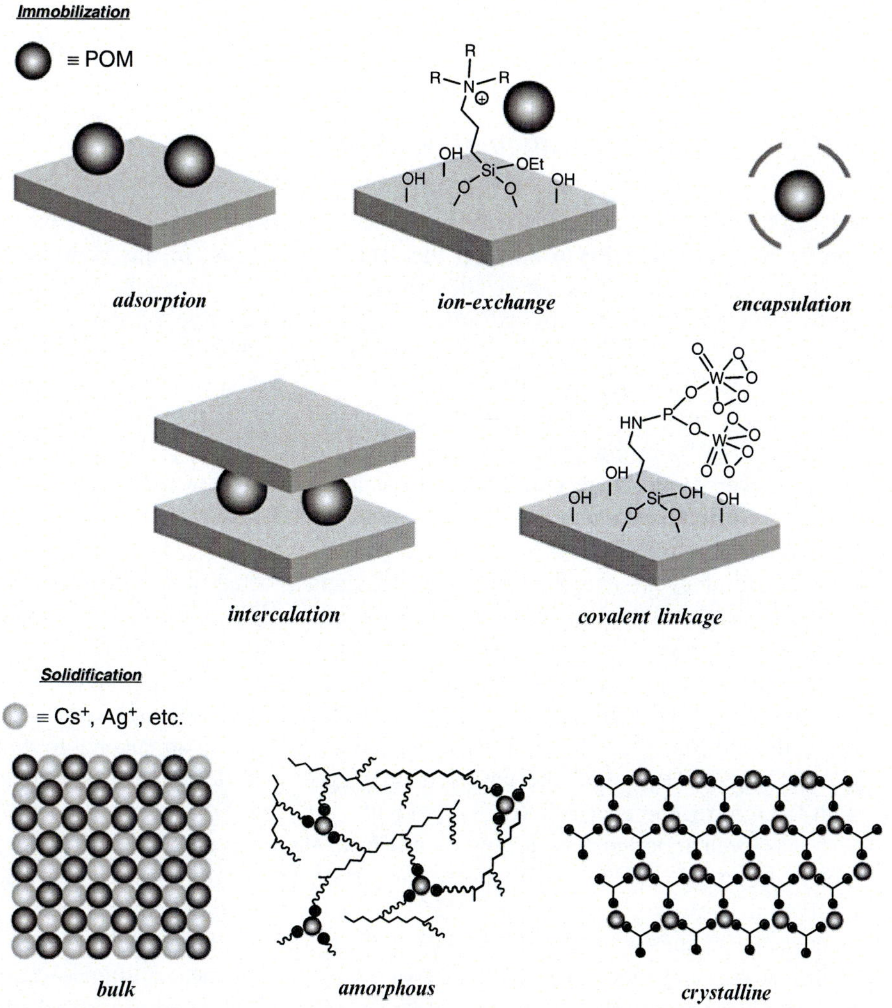

Fig. 3 Schematic representation of the strategies for heterogenization of POM-based compounds

3 Oxidation Catalysis by Multimetallic Active Sites of POM Catalysts

As mentioned in Sect. 2, many efficient homogeneous and heterogeneous oxidation systems based on POMs have been developed. Among them, some POM catalysts with multimetallic active sites show remarkable activity and selectivity in comparison with the conventional monometallic complexes. According to the proposed reaction mechanisms, the oxidation catalyses by multimetallic active sites of POMs are classified into the following four categories; (1) cooperative activation of oxidants, (2) simultaneous activation of oxidants and substrates, (3) stabilization of reaction intermediates, and (4) multielectron transfer (Fig. 1). In Sect. 3, we focus on the selective oxidations by multimetallic active sites of POM-based molecular catalysts.

3.1 Cooperative Activation of Oxidants

It is important to reveal the relationship between the structures of the active oxygen species and their activity to develop the efficient catalysts. In this context, the isolation and structural characterization of the active oxygen species provide the useful information on the oxidation chemistry. The reaction of the dinuclear peroxotungstate $[\{WO(O_2)_2\}_2(\mu\text{-}O)]^{2-}$ with H_2O_2 in CH_3CN gave the peroxo-bridging dinuclear tungsten species $[\{WO(O_2)_2\}_2(\mu\text{-}O_2)]^{2-}$ [136]. The single crystal X-ray structural analysis showed that $[\{WO(O_2)_2\}_2(\mu\text{-}O_2)]^{2-}$consisted of a $\mu\text{-}\eta^1\text{:}\eta^1\text{-}O_2{}^{2-}$ bridging group and two neutral $\{W(=O)(O_2)_2\}$ units, in which seven oxygen atoms were coordinated to a tungsten atom in a distorted pentagonal bipyramidal arrangement. A compound $[\{WO(O_2)_2\}_2(\mu\text{-}O_2)]^{2-}$ was the first structurally determined $(\mu\text{-}\eta^1\text{:}\eta^1\text{-peroxo})d^0$-dimetal complex. The stoichiometric epoxidation of cyclo-, *cis*-2-, *cis*-3-, *trans*-2-, and 1-octenes with $[\{WO(O_2)_2\}_2(\mu\text{-}O_2)]^{2-}$ produced the corresponding epoxides in 97, 100, 94, 93, and 91% yields, respectively, suggesting that $[\{WO(O_2)_2\}_2(\mu\text{-}O_2)]^{2-}$ has 1 equivalent of active oxygen species for the epoxidation. On the other hand, the stoichiometric epoxidation with $[\{WO(O_2)_2\}_2(\mu\text{-}O)]^{2-}$ hardly proceeded. The catalytic epoxidation proceeded by the reaction of $[\{WO(O_2)_2\}_2(\mu\text{-}O_2)]^{2-}$ with an alkene to form $[\{WO(O_2)_2\}_2(\mu\text{-}O)]^{2-}$ and the corresponding epoxide followed by the reaction with H_2O_2 to regenerate $[\{WO(O_2)_2\}_2(\mu\text{-}O_2)]^{2-}$ (Scheme 3).

It has been reported that Keggin- and Wells–Dawson-type mono-transition-metal-substituted POMs, $[PW_{11}O_{39}M(OH_2)]^{5-}$ ($M = Mn^{2+}$ and Co^{2+}) and $[\alpha_2\text{-}P_2W_{17}O_{61}M(Br)]^{7-}$ ($M = Mn^{3+}$, Fe^{3+}, and Co^{2+}), catalyzed the oxidations of alkenes and alkanes with PhIO (1) and (2) [137, 138]. The M atom easily becomes coordinatively unsaturated by elimination of the ligand, and the resulting polyanion is regarded as an inorganic metalloporphyrin analog. While they showed high activity and selectivity, the iodosobenzenes such as PhIO and F_5PhIO were required

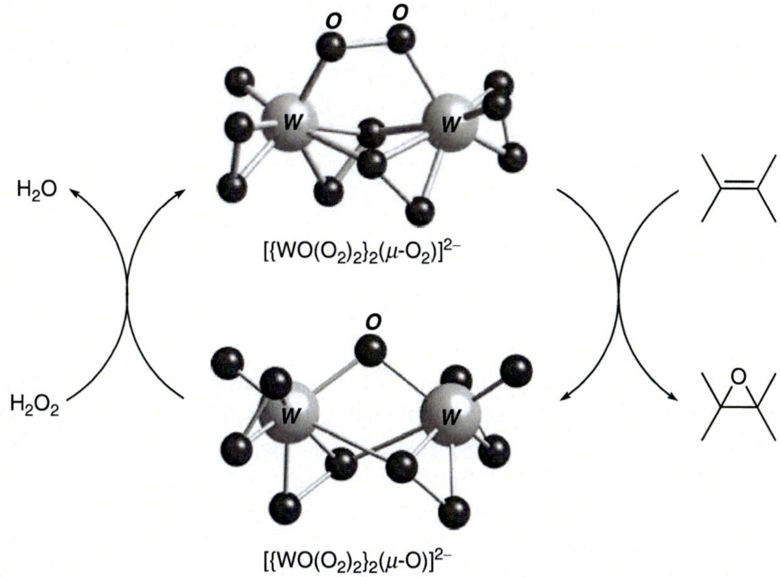

Scheme 3 Proposed mechanism for epoxidation of alkenes with H_2O_2 catalyzed by $[\{WO(O_2)_2\}_2(\mu\text{-}O)]^{2-}$ [136]

as monooxygen donors. Recent developments for the oxidation catalysis by the transition-metal-substituted POMs with multimetallic active sites can establish the use of green oxidants such as H_2O_2 and O_2. The multimetallic active sites can efficiently activate H_2O_2 and O_2, leading to the remarkable activity and selectivity, although detailed mechanistic works are still necessary in some cases [139–142].

The reaction of Keggin-type $H_3[XW_{12}O_{40}]$ ($X = P^{5+}$ and As^{5+}) with excess H_2O_2 leads to the degradation of the peroxotungstates such as $[XO_4\{WO(O_2)_2\}_4]^{3-}$ ($X = P^{5+}$ and As^{5+}), and these species has been proposed to be true catalytically active

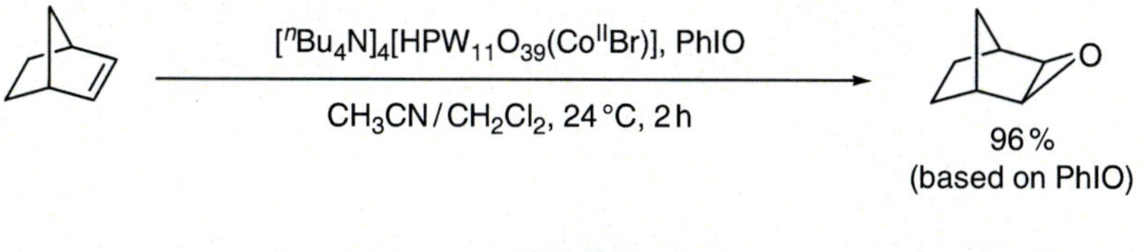

(1)

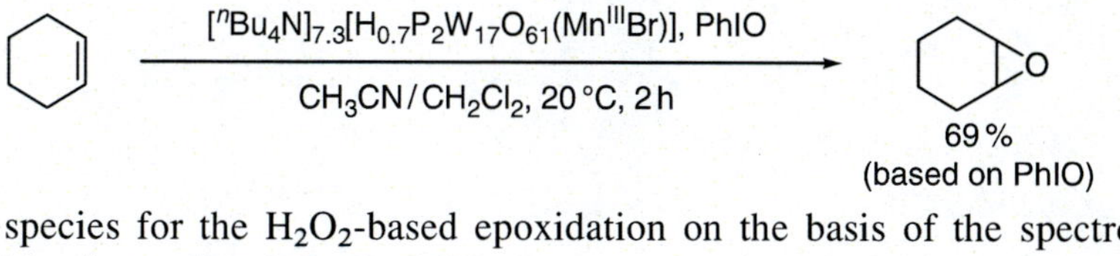

(2)

species for the H_2O_2-based epoxidation on the basis of the spectroscopic and kinetic studies [70–73, 143]. In contrast, the Ti- and Nb-containing POMs, $[PW_{11}O_{39}Ti(O_2)]^{5-}$, $[PW_{10}O_{38}Ti_2(O_2)_2]^{7-}$, $[P_2W_{12}O_{56}Nb_6(O_2)_6]^{12-}$, $[A\text{-}\beta\text{-}PW_9O_{37}Nb_3(O_2)_3]^{6-}$, $[A\text{-}\alpha\text{-}SiW_9O_{37}Nb_3(O_2)_3]^{7-}$, $[\alpha\text{-}1,2,3\text{-}H_3P_2W_{15}O_{59}Ti_3(O_2)_3]^{9-}$, and

$[H_3Nb_6O_{13}(O_2)_6]^{5-}$, have been isolated [47, 144–149]. These species have the η^2-peroxo groups coordinated to Ti and Nb atoms. The oxidation of allylic alcohols with H_2O_2 catalyzed by $[A\text{-}\beta\text{-}PW_9O_{37}Nb_3(O_2)_3]^{6-}$ gave the corresponding triols in 74–94% yields under the reflux conditions (3) [146].

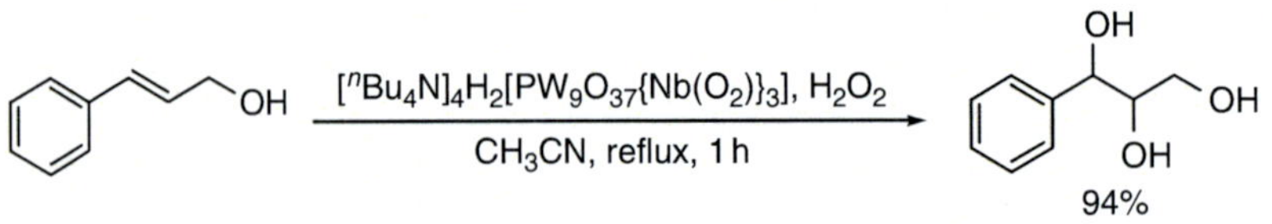

(3)

Recently, the Zr- and Hf-containing POMs with the μ-η^2:η^2-peroxo groups have been reported [150, 151]. The structures are dependent on the kinds of lacunary POM precursors. For the γ-Keggin dilacunary precursor $[\gamma\text{-}SiW_{10}O_{36}]^{8-}$, three $[\gamma\text{-}SiW_{10}O_{36}]^{8-}$ units encapsulated the unprecedented $[M_6(O_2)_6(OH)_6]^{6+}$ ($M = Zr^{4+}$ and Hf^{4+}) wheel, resulting in the trimeric species (Fig. 4a). On the other hand, for the α-Keggin monolacunary precursor $[\alpha\text{-}XW_{11}O_{39}]^{8-}$ (X = Si and Ge), two $[\alpha\text{-}XW_{11}O_{39}]^{8-}$ units encapsulated the cationic $[M_2(O_2)_2]^{4+}$ ($M = Zr^{4+}$ and Hf^{4+}) core, resulting in the dimeric species (Fig. 4b). Such a unique μ-η^2:η^2-peroxo-bridging species was also reported for the uranyl–peroxo containing 36-tungsto-8-phosphate $[Li(H_2O)K_4(H_2O)_3\{(UO_2)_4(O_2)_4(H_2O)_2\}_2(PO_3OH)_2\text{-}P_6W_{36}O_{136}]^{25-}$ [152]. Some of them were applicable to the H_2O_2-based oxidation. The dendrimer-containing POMs that were synthesized by the reaction of $[Zr_2(O_2)_2(SiW_{11}O_{39})_2]^{12-}$ with ammonium dendrons could act as a recoverable catalyst for the oxidation of sulfides to the corresponding sulfoxides and sulfones with H_2O_2 (4) [153].

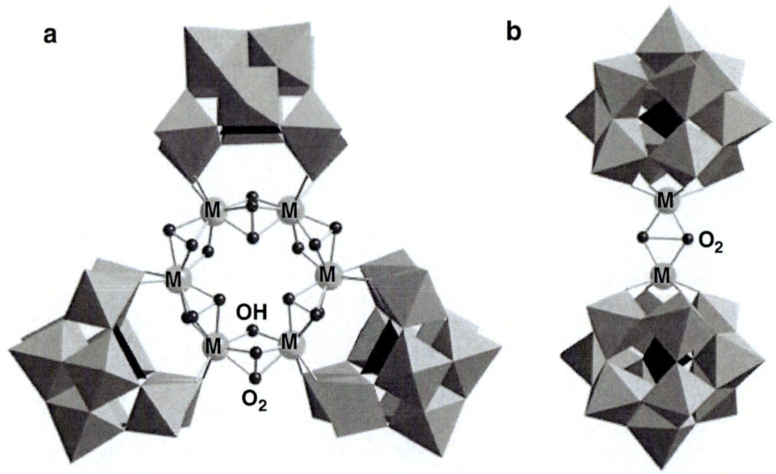

Fig. 4 Molecular structures of (**a**) $[M_6(O_2)_6(OH)_6(\gamma\text{-}SiW_{10}O_{36})_3]^{18-}$ ($M = Zr^{4+}$ and Hf^{4+}) and (**b**) $[M_2(O_2)_2(\alpha\text{-}XW_{11}O_{39})_2]^{12-}$ ($M = Zr^{4+}$ and Hf^{4+}, $X = Si^{4+}$ and Ge^{4+}) [150, 151]

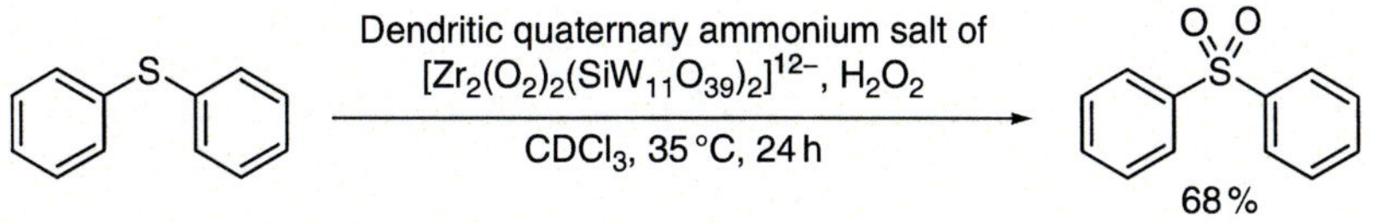

(4)

The Keggin di-vanadium-substituted silicotungstate, $[SiW_{10}O_{38}V_2(\mu\text{-}OH)_2]^{4-}$, with a $\{VO\text{-}(\mu\text{-}OH)_2\text{-}VO\}$ core could catalyze the epoxidation of various alkenes using H_2O_2 with high epoxide yields and high efficiencies of H_2O_2 utilization under very mild reaction conditions [113, 114]. Nonactivated aliphatic terminal C_3–C_{10} alkenes including propene could be transformed to the corresponding epoxide with ≥99% selectivity and ≥87% efficiency of H_2O_2 utilization. Also, the catalytic epoxidation of cyclic alkenes efficiently proceeded to afford the corresponding epoxides in high yields. The epoxidation of 3-substituted cyclohexenes such as 3-methyl-1-cyclohexene and 2-cyclohexen-1-ol showed unusual diastereoselectivities; the epoxidation diastereoselectively gave the corresponding epoxides with oxirane ring *trans* to the substituents (*anti* configuration) (5). Notably, the more accessible, but less nucleophilic double bonds in nonconjugated dienes such as *trans*-1,4-hexadiene, *R*-(+)-limonene, 7-methyl-1,6-octadiene, and 1-methyl-1, 4-cyclohexadiene were highly regioselectively epoxidized in high yields. The specific regioselectivity for the epoxidation of nonconjugated dienes probably reflects the electronic and steric characters of the active oxygen species.

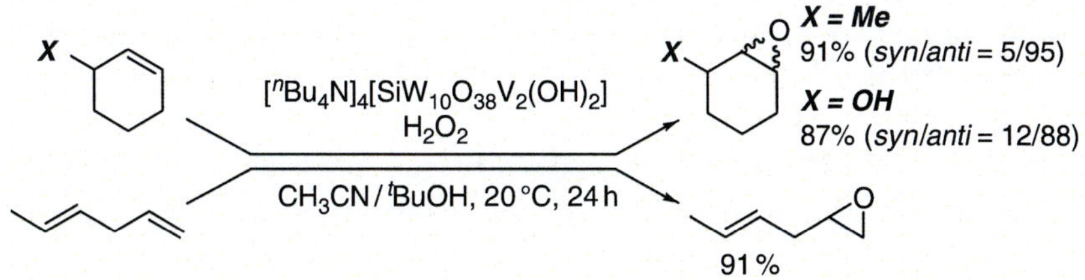

(5)

A possible reaction mechanism for the $[\gamma\text{-}SiW_{10}O_{38}V_2(\mu\text{-}OH)_2]^{4-}$-catalyzed epoxidation with H_2O_2 is shown in Scheme 4. The ^{51}V NMR, ^{183}W NMR, and CSI-MS spectroscopies show that the reaction of $[\gamma\text{-}SiW_{10}O_{38}V_2(\mu\text{-}OH)_2]^{4-}$ with H_2O_2 leads to the reversible formation of a hydroperoxo species $[\gamma\text{-}SiV_2W_{10}O_{38}(OH)(OOH)]^{4-}$. The successive dehydration of $[\gamma\text{-}SiW_{10}O_{38}V_2(OH)(OOH)]^{4-}$ forms $[\gamma\text{-}SiW_{10}O_{38}V_2(O_2)]^{4-}$, which possibly has an active oxygen species of a $\mu\text{-}\eta^2\text{:}\eta^2$-peroxo group. The kinetic and spectroscopic studies show that the present epoxidation proceeds via $[\gamma\text{-}SiW_{10}O_{38}V_2(O_2)]^{4-}$. The energy diagram of the epoxidation with density functional theory (DFT) supports the idea.

The control of the nature of the oxidant by changing the heteroatom can establish the regioselective hydroxylation of alkanes. The efficient, stereospecific, and regioselective hydroxylation of alkanes with H_2O_2 catalyzed by a di-vanadium-substituted phosphotungstate, $[\gamma\text{-}H_2PV_2W_{10}O_{40}]^{3-}$, was reported [115]. The high selectivity to alcohols, efficiency of H_2O_2 utilization, and stereospecificity were observed. The present system also showed the specific regioselectivity to the secondary alcohols for the oxidation of some cycloalkanes with both secondary

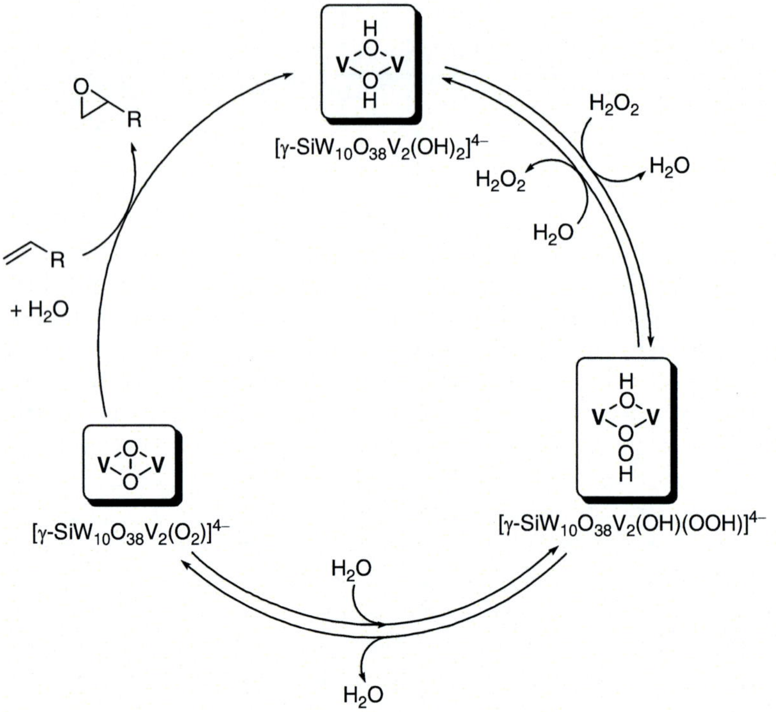

Scheme 4 Proposed mechanism for the epoxidation of alkenes with H_2O_2 catalyzed by $[\gamma\text{-SiW}_{10}O_{38}V_2(OH)_2]^{4-}$ [114]. POM frameworks are omitted for clarity

and tertiary C–H bonds. This study provides the first example of a synthetic catalyst that can achieve the specific regioselective H_2O_2-based hydroxylation of secondary C–H bonds even in the presence of more reactive tertiary C–H bonds (6).

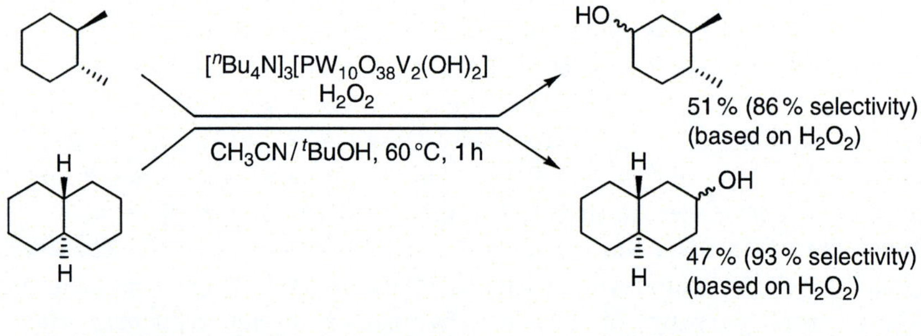

(6)

Not only the transition metals incorporated in transition-metal-substituted POMs, but also addenda atoms in POMs can activate H_2O_2. The reaction of lacunary POM $[\alpha\text{-CoW}_{11}O_{39}]^{9-}$ leads to the formation of tetraperoxo species $[\beta_3\text{-CoW}_{11}O_{35}(O_2)_4]^{10-}$ [47]. Compound $[\beta_3\text{-CoW}_{11}O_{35}(O_2)_4]^{10-}$ prepared in situ by the reaction of $[\alpha\text{-CoW}_{11}O_{39}]^{9-}$ with H_2O_2 showed the catalytic activity for the epoxidation of 2-cyclohexen-1-ol and the corresponding epoxide and ketone were obtained (7).

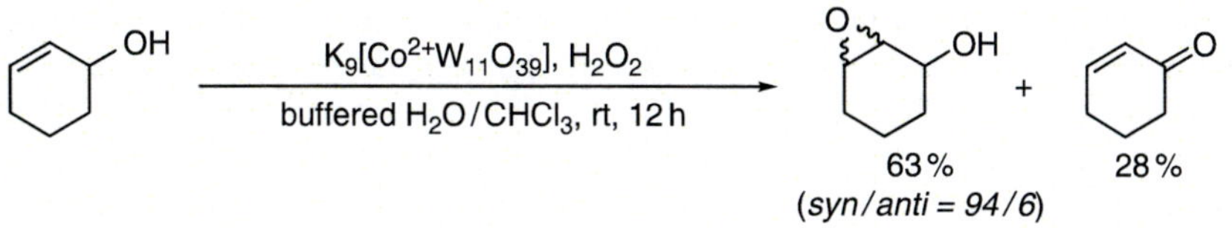

(7)

The oxo ligands at the vacant sites are basic enough to react with not only metal cations but also H^+. The tetra-*n*-butylammonium salt of a protonated divacant silicodecatungstate, $[\gamma\text{-}SiW_{10}O_{34}(H_2O)_2]^{4-}$, efficiently catalyzed oxygen-transfer reactions of various substrates including alkenes, allylic alcohols, sulfides, and organosilanes with aqueous H_2O_2 [48, 49, 101–103]. The high yields and efficiencies of H_2O_2 utilization were achieved for the epoxidation of various terminal, internal, and cyclic alkenes [typically ≥99%, (8)]. For the epoxidation of primary allylic alcohols, the corresponding epoxy alcohols were selectively obtained with small amounts of α,β-unsaturated aldehydes. Geraniol was regioselectively epoxidized at the electron-deficient allylic 2,3-double bond without formation of the 6,7-epoxy alcohol. On the other hand, geranyl acetate was selectively epoxidized at the electron-rich double bond to give the 6,7-epoxide. Various sulfides were also selectively oxidized to the corresponding sulfoxides with high efficiency of H_2O_2 utilization. In the presence of $[\gamma\text{-}SiW_{10}O_{34}(H_2O)_2]^{4-}$, various organosilanes were efficiently oxidized to corresponding silanols with H_2O_2 as an oxidant (9). The reaction of $[\gamma\text{-}SiW_{10}O_{34}(H_2O)_2]^{4-}$ with H_2O_2 led to the generation of the diperoxo species, $[\gamma\text{-}SiW_{10}O_{32}(O_2)_2]^{4-}$. The two oxo groups ($O^{2-}$) were replaced by two peroxo groups (O_2^{2-}) on the divacant lacunary site with retention of the γ-Keggin framework.

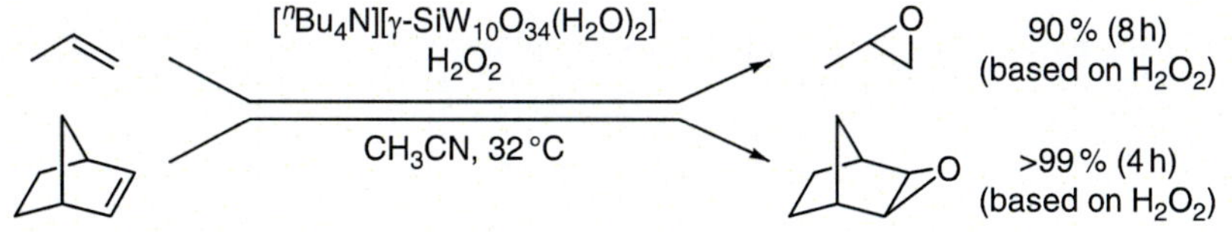

(8)

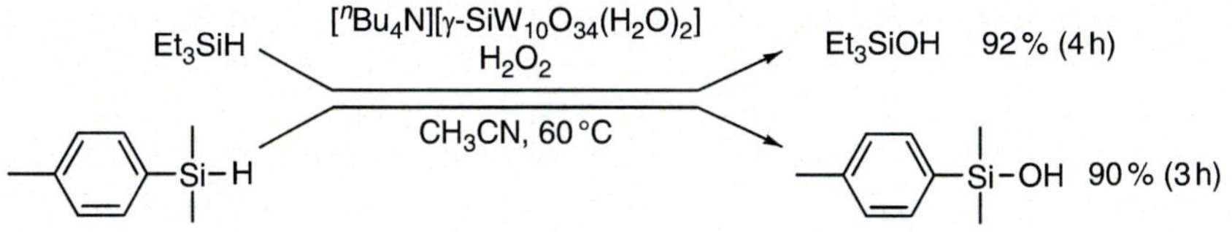

(9)

3.2 *Simultaneous Activation of Oxidants and Substrates*

Natural enzymes such as methane monooxygenases and fatty acid desaturases use the molecular recognition such as size/shape selectivity and substrate orientation to achieve the high chemo-, regio-, and stereoselectivities [154]. For the artificial catalysts, the synergistic interaction between substrates and reagents plays an important role in the control of selectivity, and a detailed information on the transition state structure in the oxygen-transfer process has been investigated.

Sandwich-type POMs, $[WZnM_2(ZnW_9O_{34})_2]^{q-}$ (M = Mn^{2+}, Ru^{3+}, Fe^{3+}, Pd^{2+}, Pt^{2+}, Zn^{2+}; q = 10–12), showed the catalytic activity for the epoxidation of chiral allylic alcohols with 30% H_2O_2 under mild conditions (ca. 20°C) in an aqueous/organic biphasic system [155]. Since the kinds of transition metals M in POMs did not affect the reactivity, chemoselectivity, and stereoselectivity for the epoxidation, a tungsten peroxo complex rather than a high-valent transition-metal-oxo species was proposed as the key intermediate. The diastereoselectivity for the epoxidation of secondary allylic alcohols showed that the epoxidation proceeds through the metal-alcoholate intermediate, and that the estimated dihedral angle between the π plane of the double bond and the hydroxy group of the allylic alcohols (C=C–C–OH) was 50–70°. A high enantioselectivity [enantiomeric excess (*ee*) 90%] was achieved with the V^{4+}=O-substituted POM, $[ZnW(VO)_2(ZnW_9O_{34})_2]^{12-}$, and the sterically demanding TADOOL-derived hydroperoxide, TADOOH ([(4*R*,5*R*)-5-[(hydroperoxydiphenyl)methyl]-2,2-dimethyl-1,3-dioxolan-4-yl]diphenylmethanol), as regenerative chiral oxygen source instead of H_2O_2 (10) [156]. The present chiral catalyst showed excellent catalytic efficiency and the turnover number (TON) reached up to 42,000.

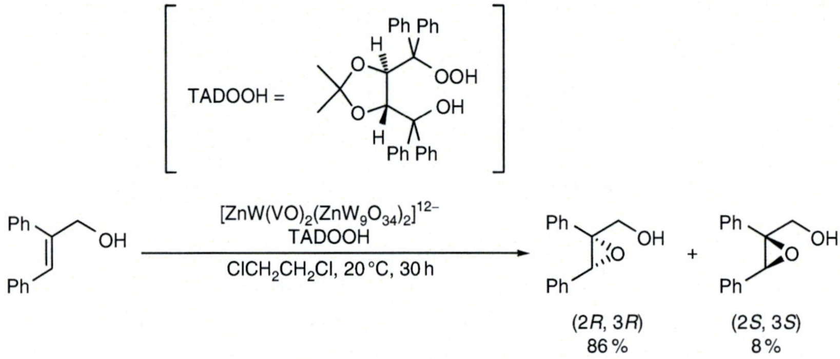

(10)

The highly chemo-, regio-, and diastereoselective and stereospecific epoxidation of various allylic alcohols with only 1 equivalent of H_2O_2 in water could be efficiently catalyzed by a dinuclear peroxotungstate, $K_2[\{WO(O_2)_2(H_2O)\}_2(\mu\text{-}O)]\cdot 2H_2O$ (11) [42, 43]. The catalyst was easily recycled with the maintenance of the catalytic performance. The catalytic reaction mechanism including the coordination of an allylic alcohol to the dinuclear peroxotungstate, the elimination of H^+ from hydroxyl group followed by the insertion of oxygen to the carbon–carbon double bond, and the regeneration of the dinuclear peroxotungstate with H_2O_2 was proposed.

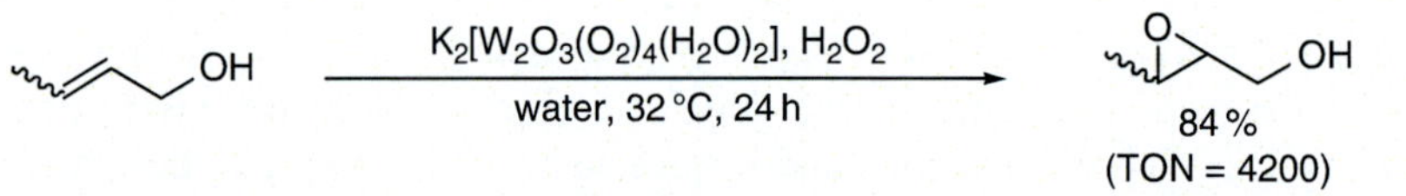

(11)

Because the possible metal-alcoholate has the longer chain in the transition state, the systems are not so efficient for the epoxidation of homoallylic alcohols [157].

However, as a result of the hydrogen-bond-mediated activation of homoallylic and allylic alcohols, the selenium-containing dinuclear tungsten species, $[SeO_4\{WO(O_2)_2\}_2]^{2-}$, showed the high catalytic activity for H_2O_2-based epoxidation of homoallylic and allylic alcohols [45]. Various kinds of homoallylic and allylic alcohols were efficiently epoxidized to the corresponding epoxy alcohols in high yields with 1 equivalent of H_2O_2 with respect to the substrates. The turnover frequency reached up to 150 h^{-1} in a 10 mmol-scale epoxidation of *cis*-3-hexen-1-ol, and this value was the highest among those reported for the transition-metal-catalyzed epoxidation of homoallylic alcohols with H_2O_2 (12). The kinetic, mechanistic, and computational studies showed that the stabilization of the transition state by the hydrogen bonding between $[SeO_4\{WO(O_2)_2\}_2]^{2-}$ and the substrates results in the high reactivity for the $[SeO_4\{WO(O_2)_2\}_2]^{2-}$-catalyzed epoxidation of homoallylic and allylic alcohols (Fig. 5). The present system could be applied to the oxidation of sulfides to the corresponding sulfoxides (13).

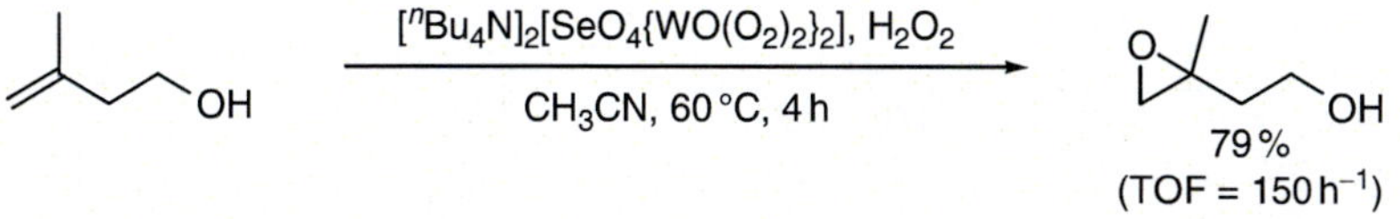

(12)

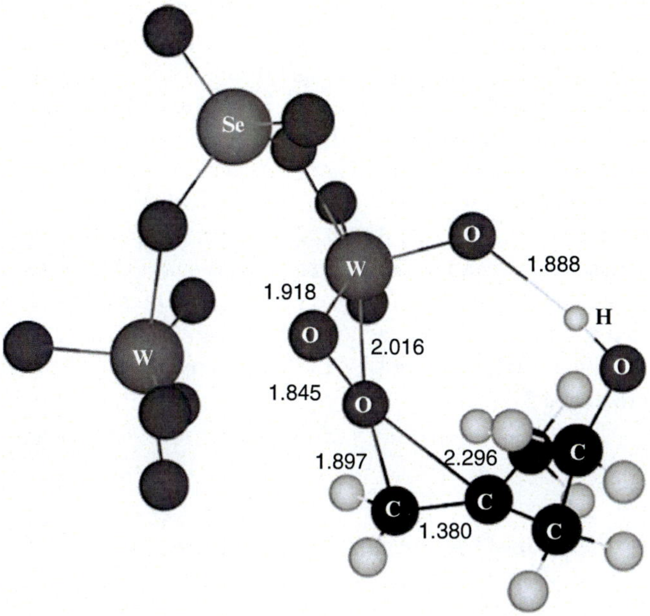

Fig. 5 Proposed transition state structure for the epoxidation of 3-methyl-3-buten-1-ol with $[SeO_4\{WO(O_2)_2\}_2]^{2-}$ in the gas phase (bond lengths in Å) [45]

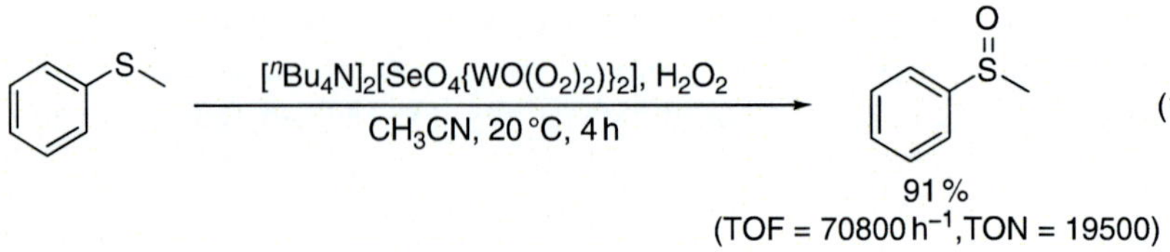

(13)

The aerobic oxidation of nitrobenzene catalyzed by $H_5[PV_2Mo_{10}O_{40}]$ proceeded regioselectively to form 2-nitrophenol in ~5% yield (14). On the other hand, a mixture of 2-, 3-, and 4-nitrophenols with a nearly statistical distribution (~1:0.9:0.45 ratio) was obtained for the aerobic oxidation of nitrobenzene in the presence of 1,1′-azobis(cyclohexanecarbonitrile) as radical initiator [158]. This study provides the first example of an aerobic regioselective hydroxylation of an arene compound. Such a unique regioselective hydroxylation resulted from an intramolecular interaction between the POM and substrate.

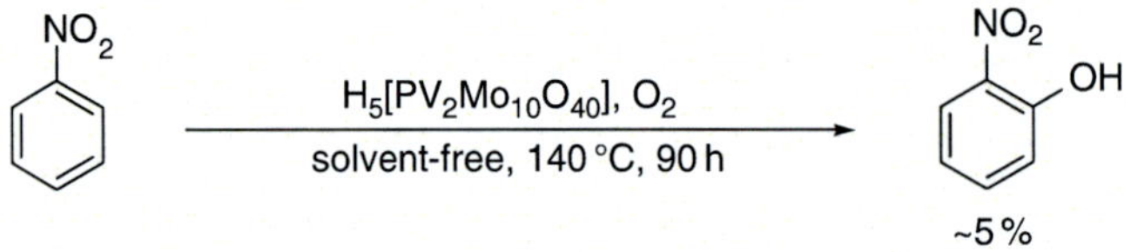

(14)

For the hybrid catalytic systems based on a metallorganic catalytic center coupled with a POM, the electronic, steric, or solubility properties of metallorganic catalysts could be modified, leading to improvement of the reactivity or a cascade of reactions. The aerobic oxidation of methane in water catalyzed by [Pt(Mebipym) Cl_2][$H_4PV_2Mo_{10}O_{40}$] (Mebipym = *N*-methy-2,2′-bipyrimidine) complex supported on SiO_2 was reported [159]. The conjugation of $[H_4PV_2Mo_{10}O_{40}]^-$ to a known Pt^{2+}–bipyrimidine complex by electrostatic interaction could facilitate the O_2 oxidation of the Pt^{2+} intermediate to a Pt^{4+} intermediate, resulting in the catalytic aerobic oxidation of methane to methanol in water. Notably, methanol was further oxidized to furnish acetaldehyde via a carbon–carbon coupling reaction (15). A Pd^{2+}-phenanthroline-$H_5[PV_2Mo_{10}O_{40}]$ catalyst prepared using an induced dipole interaction between the crown ether moiety and POM showed the catalytic activity for the Wacker oxidation of alkenes using N_2O instead of O_2 as an oxidant [160]. Although the detailed reaction mechanism is unclear, it was proposed that POM oxidized Pd^0 followed by the slow reoxidation of the reduced POM with N_2O (Scheme 5).

3.3 *Stabilization of Reaction Intermediates*

The di-copper-substituted γ-Keggin-type silicotungstate $[\gamma\text{-}H_2SiW_{10}O_{36}Cu_2(\mu\text{-}1,1\text{-}N_3)_2]^{4-}$ could act as an efficient reusable homogeneous catalyst for the aerobic oxidative alkyne homocoupling [125–127]. Various kinds of structurally diverse

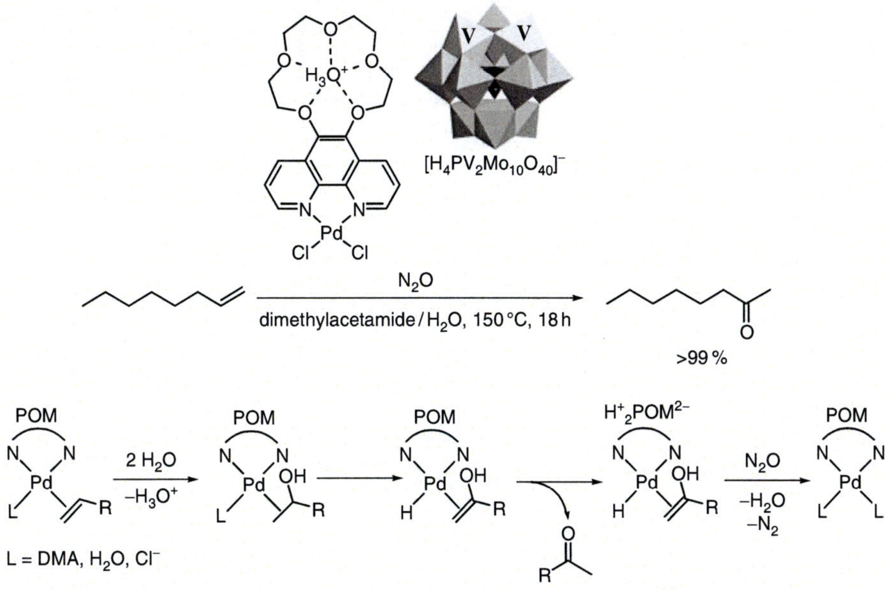

Scheme 5 Proposed mechanism for the Wacker-type oxidation of alkenes with N_2O catalyzed by metallorganic-POM hybrid catalyst [160]

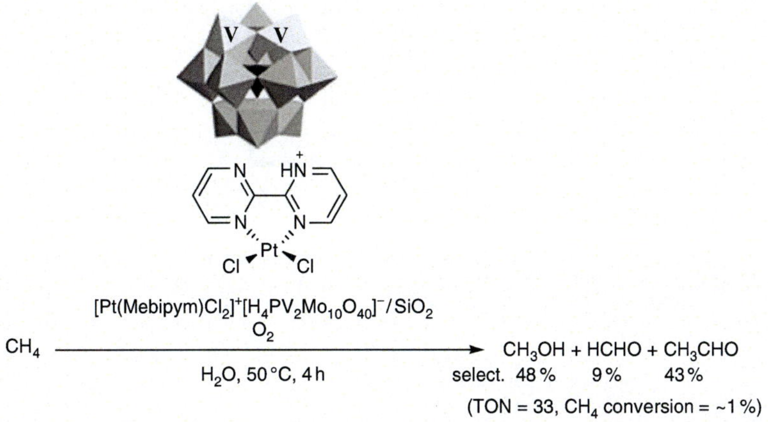

(15)

terminal alkynes including aromatic, heteroaromatic, aliphatic, double bond-containing acetylene, silylacetylene, propargylic alcohol, and propargylic amine derivatives could be converted selectively into the corresponding diynes (16). The catalytic activity of $[\gamma\text{-}H_2SiW_{10}O_{36}Cu_2(\mu\text{-}1,1\text{-}N_3)_2]^{4-}$ was much higher than those of the mono-copper-substituted silicotungstate, monomeric copper complexes, and simple copper salts, showing that the di-copper core in $[\gamma\text{-}H_2SiW_{10}O_{36}Cu_2(\mu\text{-}1,1\text{-}N_3)_2]^{4-}$ plays an important role in the present alkyne homocoupling. The reaction mechanism involving the formation of the di-copper(II)-alkynyl intermediate, reductive elimination of a diyne, and reoxidation of reduced copper species by O_2 has been proposed as shown in Scheme 6.

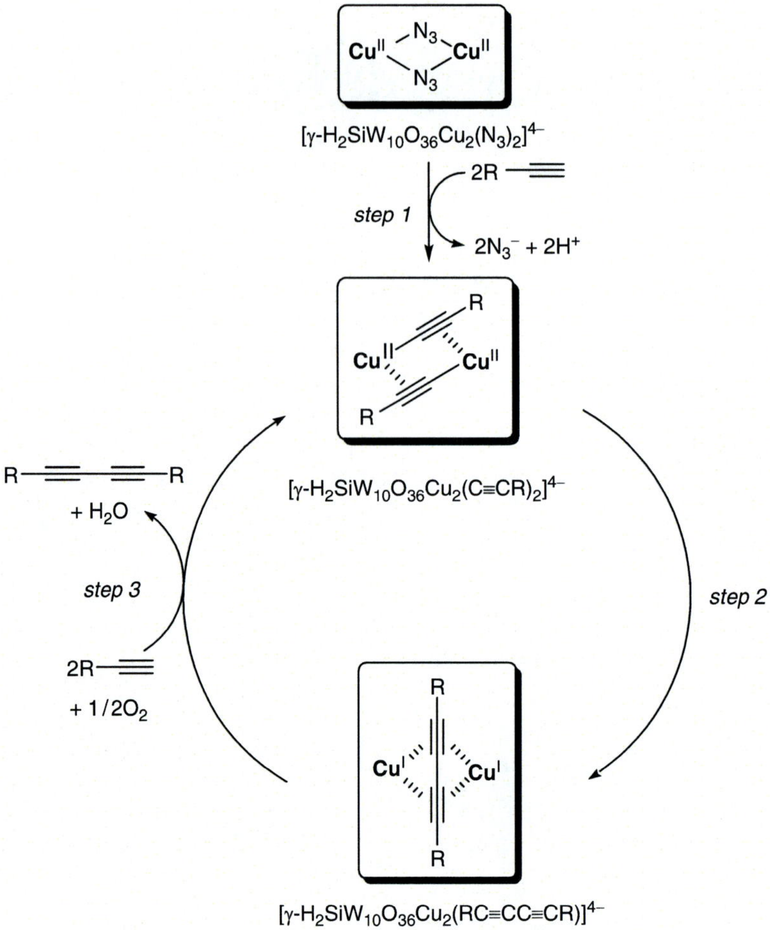

Scheme 6 Proposed mechanism for the oxidative homocoupling of alkynes catalyzed by $[\gamma\text{-}H_2SiW_{10}O_{36}Cu_2(N_3)_2]^{4-}$[126]. POM frameworks are omitted for clarity

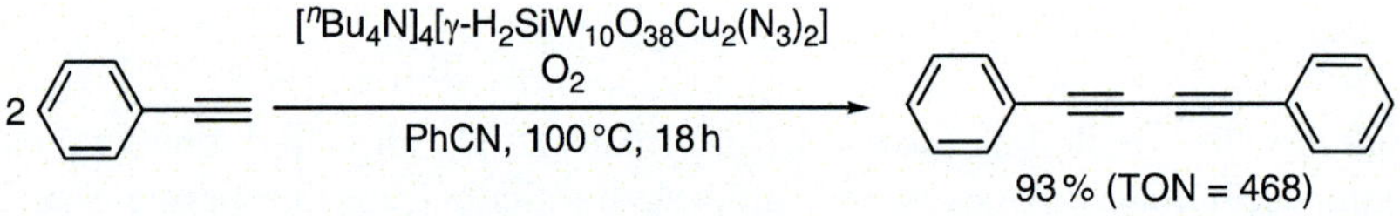

(16)

According to the reaction mechanism for the $[\gamma\text{-}H_2SiW_{10}O_{36}Cu_2(\mu\text{-}1,1\text{-}N_3)_2]^{4-}$-catalyzed oxidative homocoupling of alkyne, the reduced di-copper species can be stabilized in the absence of O_2. This di-copper-substituted γ-Keggin silicotungstate could act as an efficient precatalyst for the regioselective 1,3-dipolar cycloaddition of organic azides to alkynes without any oxidants [161, 162]. Various combinations of substrates were efficiently converted to the corresponding 1,2,3-triazole derivatives in excellent yields without any additives. The present system was applicable to a larger-scale cycloaddition of benzyl azide to phenylacetylene under solvent-free conditions (100 mmol-scale), and 21.5 g of the analytically pure corresponding triazole could be isolated (17). In this case, the turnover frequency and TON reached up to 14,800 h^{-1} and 91,500, respectively. In addition, $[\gamma\text{-}H_2SiW_{10}O_{36}Cu_2(\mu\text{-}1,1\text{-}N_3)_2]^{4-}$ could be applied to the one-pot synthesis of 1-benzyl-4-phenyl-1 *H*-1,2,3-triazole from benzyl chloride, sodium azide, and phenylacetylene [162]. The catalyst effect, kinetic, mechanistic, and computational studies show that the reduced di-copper core plays an important role in the present 1,3-dipolar cycloaddition.

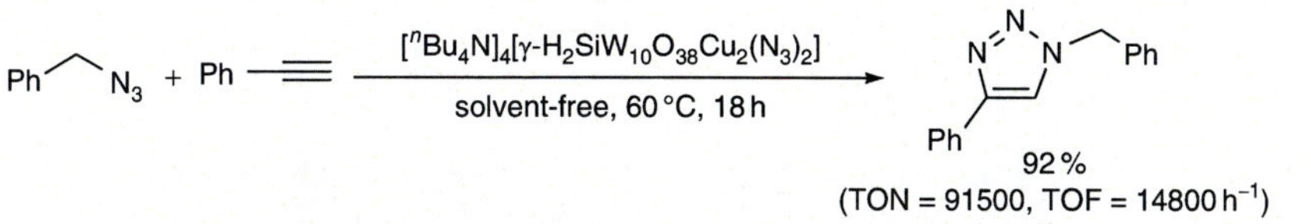

(17)

The DFT calculations showed that the possible copper–copper cooperation effect is crucial for determining the catalytic efficiency of the present cycloaddition reaction. The azide interacts with the di-copper(I)-alkynyl species containing *one* σ,π-bridging acetylide unit to form the monoalkynyl monoazido di-copper(I) intermediate. Then, the nucleophilic attack of the terminal nitrogen atom of an azide on the positively charged carbon atom of the acetylide species takes place to form the six-membered di-copper metallacycle intermediate (Fig. 6). This is the rate-limiting step for the present 1,3-dipolar cycloaddition, and the activation energy was calculated to be 59 kJ mol^{-1}. This is the first example for the remarkable enhancement of the catalytic activity by the construction of di-copper active sites.

3.4 *Multielectron Transfer*

The oxidation of water is very difficult because of its implicit complexity and requirement for the loss of four electrons and four protons from two water molecules [163–165]. The goal of artificial photosynthesis is to convert solar energy into chemical energy, with the photocatalytic oxidation of water giving O_2. However, the intermediates during the oxidation of water may degrade organic ligands [166], thus stimulating the search for all-inorganic catalysts. Recently, the major problems have been overcome by an inorganic homogeneous catalyst based on a ruthenium-containing POM $[\{Ru_4O_4(OH)_2(H_2O)_4\}(\gamma\text{-}SiW_{10}O_{36})_2]^{10-}$ that was independently synthesized and characterized by two research groups (Fig. 7) [167–172].

Hill and coworkers synthesized the mixed rubidium and potassium salt of ruthenium-containing sandwich-type POM $Rb_8K_2[\{Ru_4O_4(OH)_2(H_2O)_4\}(\gamma\text{-}SiW_{10}O_{36})_2]\cdot 25H_2O$ by the reaction of $K_8[\gamma\text{-}SiW_{10}O_{36}]$ with $RuCl_3$ in the presence of air [167]. The oxidation states of the ruthenium atoms in the POM are +4. The tetra-ruthenium-substituted POM could act as a homogeneous catalyst for the oxidation of water in phosphate buffer solution using $[Ru(bpy)_3]^{3+}$ as an oxidant (18). On the basis of the catalyst effect, mechanistic, kinetic, and computational data,

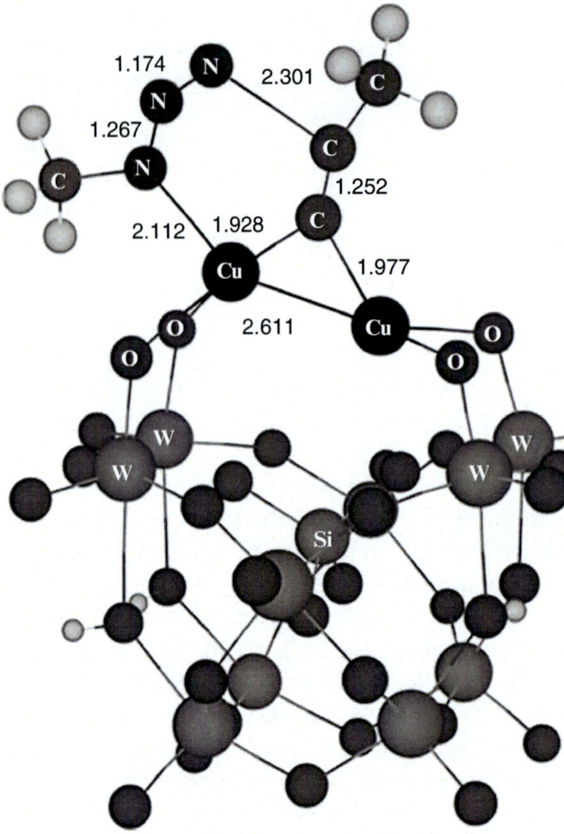

Fig. 6 Proposed transition state structure of the 1,3-dipolar cycloaddition of methylazide to methyl acetylene by the precatalyst $[\gamma\text{-}H_2SiW_{10}O_{36}Cu_2(N_3)_2]^{4-}$ in the gas phase (bond lengths in Å) [161]

a mechanism involving rapid oxidation of the catalyst in four one-electron steps followed by rate-limiting water oxidation/O_2 evolution was proposed [168]. Using the $[Ru(bpy)_3]^{3+}$ as a photosensitizer and a sacrificial electron acceptor $[S_2O_8]^{2-}$, the present system could be applied to the light-induced catalytic water oxidation (Scheme 7) [169]. The complex $[Co_4(H_2O)_2(PW_9O_{34})_2]^{10-}$ with a tetranuclear Co_4O_4 core was also a stable homogeneous catalyst for the oxidation of water with $[Ru(bpy)_3]^{3+}$ as an oxidant, and the turnover frequencies for O_2 production was $\geq 5\ s^{-1}$ at pH = 8 [170].

$$2H_2O + 4[Ru(bpy)_3]^{3+} \xrightarrow[\text{water (pH = 7.2, phosphate buffer), 24 °C}]{Rb_8K_2[\{Ru_4O_4(OH)_2(H_2O)_4\}(\gamma\text{-}SiW_{10}O_{36})_2]} O_2 + 4[Ru(bpy)_3]^{2+} + 4H^+ \quad (18)$$

69 % (based on oxidant)
(TON 39)

Bonchio and coworkers synthesized the cesium and lithium salts of similar ruthenium-containing POMs, $M_{10}[\{Ru_4O_4(OH)_2(H_2O)_4\}(\gamma\text{-}SiW_{10}O_{36})_2]$ ($M = Li^+$ and Cs^+), by the reaction of $[\gamma\text{-}SiW_{10}O_{36}]^{8-}$ with the tetra-ruthenium(IV) complex

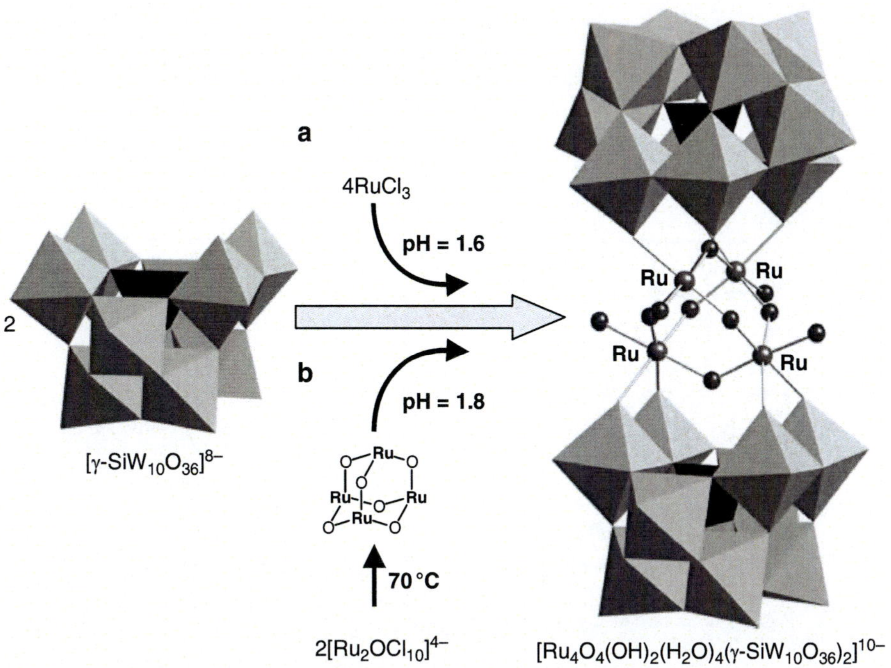

Fig. 7 Schematic representation for the synthesis of the tetra-ruthenium-substituted POM by the reaction of $[\gamma\text{-SiW}_{10}O_{36}]^{8-}$ with (**a**) $RuCl_3$ and (**b**) $[Ru_2OCl_{10}]^{4-}$ as precursors [167, 171]

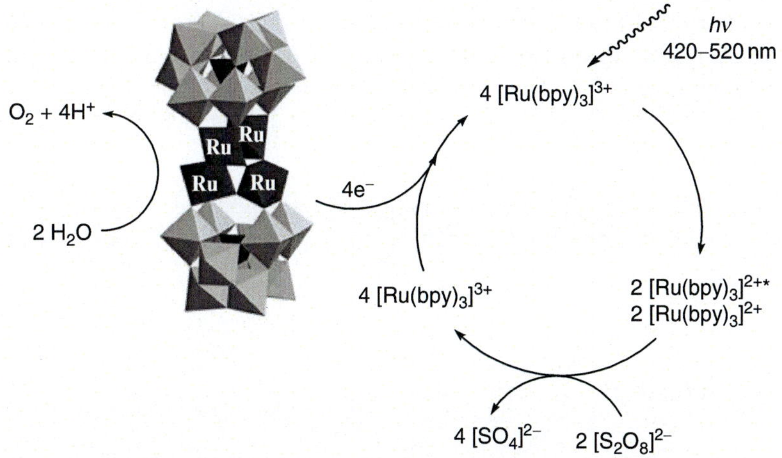

Scheme 7 Proposed mechanism for the light-induced catalytic water oxidation using the $[Ru(bpy)_3]^{3+}$ as a photosensitizer and $[S_2O_8]^{2-}$ as a sacrificial electron acceptor [169]

$[Ru_4O_6(H_2O)_n]^{4+}$ prepared in situ from $[Ru_2Cl_{10}(\mu\text{-}O)]^{4-}$ [171]. The turnover frequencies for the catalytic oxidation of water in the presence of $Li_{10}[\{Ru_4O_4(OH)_2(H_2O)_4\}(\gamma\text{-SiW}_{10}O_{36})_2]$ with $(NH_4)_2Ce(NO_3)_6$ as an oxidant are $>450\ h^{-1}$,

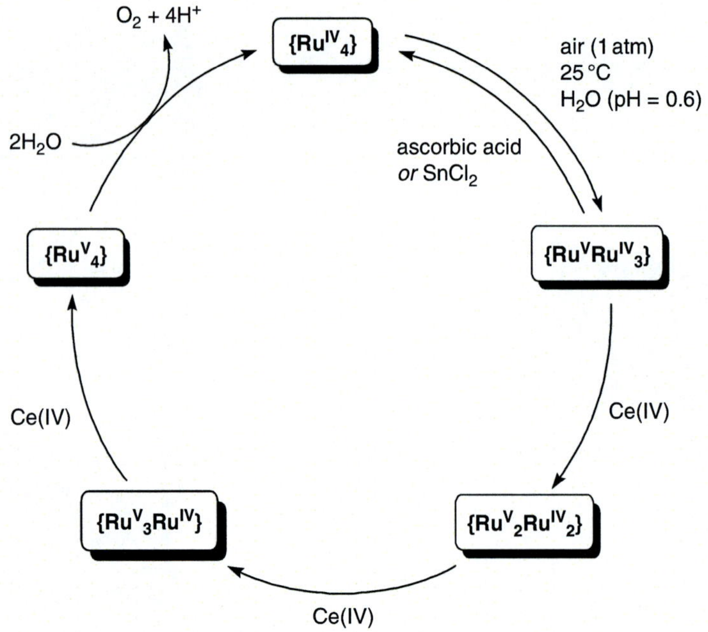

Scheme 8 Proposed mechanism for the water oxidation with Ce^{4+} as an oxidant catalyzed by $[Ru_4(H_2O)_4(\mu\text{-}O)_4(\mu\text{-}OH)_2(\gamma\text{-}SiW_{10}O_{36})_2]^{10-}$ [172]. POM frameworks are omitted for clarity

and the O_2 yield with respect to the oxidant added was 90% (19). On the basis of the UV–vis, EPR, Raman spectroscopies, and DFT calculations, a high electrophilic tetra-ruthenium(V)-hydroxo species is proposed as the competent intermediate, undergoing nucleophilic attack by an external water molecule as a key step in the formation of a new O–O bond under catalytic conditions (Scheme 8) [172].

$$2H_2O + 4Ce^{4+} \xrightarrow[\text{water (pH = 0.6), 20 °C, 2 h}]{Li_{10}[\{Ru_4O_4(OH)_2(H_2O)_4\}(\gamma\text{-}SiW_{10}O_{36})_2]} O_2 + 4Ce^{3+} + 4H^+ \quad (19)$$

90% (based on Ce^{4+})
(TOF > 450 h^{-1})

4 Conclusion

In liquid-phase oxidations, POMs can be used as efficient homogeneous and heterogeneous catalysts in combination with the environment-friendly oxidants such as H_2O_2 and O_2. Especially, the POM catalysts with multimetallic active sites show unique oxidation reactions in comparison with the conventional monometallic complexes. The activities and selectivities for the selective oxidation by multimetallic active sites of POMs can largely be enhanced by the "cooperative activation of oxidants," "simultaneous activation of oxidants and substrates," “stabilization of reaction intermediates,” and “multielectron transfer.” Recently, the chirality of POMs has received considerable attention in the fields of material sciences, catalysis, and biology [173]. However, successful stereoselective catalyses by the POM-based compounds have scarcely been reported [156, 174–178]. Various kinds of POMs have controlled the pores between particles (crystallites) [179–187], in their crystal lattices [188–196], or in the molecules [197–205], and show unique sorption and catalytic properties. Shape- and stereoselective oxidations using porous POMs are interesting future research areas.

References

1. Anastas PT, Warner JC (1998) Green chemistry: theory and practice. Oxford University Press, Oxford
2. Sheldon RA (2000) Pure Appl Chem 72:1233–1246
3. Clark JH (1999) Green Chem 1:1–8
4. Special issues on “Green Chemistry” (2002) Acc Chem Res 35:685–816
5. Sheldon RA, Kochi JK (1981) Metal catalyzed oxidations of organic compounds. Academic, New York
6. Simándi LI (2003) Advances in catalytic activation of dioxygen by metal complexes. Kluwer, Dordrecht
7. Bäckvall J-E (ed) (2004) Modern oxidation methods. Wiley-VCH, Weinheim
8. Meyer F, Limberg C (eds) (2007) Organometallic oxidation catalysis. Topics in organometallic chemistry, vol 12. Springer, Berlin
9. Mizuno N (2009) Modern heterogeneous oxidation catalysis. Wiley-VCH, Weinheim
10. Punniyamurthy T, Velusamy S, Iqbal J (2005) Chem Rev 105:2329–2363
11. Cavani F, Teles JH (2009) ChemSusChem 2:508–534
12. Shibasaki M, Yamamoto Y (2004) Multimetallic catalysts in organic synthesis. Wiley-VCH, Weinheim
13. Belloni J, Mostafavi M, Remita H, Marignier J-L, Delcourt M-O (1998) New J Chem 22:1239–1255
14. Que L Jr, Tolman WB (2002) Angew Chem Int Ed 41:1114–1137
15. Duschek A, Kirsch SF (2008) Angew Chem Int Ed 47:5703–5705
16. Lu L, Eychmüller A (2008) Acc Chem Res 41:244–253
17. Coutanceau C, Brimaud S, Lamy C, Léger J-M, Dubau L, Rousseau S, Vigier F (2008) Electrochim Acta 53:6865–6880
18. Zhong C-J, Luo J, Njoki PN, Mott D, Wanjala B, Loukrakpam R, Lim S, Wang L, Fang B, Xu Z (2008) Energy Environ Sci 1:454–466
19. Shibasaki M, Kanai M, Matsunaga S, Kumagai N (2009) Acc Chem Res 42:1117–1127
20. Bracey CL, Ellis PR, Hutchings GJ (2009) Chem Soc Rev 38:2231–2243
21. Haak RM, Wezenberg SJ, Kleij AW (2010) Chem Commun 46:2713–2723
22. Pope MT (1983) Heteropoly and isopoly oxometalates. Springer, Berlin
23. Pope MT, Müller A (2001) Polyoxometalate chemistry from topology via self-assembly to applications. Kluwer, Dordrecht
24. Yamase T, Pope MT (2002) Polyoxometalate chemistry for nano-composite design. Kluwer, New York
25. Pope MT (2004) Polyoxo anions: synthesis and structure. In: McCleverty JA, Meyer TJ (eds) Comprehensive coordination chemistry II, vol 4. Elsevier Pergamon, Amsterdam
26. Thematic issue on “Polyoxometalates” (1998) Chem Rev 98:1–389
27. Mialane P, Dolbecq A, Sécheresse F (2006) Chem Commun 3477–3485

28. Proust A, Thouvenot R, Gouzerh P (2008) Chem Commun:1837–1852
29. Long D-L, Tsunashima R, Cronin L (2010) Angew Chem Int Ed 49:1736–1758
30. Kozhevnikov IV (2002) Catalysts for fine chemical synthesis, volume 2, catalysis by polyoxometalates. Wiley, Chichester
31. Hill CL (2004) Polyoxometalates: reactivity. In: McCleverty JA, Meyer TJ (eds) Comprehensive coordination chemistry II, vol 4. Elsevier Pergamon, Amsterdam
32. Mizuno N, Kamata K, Yamaguchi K (2006) Liquid phase oxidations catalyzed by polyoxometalates. In: Richards R (ed) Surface and nanomolecular catalysis. Taylor and Francis Group, New York
33. Hill CL, Chrisina C, Prosser-McCartha M (1995) Coord Chem Rev 143:407–455
34. Okuhara T, Mizuno N, Misono M (1996) Adv Catal 41:113–252
35. Neumann R (1998) Prog Inorg Chem 47:317–370
36. Neumann R, Khenkin AM (2006) Chem Commun:2529–2538
37. Neumann R (2010) Inorg Chem 49:3594–3601
38. Attanasio D, Suber L, Thorslund K (1991) Inorg Chem 30:590–592
39. Ishii Y, Yamawaki K, Ura T, Yamada H, Yoshida T, Ogawa M (1988) J Org Chem 53: 3587–3593
40. Khenkin AM, Weiner L, Wang Y, Neumann R (2001) J Am Chem Soc 123:8531–8542
41. Lissel M, in de Wal HJ, Neumann R (1992) Tetrahedron Lett 33:1795–1798
42. Kamata K, Yamaguchi K, Hikichi S, Mizuno N (2003) Adv Synth Catal 345:1193–1196
43. Kamata K, Yamaguchi K, Mizuno N (2004) Chem Eur J 10:4728–4734
44. Sakaguchi S, Nishiyama Y, Ishii Y (1996) J Org Chem 61:5307–5311
45. Kamata K, Hirano T, Kuzuya S, Mizuno N (2009) J Am Chem Soc 131:6997–7004
46. Kamata K, Hirano T, Mizuno N (2009) Chem Commun:3958−3960
47. Server-Carrió J, Bas-Serra J, González-Núñez ME, García-Gastaldi A, Jameson GB, Baker LCW, Acerete R (1999) J Am Chem Soc 121:977–984
48. Kamata K, Yonehara K, Sumida Y, Yamaguchi K, Hikichi S, Mizuno N (2003) Science 300:964–966
49. Ishimoto R, Kamata K, Mizuno N (2009) Angew Chem Int Ed 48:8900–8904
50. Neumann R, Dahan M (1997) Nature 388:353–355
51. Neumann R, Dahan M (1998) J Am Chem Soc 120:11969–11976
52. Sloboda-Rozner D, Saha-Möller CR, Neumann R (2003) J Am Chem Soc 125:5280–5281
53. Zhang X, Anderson TM, Chen Q, Hill CL (2001) Inorg Chem 40:418–419
54. Mizuno N, Nozaki C, Kiyoto I, Misono M (1998) J Am Chem Soc 120:9267–9272
55. Weiner H, Finke RG (1999) J Am Chem Soc 121:9831–9842
56. Neumann R, Levin M (1991) J Org Chem 56:5707–5710
57. Yamaguchi K, Yoshida C, Uchida S, Mizuno N (2005) J Am Chem Soc 127:530–531
58. Maksimchuk NV, Kovalenko KA, Arzumanov SS, Chesalov YA, Melgunov MS, Stepanov AG, Fedin VP, Kholdeeva OA (2010) Inorg Chem 49:2920–2930
59. Sels BF, de Vos DE, Jacobs PA (2005) Angew Chem Int Ed 44:310–313
60. Hoegaerts D, Sels BF, de Vos DE, Verpoort F, Jacobs PA (2000) Catal Today 60:209–218
61. Phule JT, Neiwert WA, Hardcastle KI, Do BT, Hill CL (2001) J Am Chem Soc 123: 12101–12102
62. Yamada YMA, Ichinohe M, Takahashi H, Ikegami S (2001) Org Lett 3:1837–1840
63. Uchida S, Hikichi S, Akatsuka T, Tanaka T, Kawamoto R, Lesbani A, Nakagawa Y, Uehara K, Mizuno N (2007) Chem Mater 19:4694–4701
64. Hill CL (1995) Synlett:127–132
65. Hiskia A, Mylonas A, Papaconstantinou E (2001) Chem Soc Rev 30:62–69
66. Ishii Y, Yamawaki K, Yoshida T, Ura T, Ogawa M (1987) J Org Chem 52:1868–1870
67. Oguchi T, Sakata Y, Takeuchi N, Kaneda K, Ishii Y, Ogawa M (1989) Chem Lett:2053–2056
68. Sakaue S, Sakata Y, Nishiyama Y, Ishii Y (1992) Chem Lett:289–292
69. Ishii Y, Sakata Y (1990) J Org Chem 55:5545–5547

70. Aubry C, Chottard G, Platzer N, Brégeault J-M, Thouvenot R, Chauveau F, Huet C, Ledon H (1991) Inorg Chem 30:4409–4415
71. Dengel AC, Griffith WP, Parkin BC (1993) J Chem Soc Dalton Trans:2683–2688
72. Salles L, Aubry C, Thouvenot R, Robert F, Dorémieux-Morin C, Chottard G, Ledon H, Jeannin Y, Brégeault J-M (1994) Inorg Chem 33:871–878
73. Duncan DC, Chambers RC, Hecht E, Hill CL (1995) J Am Chem Soc 117:681–691
74. Matveev KI (1977) Kinet Katal 18:862–867
75. Khenkin AM, Rosenberger A, Neumann R (1999) J Catal 182:82–91
76. Neumann R, Lissel M (1989) J Org Chem 54:4607–4610
77. Atlamsani A, Brégeault J-M, Ziyad M (1993) J Org Chem 58:5663–5665
78. Hamamoto M, Nakayama K, Nishiyama Y, Ishii Y (1993) J Org Chem 58:6421–6425
79. Neumann R, Khenkin AM, Vigdergauz I (2000) Chem Eur J 6:875–882
80. Monflier E, Blouet E, Barbaux Y, Mortreux A (1994) Angew Chem Int Ed Engl 33: 2100–2102
81. Passoni LC, Cruz AT, Buffon R, Shuchardt U (1997) J Mol Catal A 120:117–123
82. Yokota T, Tani M, Sakaguchi S, Ishii Y (2003) J Am Chem Soc 125:1476–1477
83. Yokota T, Sakaguchi S, Ishii Y (2002) Adv Synth Catal 344:849–854
84. Khenkin AM, Weiner L, Neumann R (2005) J Am Chem Soc 127:9988–9989
85. Kozhevnikov IV, Matveev KI (1983) Appl Catal 5:135–150
86. Harrup MK, Hill CL (1996) J Mol Catal A 106:57–66
87. Weinstock IA, Barbuzzi EMG, Wemple MW, Cowan JJ, Reiner RS, Sonnen DM, Heintz RA, Bond JS, Hill CL (2001) Nature 414:191–195
88. Niederhoffer EC, Timmons JH, Martell AE (1984) Chem Rev 84:137–207
89. Thematic issue on "Metal-Dioxygen Complexes" (1994) Chem Rev 94:567–856
90. Thematic issue on "Bioinorganic Enzymology" (1996) Chem Rev 96:2237–3042
91. Thematic issue on "Biomimetic Inorganic Chemistry" (2004) Chem Rev 104:347–1200
92. Dickman MH, Pope MT (1994) Chem Rev 94:569–584
93. Sergienko VS (2008) Crystallogr Rep 53:18–46
94. Venturello C, D'Aloisio R, Bart JCJ, Ricci M (1985) J Mol Catal 32:107–110
95. Liu Y, Murata K, Inaba M (2004) Chem Commun:582–583
96. Kamata K, Ishimoto R, Hirano T, Kuzuya S, Uehara K, Mizuno N (2010) Inorg Chem 49: 2471–2478
97. Kamata K, Hirano T, Ishimoto R, Mizuno N (2010) Dalton Trans 39:5509–5518
98. Schwegler M, Floor M, van Bekkum H (1988) Tetrahedron Lett 29:823–826
99. Watanabe Y, Yamamoto K, Tatsumi T (1999) J Mol Catal A 145:281–289
100. Schroden RC, Blansford CF, Melde BJ, Johnson BJS, Stein A (2001) Chem Mater 3: 1074–1081
101. Kamata K, Nakagawa Y, Yamaguchi K, Mizuno N (2004) J Catal 224:224–228
102. Mizuno N, Yamaguchi K, Kamata K (2005) Coord Chem Rev 249:1944–1956
103. Kamata K, Kotani M, Yamaguchi K, Hikichi S, Mizuno N (2007) Chem Eur J 13:639–648
104. Phan TD, Kinch MA, Barker JE, Ren T (2005) Tetrahedron Lett 46:397–400
105. Carraro M, Sandei L, Sartorel A, Scorrano G, Bonchio M (2006) Org Lett 8:3671–3674
106. Berardi S, Bonchio M, Carraro M, Conte V, Sartorel A, Scorrano G (2007) J Org Chem 72:8954–8957
107. Musaev DG, Morokuma K, Geletii YV, Hill CL (2004) Inorg Chem 43:7702–7708
108. Prabhakar R, Morokuma K, Hill CL, Musaev DG (2006) Inorg Chem 45:5703–5709
109. Sartorel A, Carraro M, Bagno A, Scorrano G, Bonchio M (2007) Angew Chem Int Ed 46: 3255–3258
110. Bar-Nahum I, Cohen H, Neumann R (2003) Inorg Chem 42:3677–3684
111. Bar-Nahum I, Neumann R (2003) Chem Commun:2690–2691
112. Bar-Nahum I, Narasimhulu KV, Weiner L, Neumann R (2005) Inorg Chem 44:4900–4902
113. Nakagawa Y, Kamata K, Kotani M, Yamaguchi K, Mizuno N (2005) Angew Chem Int Ed 44:5136–5141

114. Nakagawa Y, Mizuno N (2007) Inorg Chem 46:1727–1736
115. Kamata K, Yonehara K, Nakagawa Y, Uehara K, Mizuno N (2010) Nat Chem 2:478–483
116. Bösing M, Nöh A, Loose I, Krebs B (1998) J Am Chem Soc 120:7252–7259
117. Ritorto MD, Anderson TM, Neiwert WA, Hill CL (2004) Inorg Chem 43:44–49
118. Neumann R, Gara M (1994) J Am Chem Soc 116:5509–5510
119. Neumann R, Gara M (1995) J Am Chem Soc 117:5066–5074
120. Anderson TM, Zhang X, Hardcastle KI, Hill CL (2002) Inorg Chem 41:2477–2488
121. Zhang X, Chen Q, Duncan DC, Campana CF, Hill CL (1997) Inorg Chem 36:4208–4215
122. Mizuno N, Nozaki C, Kiyoto I, Misono M (1998) J Am Chem Soc 120:9267–9272
123. Nishiyama Y, Nakagawa Y, Mizuno N (2001) Angew Chem Int Ed 40:3639–3641
124. Weiner H, Finke RG (1999) J Am Chem Soc 121:9831–9842
125. Kamata K, Yamaguchi S, Kotani M, Yamaguchi K, Mizuno N (2008) Angew Chem Int Ed 47:2407–2410
126. Yamaguchi K, Kamata K, Yamaguchi S, Kotani M, Mizuno N (2008) J Catal 258:121–130
127. Mizuno N, Kamata K, Nakagawa Y, Oishi T, Yamaguchi K (2010) Catal Today. 17:1261–1267
128. Sloboda-Rozner D, Witte P, Alsters PL, Neumann R (2004) Adv Synth Catal 346:339–345
129. Vasylyev MV, Neumann R (2004) J Am Chem Soc 126:884–890
130. Haimov A, Neumann R (2006) J Am Chem Soc 128:15697–15700
131. Yamaguchi K, Mizuno N (2002) New J Chem 26:972–974
132. Anderson TM, Neiwert WA, Kirk ML, Piccoli PMB, Schultz AJ, Koetzle TF, Musaev DG, Morokuma K, Cao R, Hill CL (2004) Science 306:2074–2077
133. Anderson TM, Cao R, Slonkina E, Hedman B, Hodgson KO, Hardcastle KI, Neiwert WA, Wu S, Kirk ML, Knottenbelt S, Depperman EC, Keita B, Nadjo L, Musaev DG, Morokuma K, Hill CL (2005) J Am Chem Soc 127:11948–11949
134. Cao R, Anderson TM, Piccoli PMB, Schultz AJ, Koetzle TF, Geletii YV, Slonkina E, Hedman B, Hodgson KO, Hardcastle KI, Fang X, Kirk ML, Knottenbelt S, Kögerler P, Musaev DG, Morokuma K, Takahashi M, Hill CL (2007) J Am Chem Soc 129:11118–11133
135. Hagen J (1999) Industrial catalysis: a practical approach. Wiley-VCH, Weinheim
136. Kamata K, Kuzuya S, Uehara K, Yamaguchi S, Mizuno N (2007) Inorg Chem 46:3768–3774
137. Hill CL, Brown RB Jr (1986) J Am Chem Soc 108:536–538
138. Mansuy D, Bartoli J-F, Battioni P, Lyon DK, Finke RG (1991) J Am Chem Soc 113: 7222–7226
139. Yin C-X, Finke RG (2005) Inorg Chem 44:4175–4188
140. Morris AM, Anderson OP, Finke RG (2009) Inorg Chem 48:4411–4420
141. Yin C-X, Finke RG (2005) J Am Chem Soc 127:9003–9013
142. Botar B, Geletii YV, Kögerler P, Musaev DG, Morokuma K, Weinstock IA, Hill CL (2006) J Am Chem Soc 128:11268–11277
143. Piquemal J-Y, Salles L, Chottard G, Herson P, Ahcine C, Brégeault J-M (2006) Eur J Inorg Chem:939–947
144. Yamase T, Ozeki T, Motomura S (1992) Bull Chem Soc Jpn 65:1453–1459
145. Judd DA, Chen Q, Campana CF, Hill CL (1997) J Am Chem Soc 119:5461–5462
146. Harrup MK, Kim G-S, Zeng H, Johnson RP, VanDerveer D, Hill CL (1998) Inorg Chem 37:5550–5556
147. Kim G-S, Zeng H, Hill CL (2003) Bull Korean Chem Soc 24:1005–1008
148. Sakai Y, Kitakoga Y, Hayashi K, Yoza K, Nomiya K (2004) Eur J Inorg Chem:4646–4652
149. Ohlin CA, Villa EM, Fettinger JC, Casey WH (2008) Angew Chem Int Ed 47:8251–8254
150. Bassil BS, Mal SS, Dickman MH, Kortz U, Oelrich H, Walder L (2008) J Am Chem Soc 130:6696–6697
151. Mal SS, Nsouli NH, Carraro M, Sartorel A, Scorrano G, Oelrich H, Walder L, Bonchio M, Kortz U (2010) Inorg Chem 49:7–9
152. Mal SS, Dickman MH, Kortz U (2008) Chem Eur J 14:9851–9855
153. Jahier C, Mal SS, Kortz U, Nlate S (2010) Eur J Inorg Chem:1559–1566

154. Mas-Ballesté R, Que L Jr (2006) Science 312:1885–1886
155. Adam W, Alsters PL, Neumann R, Saha-Möller CR, Sloboda-Rozner D, Zhang R (2003) J Org Chem 68:1721–1728
156. Adam W, Alsters PL, Neumann R, Saha-Möller CR, Seebach D, Beck AK, Zhang R (2003) J Org Chem 68:8222–8231
157. Li Z, Yamamoto H (2010) J Am Chem Soc 132:7878–7880
158. Khenkin AM, Weiner L, Neumann R (2005) J Am Chem Soc 127:9988–9989
159. Bar-Nahum I, Khenkin AM, Neumann R (2004) J Am Chem Soc 126:10236–10237
160. Ettedgui J, Neumann R (2009) J Am Chem Soc 131:4–5
161. Kamata K, Nakagawa Y, Yamaguchi K, Mizuno N (2008) J Am Chem Soc 130: 15304–15310
162. Yamaguchi K, Kotani M, Kamata K, Mizuno N (2008) Chem Lett 37:1258–1259
163. Meyer TJ (2008) Nature 451:778–779
164. Kalyanasundaram K, Graetzel M (2010) Curr Opin Biotechnol 21:298–310
165. Lewis NS, Nocera DG (2006) Proc Natl Acad Sci USA 103:15729–15735
166. Rüttlinger W, Dismukes GC (1997) Chem Rev 97:1–24
167. Geletii YV, Botar B, Kögerler P, Hillesheim DA, Musaev DG, Hill CL (2008) Angew Chem Int Ed 47:3896–3899
168. Geletii YV, Besson C, Hou Y, Yin Q, Musaev DG, Quiñonero D, Cao R, Hardcastle KI, Proust A, Kögerler P, Hill CL (2009) J Am Chem Soc 131:17360–17370
169. Geletii YV, Huang Z, Hou Y, Musaev DG, Lian T, Hill CL (2009) J Am Chem Soc 131:7522–7523
170. Yin Q, Tan JM, Besson C, Geletii YV, Musaev DG, Kuznetsov AE, Luo Z, Hardcastle KI, Hill CL (2010) Science 328:342–345
171. Sartorel A, Carraro M, Scorrano G, De Zorzi R, Geremia S, McDaniel ND, Bernhard S, Bonchio M (2008) J Am Chem Soc 130:5006–5007
172. Sartorel A, Miró P, Salvadori E, Romain S, Carraro M, Scorrano G, Di Valentin M, Llobet A, Bo C, Bonchio M (2009) J Am Chem Soc 131:16051–16053
173. Hasenknopf B, Micoine K, Lacôte E, Thorimbert S, Malacria M, Thouvenot R (2008) Eur J Inorg Chem:5001–5013
174. Carraro M, Modugno G, Sartorel A, Scorrano G, Bonchio M (2009) Eur J Inorg Chem:5164–5174
175. Li J, Hu S, Luo S, Cheng J-P (2009) Eur J Org Chem:132–140
176. Adam W, Alsters PL, Neumann R, Saha-Möller CR, Seebach D, Zhang R (2003) Org Lett 5:725–728
177. Luo S, Li J, Xu H, Zhang L, Cheng J-P (2007) Org Lett 9:3675–3678
178. Tan H, Li Y, Chen W, Liu D, Su Z, Lu Y, Wang E (2009) Chem Eur J 15:10940–10947
179. Okuhara T (2002) Chem Rev 102:3641–3666
180. Okuhara T, Watanabe H, Nishimura T, Inumaru K, Misono M (2000) Chem Mater 12: 2230–2238
181. Okuhara T, Mizuno N, Misono M (2001) Appl Catal A 222:63–77
182. Okuhara T (2003) Appl Catal A 256:213–224
183. Yoshinaga Y, Seki K, Nakato T, Okuhara T (1997) Angew Chem Int Ed 36:2833–2835
184. Ito T, Inumaru K, Misono M (2001) Chem Mater 13:824–831
185. Parent MA, Moffat JB (1996) Langmuir 12:3733–3739
186. Volkmer D, Bredenkötter B, Tellenbröker J, Kögerler P, Kurth DG, Lehmann P, Schnablegger H, Schwahn D, Pipenbrink M, Krebs B (2002) J Am Chem Soc 124:10489–10496
187. Kang Z, Wang E, Jiang M, Lian S, Li Y, Hu C (2003) Eur J Inorg Chem:370–376
188. Khan MI, Yohannes E, Powell D (1999) Inorg Chem 38:212–213
189. Khan MI, Yohannes E, Doedens RJ (1999) Angew Chem Int Ed 38:1292–1294
190. Hagrman D, Hagrman PJ, Zubieta J (1999) Angew Chem Int Ed 38:3165–3168
191. Son JH, Choi H, Kwon YU (2000) J Am Chem Soc 122:7432–7433

192. Schmitt W, Baissa E, Mandel A, Anson CE, Powell AK (2001) Angew Chem Int Ed 40: 3577–3581
193. Uchida S, Mizuno N (2007) Coord Chem Rev 251:2537–2546
194. Mizuno N, Uchida S (2006) Chem Lett 35:688–693
195. Mizuno N, Uchida S, Uehara K (2009) Pure Appl Chem 81:2369–2376
196. Ishii Y, Takenaka Y, Konishi K (2004) Angew Chem Int Ed 43:2702–2705
197. Müller A, Krickemeyer E, Bögge H, Schmidtmann M, Beugholt C, Das SK, Peters F (1999) Chem Eur J 5:1496–1502
198. Müller A, Shah SQN, Bögge H, Schmidtmann M (1999) Nature 397:48–50
199. Müller A, Krickemeyer E, Bögge H, Schmidtmann M, Botar B, Talismanova MO (2003) Angew Chem Int Ed 42:2085–2090
200. Müller A, Krickemeyer E, Bögge H, Schmidtmann M, Körgerler P, Rosu C, Bockmann E (2001) Angew Chem Int Ed 40:4034–4037
201. Tianbo L, Diemann E, Li H, Dress AWM, Müller A (2003) Nature 426:59–62
202. Volkmer D, Chesne AD, Kurth DG, Schnablegger H, Lehmann P, Koop MJ, Müller A (2000) J Am Chem Soc 122:1995–1998
203. Salignac B, Riedel S, Dolbecq A, Sécheresse F, Cadot E (2000) J Am Chem Soc 122: 10381–10389
204. Cadot E, Marrot J, Sécheresse F (2001) Angew Chem Int Ed 40:774–777
205. Contant R, Tézé A (1985) Inorg Chem 24:4610–4614

Top Organomet Chem (2011) 37: 161–183
DOI: 10.1007/3418_2011_4

Published online: 12 March 2011

Bifunctional Acid Catalysts for Organic Synthesis

Pingfan Li and Hisashi Yamamoto

Abstract The concept of bifunctional acid catalysis is very helpful for inventing new catalytic asymmetric reactions. Compared with single functional acid catalysts, cooperative effect of two acid components has the potential to fine tune the reactivity as well as the selectivity of desired reaction pathways. This chapter focuses on some representative examples on the recent developments of bifunctional acid catalysis, including combined acid catalysis and other cooperative acid catalysis.

Keywords Asymmetric synthesis · Bifunctional acid catalysis · Combined acid catalysis · Cooperative acid catalysis · Designer acid

Contents

P. Li and H. Yamamoto (✉)
Department of Chemistry, The University of Chicago, Chicago, IL, USA
e-mail: yamamoto@uchicago.edu

1 Introduction

Designer acids are acid catalysts tailored through creative imagination, and optimized through careful engineering, which are expected to show high reactivity, selectivity, versatility, and robustness for various synthetic applications [1, 2].

In this rapidly growing field of asymmetric catalysis [3, 4], the use of chiral Lewis acid catalysts has been well appreciated by us during the past three decades [5–7]. Although we treat transition-metal catalysis separately from Lewis acid catalysis, it should be noted that, as long as electron pair donors/acceptors are involved, the interactions between transition metals and corresponding substrates are always Lewis acid/Lewis base interactions; and thus, any electron pair acceptor catalyst initiated asymmetric reaction could be regarded as chiral Lewis acid catalyzed reaction in its broadest sense.

All these acid catalysts are ultimately consisted of two components: a partially cationic center atom(s), for example hydrogen, boron, silicon, transition/main-group metal cation, tetraalkylammonium cation, and so on; and a partially anionic ligand (s), with space-filling backbones and coordinating atoms such as oxygen, nitrogen, sulfur, phosphine, and so on. A lot of efforts have been made to find the perfect ligand backbones for achieving highly enantioselective transformations, which led to the recognition of several privileged ligand structures [8]. The concept of combined acid catalysis, on the contrary, could also be regarded as an efficient way for designing novel ligand structures, in which the intermolecular or intramolecular coordination of Lewis acidic and Lewis basic sites of these combined acid catalysts actually build up novel ligand structures in the form of inorganic–organic hybrids (Fig. 1). As will be discussed in this chapter, combined acid catalysts could achieve unprecedented reactivities and selectivities compared with their single functional counterparts.

Beyond the concept of combined acid catalysis, there are also many other bifunctional acid catalysts interacting with nucleophiles and electrophiles simultaneously, and thus benefiting through such cooperative effect. Selected examples on Lewis acid/hydrogen bonding cooperative catalysis and Lewis acid/transition-metal

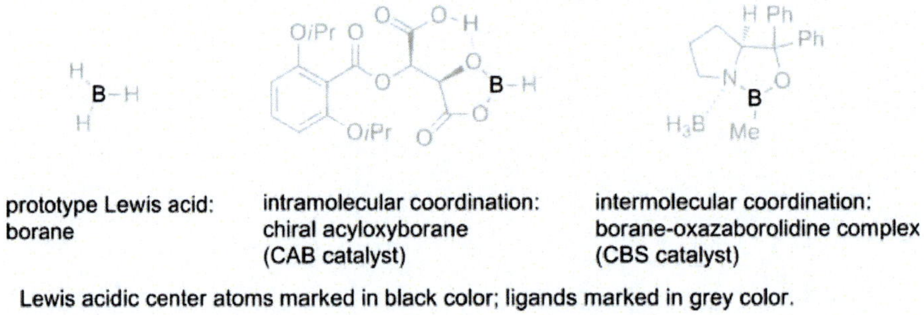

Fig. 1 Structures of a prototype borane Lewis acid catalyst and borane-based combined acid catalysts

cooperative catalysis will be discussed. Readers may also find the review papers on cooperative effect of Lewis acids with transition metals [9], as well as organocatalysts with transition metals [10] quite interesting.

Our literature selection on bifunctional acid catalysis will be focused on publications in the last decade, while seminal papers of historical importance, as well as some related references, are also included.

2 Combined Acid Catalysis

Combined acid catalysts are bifunctional acid catalysts consisted by at least two acid components, in which one acid component's cationic center coordinate to the other one's anionic ligand, and thus altering the reactivity and selectivity. We have suggested the classification of combined acid catalysts into four general categories: Lewis acid-assisted Lewis acid (LLA), Brønsted acid-assisted Lewis acid (BLA), Lewis acid-assisted Brønsted acid (LBA), and Brønsted acid-assisted Brønsted acid (BBA) [1]. This classification is used solely for simplification of our discussion. As we have noted above, Brønsted acid, or proton-based acid, is actually a unique type of Lewis acid. Introducing the concept of combined acid catalysis is aim at encouraging and enhancing the success rate of our future research to uncover novel designer acid systems for asymmetric synthesis.

2.1 *Oxazaborolidine-Based Catalysts*

The pioneering work from Itsuno group [11–14] on stoichiometric 1,2-aminoalcohol-borane complex-mediated borane reduction of ketones led to the discovery of well-defined oxazaborolidine catalyzed asymmetric reduction by Corey and coworkers [15–18]. Known as Corey–Bakshi–Shibata reaction, or CBS reduction, this reaction has become a standard method for making chiral secondary alcohols for complex molecule synthesis [19]. The generally accepted mechanism of this reaction is shown in Fig. 2. Coordination of the electrophilic reductant BH_3 to the nitrogen atom of

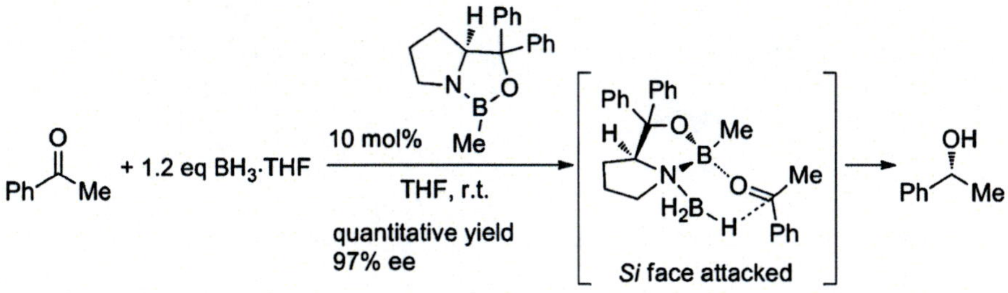

Fig. 2 Stereochemical rational for CBS reduction

oxazaborolidine not only enhanced the nucleophilicity of the hydride, but also enhanced the Lewis acidity of the borane atom of oxazaborolidine (Lewis acid activation of Lewis acid, LLA), which coordinates to the less sterically hindered/more basic lone pair electrons of the ketone substrate's carbonyl group. Hydride reduction then proceeds through a six-membered ring transition state to give the observed major enantioenriched alcohol product.

Oxazaborolidines themselves are usually not active enough as Lewis acid catalyst; unless a Lewis acidic reactant is involved (for example, in CBS reduction), or additional electron withdrawing group is attached [20]. A significant breakthrough in this chemistry came as the Corey group reported [21, 22] that when a proline derived oxazaborolidine was used together with a Brønsted acid component such as triflic acid (TfOH) or bistriflimide (Tf_2NH), these combined BLA catalysts can promote highly selective Diels-Alder reactions (Fig. 3). Unsaturated aldehydes, ketones, esters, and quinones can all be used in enantioselective intermolecular Diels–Alder reactions with various dienes [23–25]. Corresponding intramolecular [26], as well as transannular [27], Diels–Alder reactions were also shown to proceed very smoothly. Many other reaction types are also applicable with this exceptionally general catalyst, including cyanosilylation of aldehydes [28] and ketones [29], [2+2] cycloaddition between ketene and aldehydes [30], and Michael additions [31]. The observed facial selectivities can be well explained by the pretransition state assembly models as shown in Fig. 3. Weak hydrogen bonding between C–H

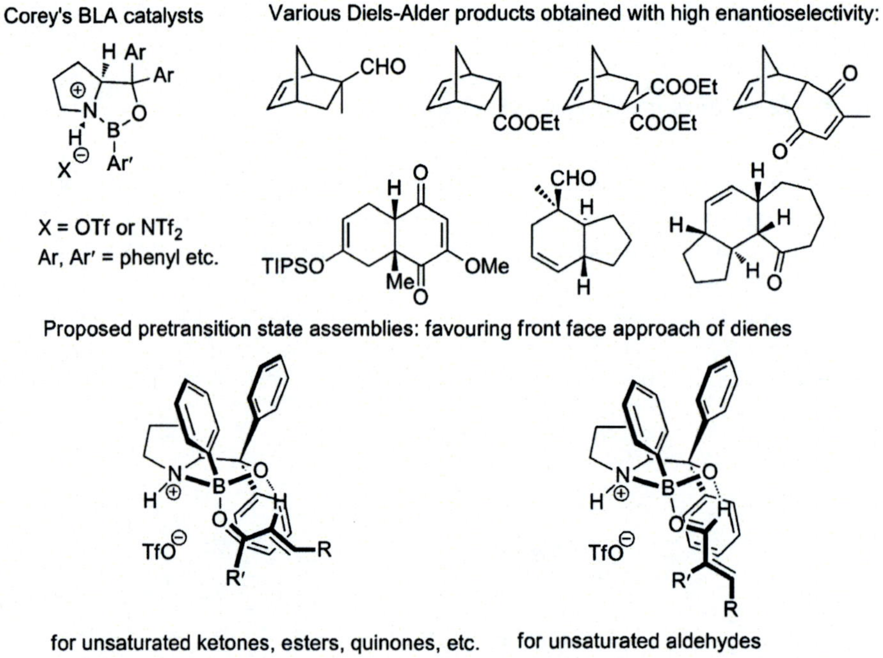

Fig. 3 Application of strong Brønsted acid-activated oxazaborolidine catalysts, and Corey's pretransition state assembly models

moieties and oxygen atom from oxazaborolidines [32, 33], as well as π–π interaction between the unsaturated dienophile substrates and phenyl rings from oxazaborolidines, are believed to be crucial for shielding one of the enantiotopic faces and achieving high degree of stereocontrol. The beauty of this model lies in the fact that it has been uniformly successful so far to predict various products' absolute stereochemistry. It has also been supported by the computational studies from the Houk group [34].

The Corey group has utilized these catalysts for a number of natural product synthesis, including estrone [35, 36], desogestrel [35], oseltamivir (Tamiflu) [37], dolabellatrienone and dolabellatriene [38], aflatoxin B_2 [39], laurenditerpenol [40], and various others [41].

A more recent contribution from the Corey group deserves special appreciation. These above-mentioned BLA catalysts can be viewed as borane Lewis acids with protonated secondary amine as part of their ligands, while combination with Brønsted acid is responsible for the enhanced Lewis acidity. A related oxazaborolidine catalyst with cationic tertiary amine as part of its ligand can also be used in enantioselective Diels–Alder reaction (Fig. 4) [42]. This single functional acid catalyst was shown to give similar or superior performance compared with related combined acid catalysts, though its preparation seems to be more difficult.

In 2005, we developed a new type of LLA catalyst system with a valine-derived oxazaborolidine and tin(IV) chloride for enantioselective Diels–Alder reaction (Fig. 5) [43]. Remarkably, $SnCl_4$ can be used in excessive catalyst loading without compromising enantioselectivity, which means that the achiral Lewis acid mediated racemic pathway should be significantly slower than the LLA catalyzed enantioselective pathway. When excessive $SnCl_4$ were used, it can neutralize incidental Lewis base impurities (e.g., moisture introduced during reaction set-up). In fact, the reaction can even proceed smoothly when various Lewis bases, such as water and amines, were added in, making this reaction system highly robust. The robustness of this catalyst system has set an example and high bar for future development of acid catalysis.

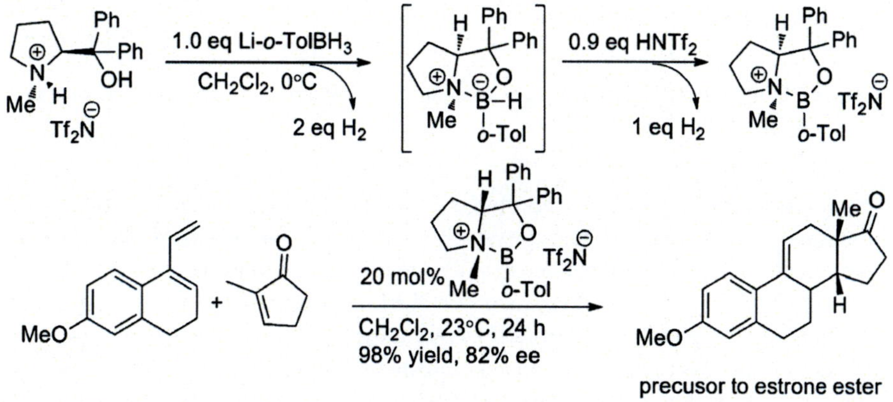

Fig. 4 Preparation and application of a cationic oxazaborolidine catalyst

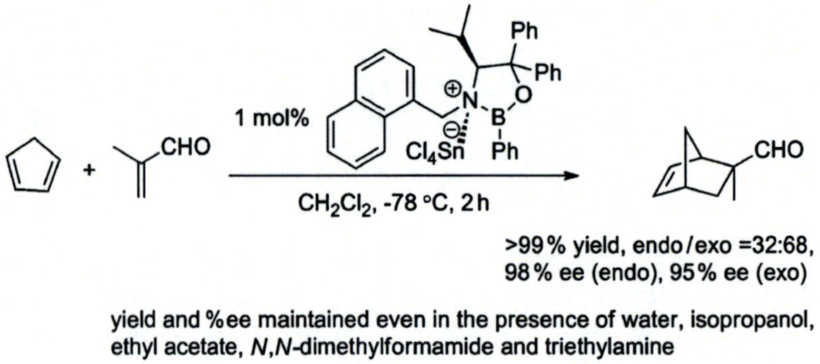

Fig. 5 $SnCl_4$ activated oxazaborolidine catalyst for Lewis base tolerant asymmetric Diels–Alder reaction

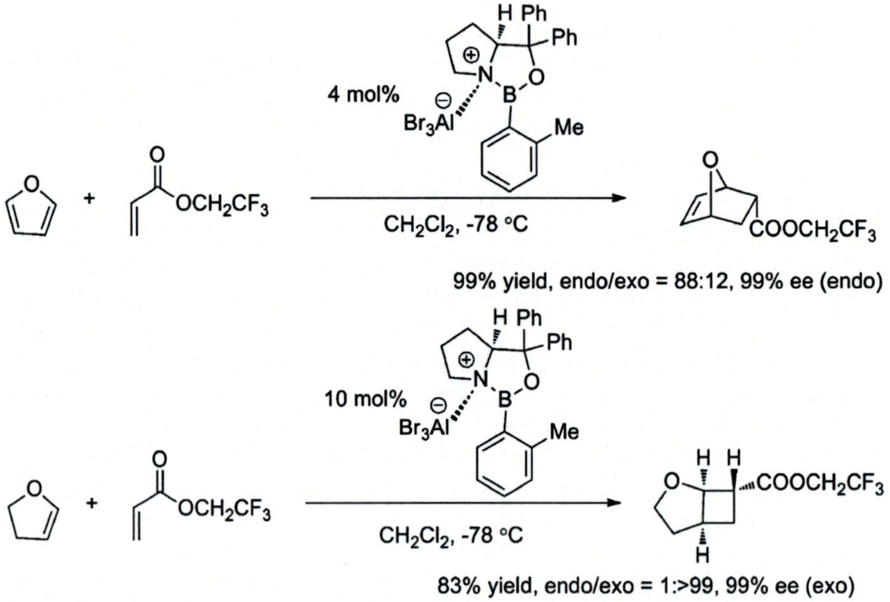

Fig. 6 $AlBr_3$ activated oxazaborolidine catalyst for asymmetric furan Diels–Alder reaction and [2+2] cycloaddition

In 2007, the Corey group reported the use of $AlBr_3$ as Lewis acid activator for a number of oxazaborolidine catalyzed reactions (Fig. 6) [44, 45]. In this study, complete complexation between $AlBr_3$ and oxazaborolidine was observed by ^{1}H NMR, which might be the reason for the higher reactivity of this LLA catalyst compared with previously reported BLA catalysts.

During our synthetic study on a Robinson annulation approach toward platensimycin, we used a carbon based Brønsted acid to activate the valine derived oxazaborolidine catalyst to prepare a key chiral building block [46]. This new

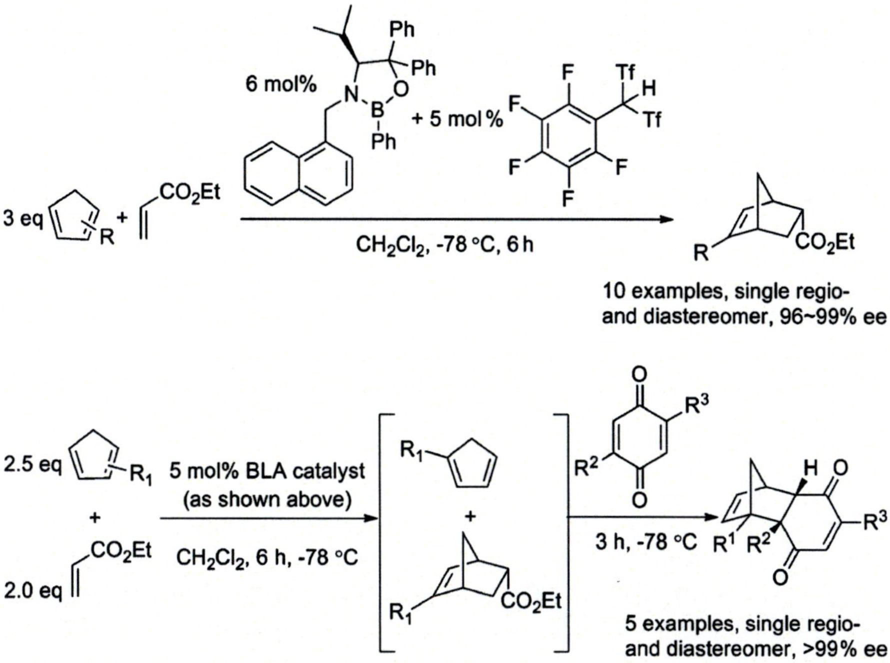

Fig. 7 Carbon-based Brønsted acid activated oxazaborolidine catalyst for asymmetric Diels–Alder reactions

BLA catalyst is effective for highly enantio-, diastereo-, and regioselective Diels-Alder reaction between methyl acrylate and substituted cyclopentadienes (Fig. 7) [47]. The better performance of carbon based Brønsted acid compared with TfOH or Tf_2NH were thought to be resulting from the lower nucleophilicity of its sterically bulky counteranion, leaving the borane atom more Lewis acidic. While these cyclopentadiene substrates exist as mixture of 1- and 2- substituted isomers, only 2-substituted cyclopentadienes react with methylacrylate at low temperature. In addition, after consumption of all 2-substituted cyclopentadienes, the remaining 1-substituted cyclopentadienes can then participate in Diels-Alder reaction with newly added quinone substrates which are more reactive dienophiles, giving otherwise difficult to access Diels–Alder products with two adjacent all-carbon quaternary stereocenters.

Later, we further extended the application of this type of BLA catalyst for highly selective Diels-Alder reaction of alkynyl ketone substrates (Fig. 8) [48]. In this case, dienophiles lacking the typical hydrogen bond donor moiety required by Corey's pretransition state assembly model can also give rise to high enantioselectivities. These results echoed the previous study by Paddon-Row and co-workers on related cationic oxazaborolidine catalyzed Diels-Alder reaction without involvement of hydrogen bonding, in which other stereoelectronic effects may contribute to the preorganization of Lewis acid/carbonyl coordination [49]. Considering the

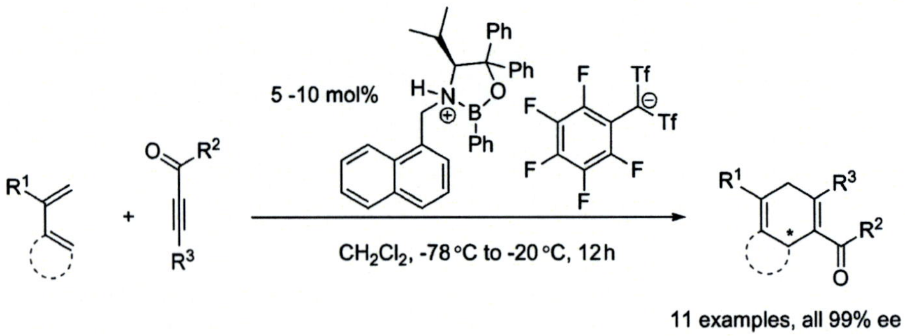

Fig. 8 Carbon-based Brønsted acid activated oxazaborolidine catalyst for asymmetric Diels–Alder reaction of alkynyl ketones

endo/exo approach of diene substrate toward alkynyl ketone dienophile can produce enantiomers, the remarkably good enantioselectivities highlighted the superiority of this catalyst system. Presumably, the extremely large $(C_6F_5)CTf_2^-$ counteranion may extend the asymmetric induction good enough for achieving nice results for this challenging reaction.

2.2 *BINOL-Aluminum Complex Based Catalysts*

In 1988, our group developed the use of BINOL-aluminum complex to catalyze highly enantioselective hetero-Diels-Alder reaction between Danishefsky type dienes and aldehydes (Fig. 9) [50]. Since then, a lot of catalytic asymmetric reactions utilizing this type of mononucleus aluminum catalysts have been developed [6].

In 2001, we extended this chemistry to the use of multinucleus BINOL-aluminum catalysts, which could be regarded as LLA catalysts, for a number of Diels-Alder reactions (Fig. 10) [51]. These complex structures have been elucidated on the basis of NMR studies, as well as gas generation measurements. Such kind of aluminoxide catalysts can be viewed as chiral versions of methyl aluminoxane (MAO), a very powerful Lewis acid co-catalyst for cationic polymerization of olefins. We thus believed further extension of this chemistry should be very promising.

During our recent study on utilizing tropone as diene for platencin synthesis [52], many other types of Lewis acid catalysts including the above mentioned oxazaborolidine catalysts all failed to promote the desired inverse-electron-demand Diels–Alder reaction. Fortunately, the dinucleus BINOL-aluminum catalysts were found to give very nice yields and enantioselectivities (Fig. 11) [53]. Activation mode in this reaction can be either intramolecular LLA assembly or double aluminum coordination to carbonyl substrate. Future research efforts are directed toward more applications with this type of catalyst, and more detailed understanding of the nature of this type of catalysts.

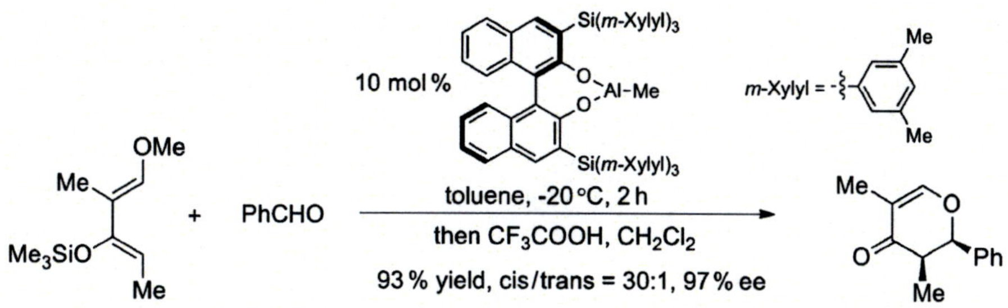

Fig. 9 Mononucleus BINOL-aluminum complex catalyzed hetero-Diels–Alder reaction

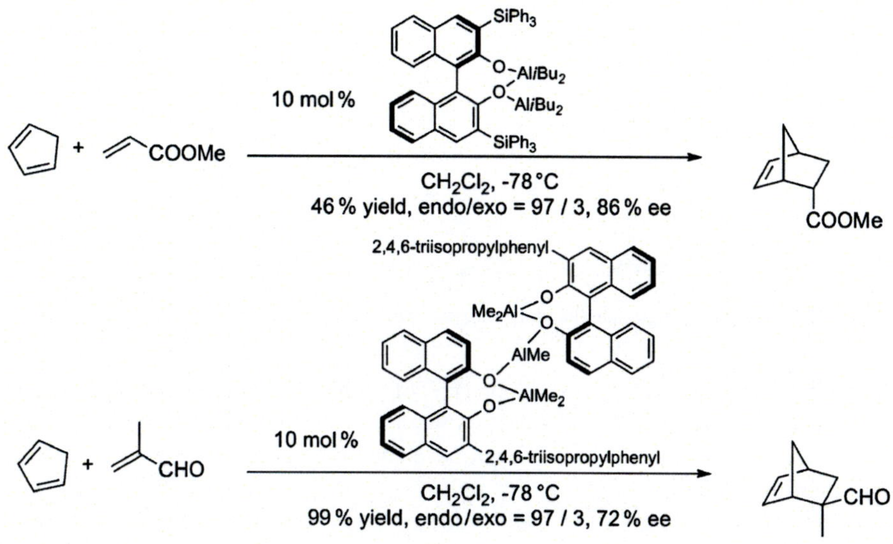

Fig. 10 Multinucleus BINOL-aluminum complexes catalyzed Diels–Alder reactions

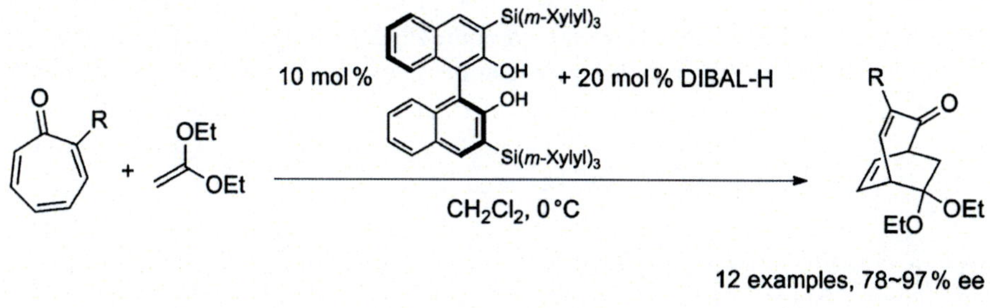

Fig. 11 Dinucleus BINOL-aluminum complex catalyzed tropone Diels–Alder reaction

2.3 *Brønsted Acid Catalyzed Asymmetric Protonation Reactions*

LBA catalyst prepared from tin(IV) chloride and chiral BINOL can be used as stoichiometric reagent for enantioselective protonation of silyl enol ethers [54].

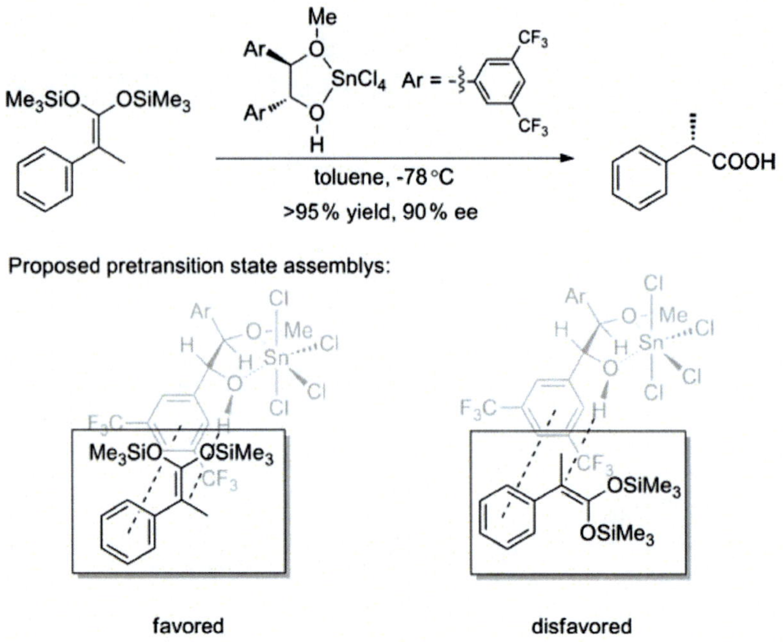

Fig. 12 $SnCl_4$-based LBA-mediated asymmetric protonation reaction

In 2003, our group developed a new type of protonation reagent involving mono-alkylated 1,2-diarylethane-1,2-diols as ligands (Fig. 12) [55]. A model for understanding the enantioselectivity was proposed based on the crystallographic structure of this LBA catalyst.

In 2008, we reported the use of chiral *N*-trifyl thiophosphoimide to catalyze enantioselective protonation of silyl enol ethers with various achiral proton sources (Fig. 13) [56]. It was found that neither the achiral acids nor stoichiometric amount of chiral catalyst alone can protonate the silyl enol ether substrate under such reaction conditions. We believe the combined BBA catalyst, which is an oxonium cation with chiral thiophosphoimide counteranion, is the reactive species for this protonation reaction. On the other hand, since the extremely bulky chiral counter anion cannot accomplish the desilylation step, presence of achiral proton source for catalyst regeneration turns out to be essential.

2.4 BINOL Catalyzed Asymmetric Organoborane Addition Reactions

In 2006, Schaus and co-workers reported the BINOL derivatives catalyzed asymmetric allylboration of ketones. 3,3′-dibromo substituted BINOL was found to be very effective catalyst for allylboration of a wide variety of ketone substrates

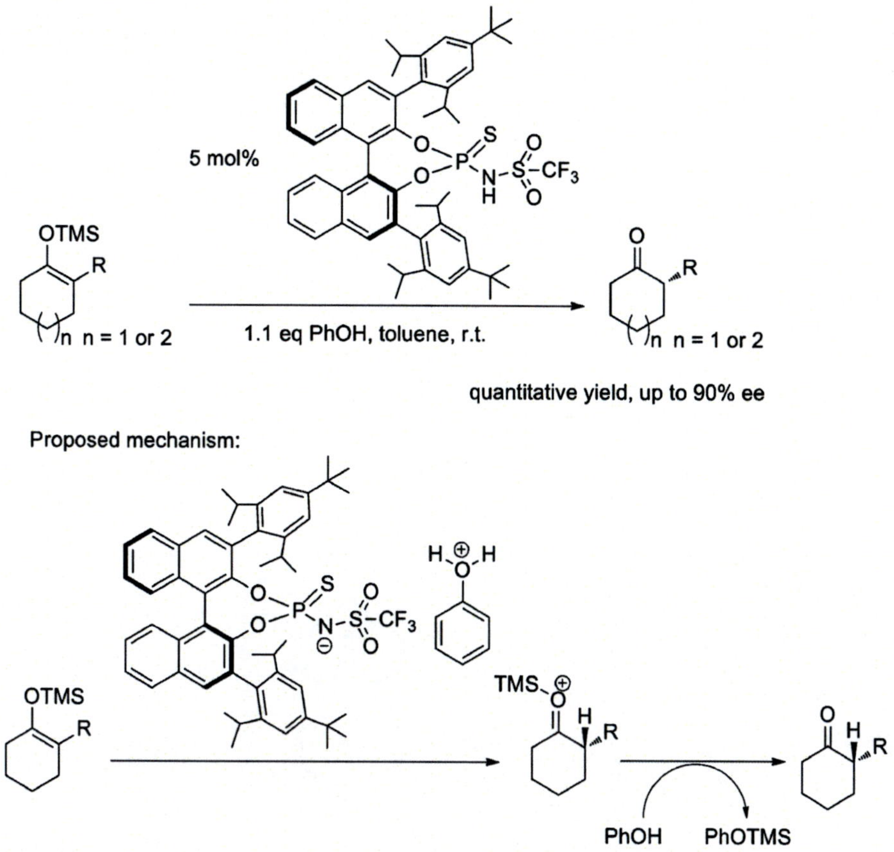

Fig. 13 Thiophosphoimide-based BBA catalyzed asymmetric protonation reaction

(Fig. 14) [57]. In the proposed mechanism, ligand exchange between one of allylboronate's alkoxy groups with the BINOL catalyst can lead to intramolecular hydrogen bonding, which enhanced the Lewis acidity of the borane atom and reactivity of allylboronate. More recently, insights from kinetic studies helped further improvement of this reaction: addition of *t*-BuOH could accelerate ligand exchange and the overall reaction rate. This allowed reduction of catalyst loading from 15 to 2 mol% for the room temperature reaction involving more robust cyclic boronates at room temperature reaction conditions [58].

The Schaus group also reported related reactions involving asymmetric allylboration of acyl imines [59], asymmetric three component Petasis condensation reaction of secondary amines/glyoxylates/alkenyl boronates [60], as well as addition of aryl, vinyl, and alkynyl boronates to acyl imines (Fig. 15) [61]. In the later reaction, a two point coordination transition state was proposed to account the observed facial selectivity.

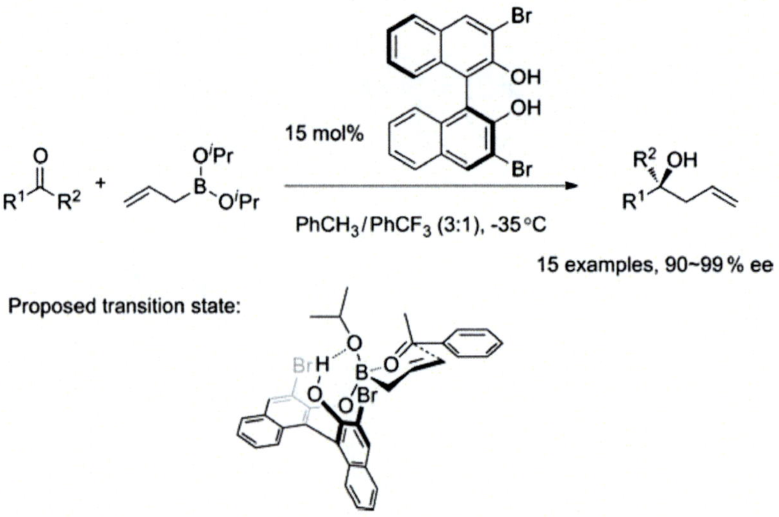

Fig. 14 BINOL catalyzed asymmetric allylboration of ketones

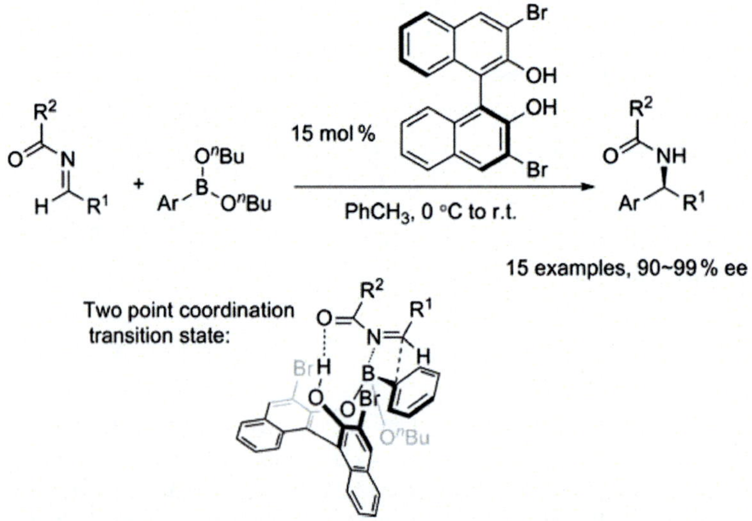

Fig. 15 BINOL catalyzed asymmetric addition of boronates to acyl imines

2.5 *Lewis Acid/Chiral Phosphoric Acid Combination*

Since the seminal reports from the Akiyama group [62] and the Terada group [63] in 2004, chiral phosphoric acid catalysis has become a very exciting field in the past few years [64–66]. Our group has made some efforts to increase the acidity of this type of catalysts through modifying chiral phosphoric acid structure to corresponding chiral *N*-triflyl phosphoramide [67, 68] or *N*-triflyl thio- and

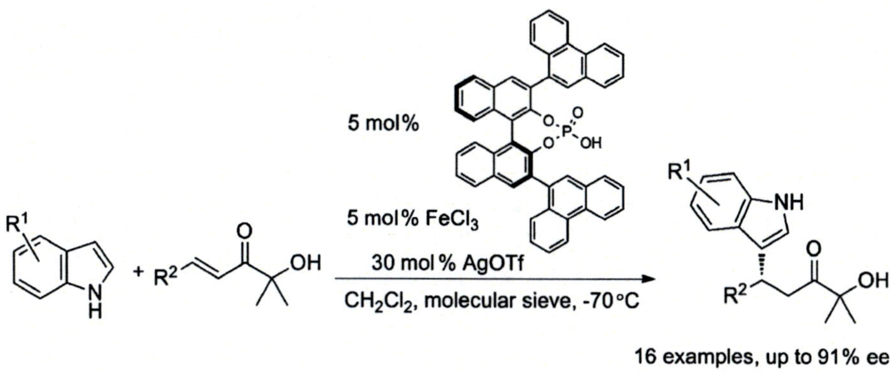

Fig. 16 Iron activation of chiral phosphoric acid for Fridel–Crafts reaction

seleno-phosphoramide [56, 69]. Another interesting approach for enhancing reactivities of chiral phosphoric acid catalysis has appeared recently from two independent studies, in which Lewis acid activation of Brønsted acid (LBA) consideration is the key to success.

In 2010, Xia, Huang and co-workers reported an iron-assisted chiral phosphoric acid catalyst for the Fridel–Crafts reaction between indole and β-aryl α′-hydroxy enones (Fig. 16) [70]. Due to the evidence from electron spray ionization mass spectrometry (ESI-MS) analysis for the presence of iron phosphate salt, an iron/proton dual activation mechanism was proposed to account the observed reactivity/selectivity trends. Addition of a third acid component, silver triflate, was shown to be beneficial for the enantioselectivity, presumably by exchanging chloride to the less coordinating triflate anion.

Also in 2010, Luo and co-workers developed another combined acid catalysis system involving chiral phosphoric acid and magnesium fluoride for asymmetric Fridel–Crafts reaction of unsaturated ketoesters (Fig. 17) [71]. These two acid components were found to be inactive for this reaction when used individually, while the combined use of this Lewis acid-assisted Brønsted acid system gave good yields and excellent enantioselectivities for a variety of substrates. Other Lewis acid additives, such as $MgSO_4$, $InBr_3$, $Bi(OTf)_3$, $Zn(OTf)_2$, $Ni(OTf)_2$, and $Cu(OTf)_2$, can also promote this reaction, though with lower enantioselectivities. We believe more studies along this line could extend the chiral phosphoric acid chemistry to an even more interesting level.

2.6 *Brønsted Acid/Chiral (Thio)urea Combination*

More recently, new developments of chiral hydrogen bond donor catalysts [72, 73], especially the urea/thiourea type catalysts, have led to the realization of the combined use of achiral Brønsted acid catalysts and chiral thiourea catalysts. Representative examples are summarized herein.

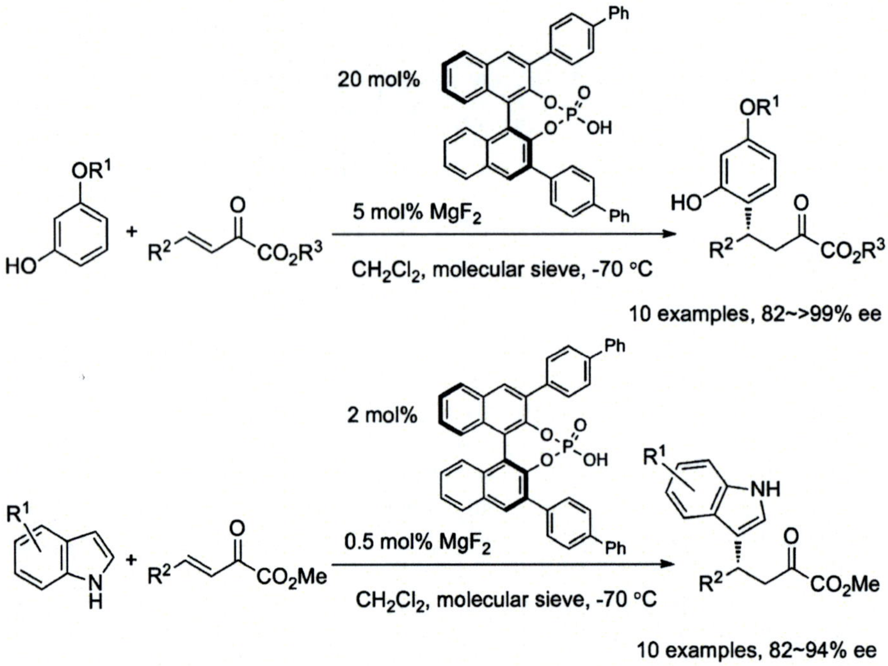

Fig. 17 Magnesium activation of chiral phosphoric acid for Fridel–Crafts reaction

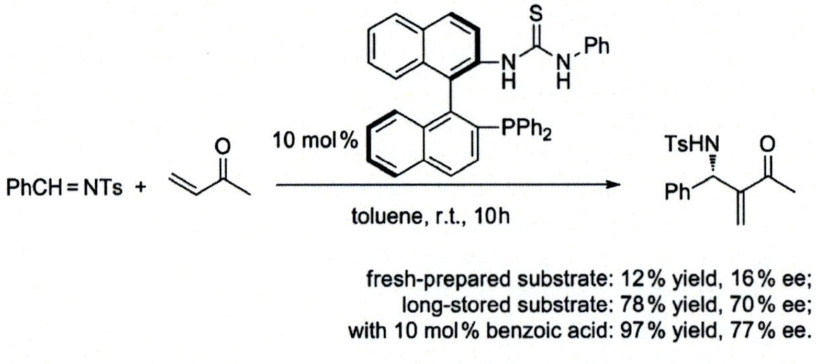

Fig. 18 Cooperative catalysis of chiral thiourea and Brønsted acid for aza-Morita–Baylis–Hillman reaction

The cooperative effect of Brønsted acid catalysts and thiourea catalysts has been noticed by the Shi group (Fig. 18) [74]. In their 2007 paper on chiral thiourea-phosphine catalyzed asymmetric aza-Morita-Baylis-Hillman reaction, Shi and co-workers described that when they used a freshly prepared *N*-benzylidene-4-methylbenzenesulfonamide substrate, much lower yield and enantioselectivity of the aza-Morita-Baylis-Hillman product was obtained compared with their initial result when using a long-stored substrate. They subsequently found that the long-stored substrate contained a small amount of 4-methylbenzenesulfonamide and

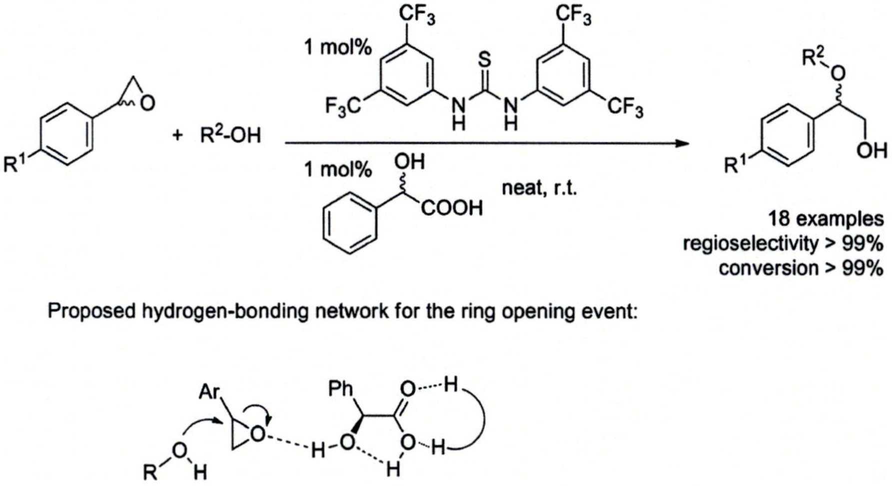

Fig. 19 Cooperative catalysis of achiral thiourea and Brønsted acid for styrene oxide ring opening

benzoic acid, while benzoic acid indeed accelerated the reaction of the freshly prepared substrate. Careful ^{31}P NMR studies on the cooperative effect between benzoic acid and chiral thiourea-phosphine catalyst were then undertaken, which provided nice insights for this acid–base bifunctional catalysis reaction: appropriate amount of benzoic acid could help proton-transfer of some key intermediates in the catalytic cycle, a task not attainable by the thiourea moiety alone.

In 2008, the Schreiner group also found that a mandelic acid impurity was actually accelerating their achiral thiourea catalyzed styrene oxide ring opening reaction (Fig. 19) [75]. Interaction between mandelic acid and thiourea was found to be very strong that the complexation energy could not be determined through conventional NMR titration. Computational studies, on the contrary, showed significant stability of ternary complex (as shown in Fig. 19) over corresponding binary complexes.

In 2009, the Jacobsen group reported a highly enantioselective Pictet–Spengler reaction starting from tryptamine and aldehyde precursors (Fig. 20) [76]. Cooperative catalysis of chiral thiourea ligand and benzoic acid accomplished the initial imine formation, and then asymmetric cyclization in one pot. The products' chirality was clearly generated through interactions between the ion pair between iminium cation and chiral thiourea bonded benzoate anion. The use of stronger achiral acids such as hydrogen chloride and methanesulfonic acid gave product in lower yield and ee, probably due to catalyst inhibition through protonation of the Lewis basic product.

In 2010, the Jacobsen group further advanced this chemistry to the combined use of strong achiral acid components with chiral urea and thiourea catalysts for Povarov reaction (Fig. 21) [77]. Different types of electron-rich alkenes, such as 2,3-dihydrofuran, various enamides and enecarbamates, can be used for the Povarov reaction with imine substrates. Detailed kinetic, as well as computational studies, led to the picture of inducing high enantioselectivity with protio-iminium

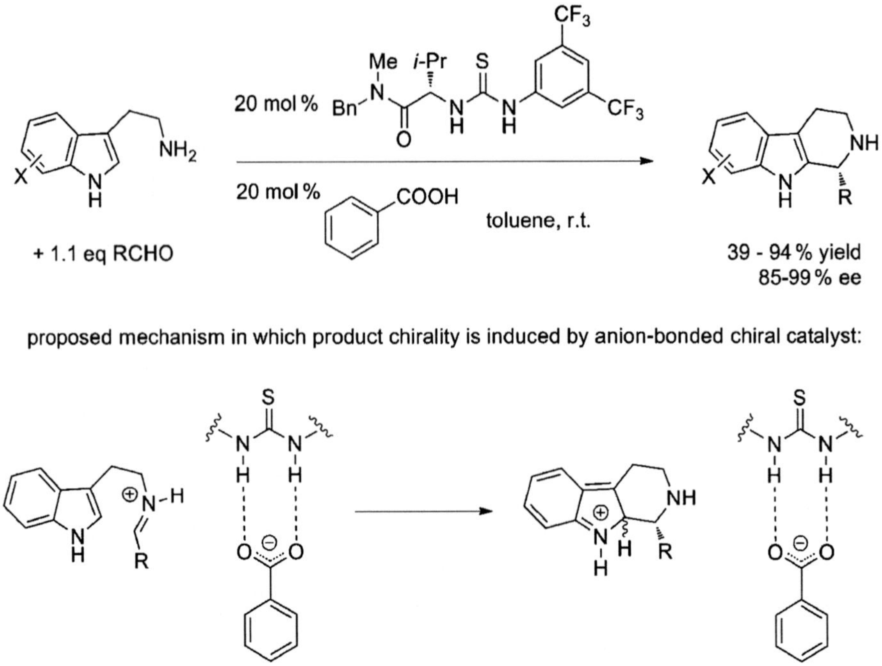

Fig. 20 Cooperative catalysis of chiral thiourea and Brønsted acid for Pictet–Spengler reaction

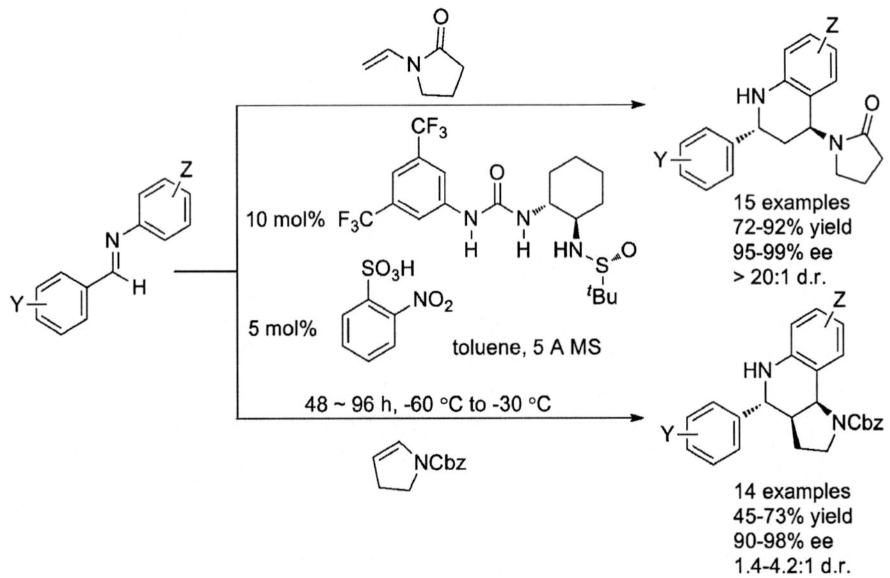

Fig. 21 Cooperative catalysis of chiral thiourea and Brønsted acid for Povarov reaction

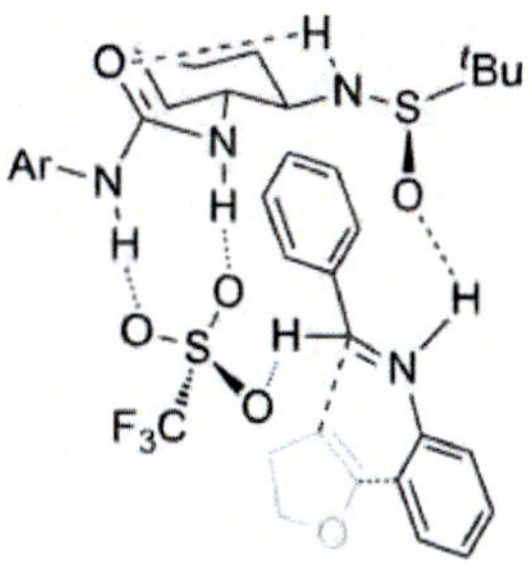

Fig. 22 A hydrogen bonding network model derived from kinetic and computational stuides

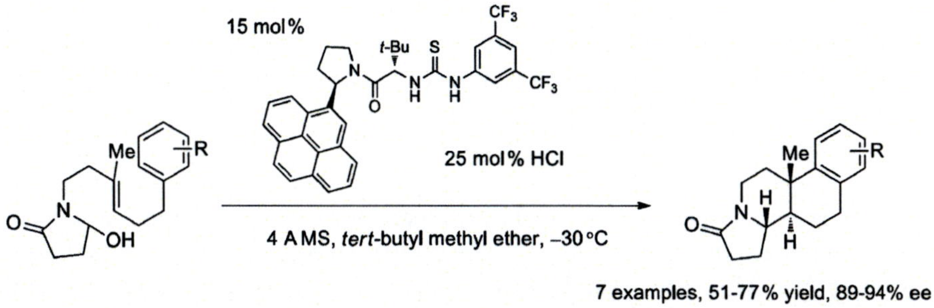

Fig. 23 Cooperative catalysis of chiral thiourea and Brønsted acid for cationic cyclization

ions through a network of noncovalent interactions (Fig. 22). The optimized chiral urea catalyst effectively suppressed the intrinsic reactivity of sulfonic acid catalyzed background reaction, and thus led to the realization of high enantioselectivity.

Also in 2010, the Jacobsen group reported the combined use of hydrochloric acid and thiourea catalysts for the highly enantioselective cationic double cyclization reaction initiating from an *N*-acyliminium ion (Fig. 23) [78]. The catalyst optimization research led to the identification of a pyrenyl-substituted thiourea structure, which seems to indicate the importance of cationic-π interaction during this reaction.

Retrospectively, these new developments from the Jacobsen group were built upon studies of counteranion-binding chiral thiourea catalysis [79–88]. Mechanistic studies, both experimentally and computationally [89, 90], greatly improved their understanding of these reactions, which ultimately led to new catalyst design. Given the versatile ability of urea and thiourea catalysts in anion binding interactions, this strategy could pave the way for future development of new synthetically useful asymmetric transformations involving cationic intermediates.

3 Beyond Combined Acid Catalysis

The remaining part of this chapter will highlight some recent examples in which bifunctional acid catalysis does not involve combined acid catalysis, namely, two acid components (either Lewis acid, hydrogen bonding, or transition metal) are coordinating with the substrates (both electrophile and nucleophile) independently, rather than coordinating with each other.

3.1 Lewis Acid/Hydrogen Bonding Cooperative Catalysis

During their investigation of bis(oxazoline) and bis(thiazoline) ligands catalysis, Du, Xu and co-workers developed a class of C_2-symmetrical tridentate bis(oxazoline)/bis(thiazoline) ligands containing diphenylamine units, which are applicable for a number of asymmetric reactions [91–98]. The importance of the NH group was clearly demonstrated by structure–reactivity relationship study with a ligand bearing CH_2 group, which gave much lower enantioselectivity [92]. The proposed transition state assemblies for these reactions involve simultaneous metal centered Lewis acid activation of electrophiles, and NH-based hydrogen bonding with nucleophiles (Fig. 24) [92, 95]. Although the later interaction might diminish the substrate's nucleophilicity, it might be compensated by entropy gains, and could be very important in achieving highly ordered transition state assembly and higher enantioselectivity. For example, in the zinc catalyzed asymmetric Friedel–Crafts alkylation of indole with β-nitrostyrene, NH•••π interaction between the catalyst

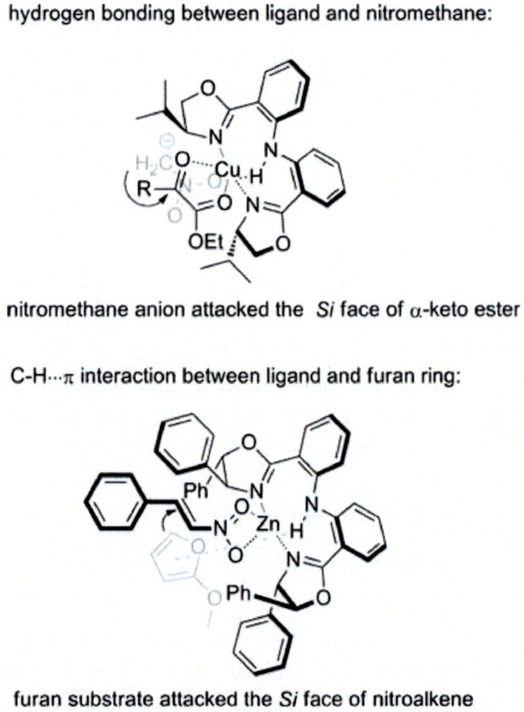

Fig. 24 Hydrogen bonding models proposed for reactions involving diphenylamine-based ligands

and indole ring was proposed [94]. Structure reactivity relationship studies on such ligands were also carried out to elucidate the importance of the proposed NH effect [97]. While enhancement of NH acidity by electronic tuning of backbone phenyl rings did not improve enantioselectivity, we are curious to know whether decreasing N–H acidity might diminish the enantioselectivity.

Similar hypothesis of this kind of cooperative catalysis mechanism has been proposed by Matsunaga, Shibasaki, and co-workers in their heterobimetallic gallium/ytterbium-Schiff base complex catalyzed asymmetric α-addition of isocyanides to aldehydes [99]. Research along this line will be reviewed in another chapter of this book [100].

3.2 *Lewis Acid/Transition-Metal Cooperative Catalysis*

Many rhodium(II) complexes are excellent catalysts for metal-carbenoid-mediated enantioselective C–H insertion reactions [101]. In 2002, computational studies by Nakamura and co-workers suggested the dirhodium tetracarboxylate catalyzed diazo compounds insertion reaction to alkanes' C–H bonds proceed through a three-centered hydride-transfer-like transition state (Fig. 25) [102]. Only one rhodium atom of the catalyst is involved in the formation of rhodium carbene intermediate, while the other rhodium atom served as a mobile ligand, which enhanced the electrophilicity of the first one and facilitate the cleavage of rhodium–carbon bond. In this case, the metal–metal bond constitutes a special example of Lewis acid activation of Lewis acidic transition-metal catalyst.

In 2008, the White group reported a new approach for enantioselective allylic oxidation of C–H bonds, namely chiral Lewis acid activation of achiral transition-metal catalyzed reaction (Fig. 26) [103]. They proposed the use of a chiral

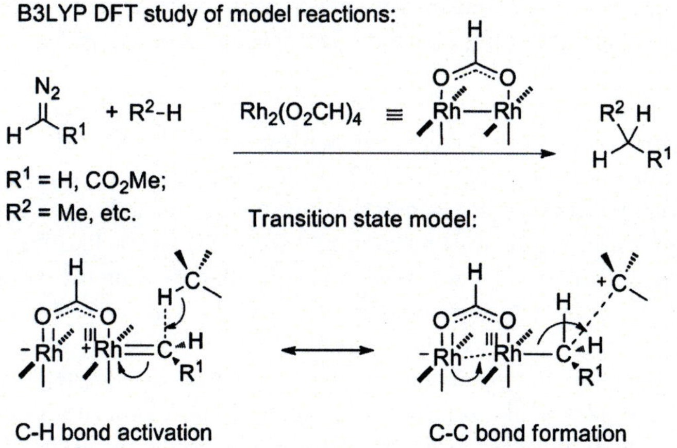

Fig. 25 Dirhodium complex catalysis model derived from computational studies

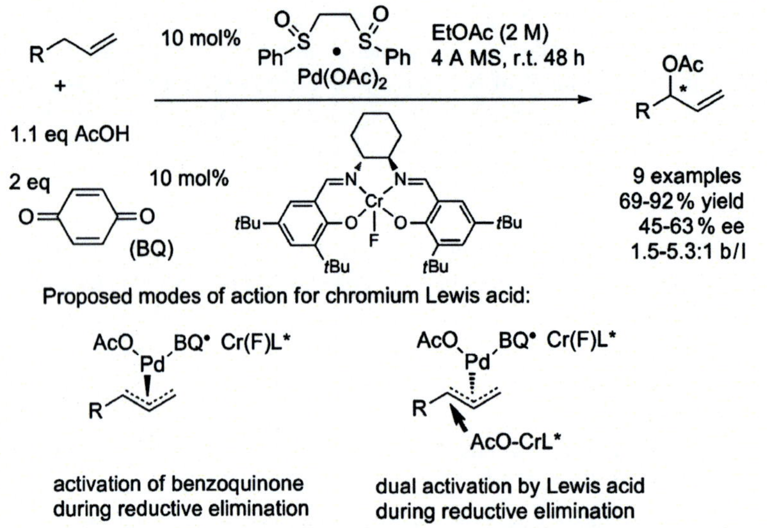

Fig. 26 Chiral Lewis acid activation of transition-metal catalyzed reaction

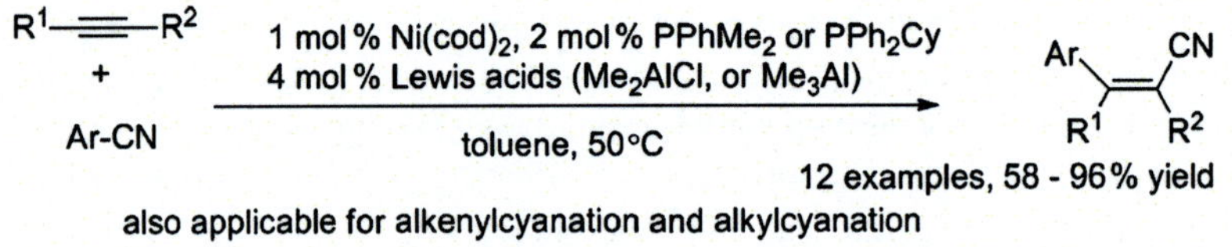

Fig. 27 Achiral Lewis acid activation of achiral transition-metal catalyzed reaction

chromium-salen complex-based Lewis acid catalyst to coordinate with the benzoquinone and eventually obtained modest asymmetric induction (up to 63% ee). Further development of this concept could open up a new door for catalyst design and development [104].

Another important strategy in bifunctional acid catalyst design is the use of achiral Lewis acid to activate chiral transition-metal catalyzed reaction. Nakao, Hiyama, and co-workers reported in 2007 very general nickel catalyzed carbocyanation reactions of alkynes in which various Lewis acid catalysts (Me_3Al, Me_2AlCl, BPh_3) significantly enhanced reaction rate (Fig. 27) [105]. In 2008, the Jacobsen group reported a nickel catalyzed intramolecular asymmetric arylcyanation of unactivated olefins with high enantioselectivities (Fig. 28) [106]. Tangphos was found to be the optimal phosphine ligand for this nickel catalyzed reaction, while addition of BPh_3 as Lewis acid catalyst greatly enhanced the yield.

More examples of these types of cooperative use of transition metal catalysts with Lewis acid catalysts can be found in a recent review [9]. And we expect many more asymmetric reactions of this type could be developed in the near future.

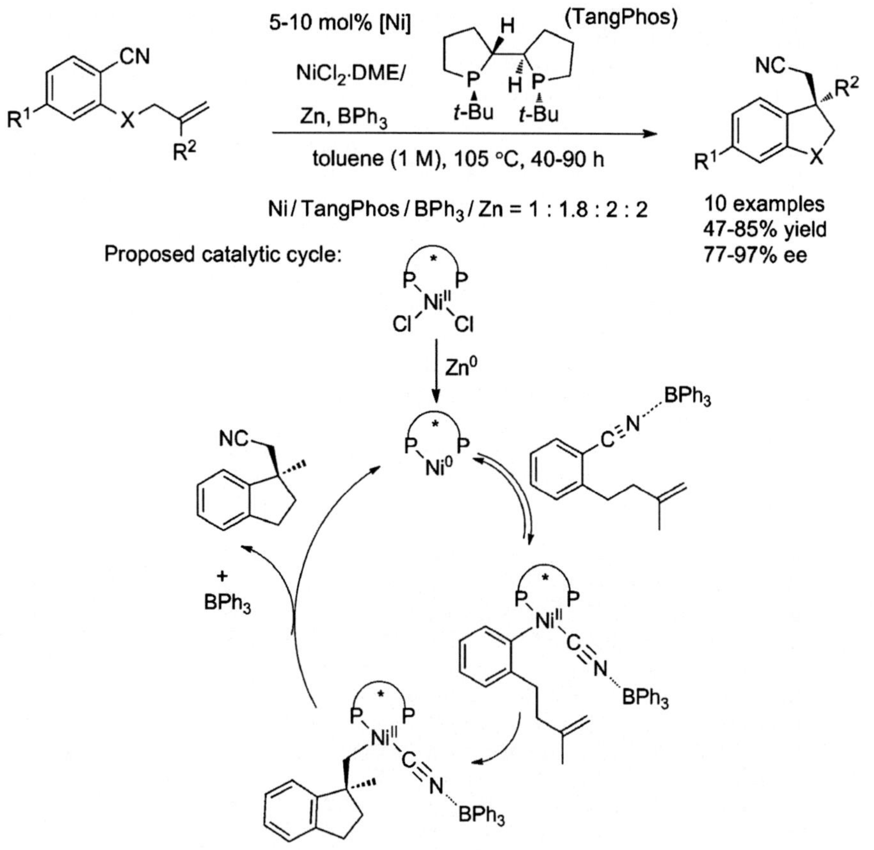

Fig. 28 Achiral Lewis acid activation of chiral transition-metal catalyzed reaction

Acknowledgments This work was supported by NIH (5R01GM068433-06), NSF (CHE-0717618), and a Eugene Olshansky Memorial Scholarship (to P.L.).

References

1. Yamamoto H, Futatsugi K (2005) Angew Chem Int Ed 44:1924
2. Yamamoto H, Futatsugi K (2008) Recent advance of "combined acid" strategy for asymmetric catalysis. In: Yamamoto H, Ishihara K (eds) Acid catalysis in modern organic synthesis. Wiley-VCH, Weinheim
3. Jacobsen EN, Pfaltz A, Yamamoto H (eds) (1999) Comprehensive asymmetric catalysis. Springer, Berlin
4. Ojima I (ed) (2010) Catalytic asymmetric synthesis, 3rd edn. Wiley, Hoboken
5. Yamamoto H (ed) (1999) Lewis acid reagents: a practical approach. Oxford University Press, Oxford
6. Yamamoto H (ed) (2000) Lewis acids in organic synthesis. Wiley-VCH, Weinheim

7. Yamamoto H, Ishihara K (eds) (2008) Acid catalysis in modern organic synthesis. Wiley-VCH, Weinheim
8. Yoon TP, Jacobsen EN (2003) Science 299:1691
9. Wang C, Xi Z (2007) Chem Soc Rev 36:1395
10. Zhong C, Shi X (2010) European J Org Chem 2999
11. Hirao A, Itsuno S, Nakahama S, Yamazaki N (1981) J Chem Soc Chem Commun 315
12. Itsuno S, Hirao A, Nakahama S, Yamazaki N (1983) J Chem Soc Perkin Trans 1 1673
13. Itsuno S, Ito K, Hirao A, Nakahama S (1983) J Chem Soc Chem Commun 469
14. Itsuno S, Nakano M, Miyazaki K, Masuda H, Ito K, Hirao A, Nakahama S (1985) J Chem Soc Perkin Trans 1 2039
15. Corey EJ, Bakshi RK, Shibata S (1987) J Am Chem Soc 109:5551
16. Corey EJ, Bakshi RK, Shibata S, Chen C-P, Singh VK (1987) J Am Chem Soc 109:7925
17. Corey EJ, Shibata S, Bakshi RK (1988) J Org Chem 53:2861
18. Corey EJ (1990) Pure Appl Chem 62:1209
19. Corey EJ, Helal CJ (1998) Angew Chem Int Ed 37:1986
20. Corey EJ, Loh T-P (1991) J Am Chem Soc 113:8966
21. Corey EJ, Shibata T, Lee TW (2002) J Am Chem Soc 124:3808
22. Corey EJ (2009) Angew Chem Int Ed 48:2100
23. Ryu DH, Lee TW, Corey EJ (2002) J Am Chem Soc 124:9992
24. Ryu DH, Corey EJ (2003) J Am Chem Soc 125:6388
25. Ryu DH, Zhou G, Corey EJ (2004) J Am Chem Soc 126:4800
26. Zhou G, Hu Q-Y, Corey EJ (2003) Org Lett 5:3979
27. Balskus EP, Jacobsen EN (2007) Science 317:1736
28. Ryu DH, Corey EJ (2004) J Am Chem Soc 126:8106
29. Ryu DH, Corey EJ (2005) J Am Chem Soc 127:5384
30. Gnanadesikan V, Corey EJ (2006) Org Lett 8:4943
31. Liu D, Hong S, Corey EJ (2006) J Am Chem Soc 128:8160
32. Corey EJ, Rhode JJ (1997) Tetrahedron Lett 38:37
33. Corey EJ, Lee TW (2001) Chem Commun 1321
34. Paddon-Row MN, Anderson CD, Houk KN (2009) J Org Chem 74:861
35. Hu Q-Y, Rege PD, Corey EJ (2004) J Am Chem Soc 126:5984
36. Yeung Y-Y, Chein R-J, Corey EJ (2007) J Am Chem Soc 129:10346
37. Yeung Y-Y, Hong S, Corey EJ (2006) J Am Chem Soc 128:6310
38. Snyder SA, Corey EJ (2006) J Am Chem Soc 128:740
39. Zhou G, Corey EJ (2005) J Am Chem Soc 127:11958
40. Mukherjee S, Scopton AP, Corey EJ (2010) Org Lett 12:1836
41. Hu Q-Y, Zhou G, Corey EJ (2004) J Am Chem Soc 126:13708
42. Canales E, Corey EJ (2008) Org Lett 10:3271
43. Futatsugi K, Yamamoto H (2005) Angew Chem Int Ed 44:1484
44. Liu D, Canales E, Corey EJ (2007) J Am Chem Soc 129:1498
45. Canales E, Corey EJ (2007) J Am Chem Soc 129:12686
46. Li P, Payette JN, Yamamoto H (2007) J Am Chem Soc 129:9534
47. Payette JN, Yamamoto H (2007) J Am Chem Soc 129:9536
48. Payette JN, Yamamoto H (2009) Angew Chem Int Ed 48:8060
49. Paddon-Row MN, Kwan LCH, Willis AC, Sherburn MS (2008) Angew Chem Int Ed 47:7013
50. Maruoka K, Itoh T, Shirasaka T, Yamamoto H (1988) J Am Chem Soc 110:310
51. Ishihara K, Kobayashi J, Inanaga K, Yamamoto H (2001) Synlett 394
52. Li P, Yamamoto H (2010) Chem Commun. doi:10.1039/c0cc01619e
53. Li P, Yamamoto H (2009) J Am Chem Soc 131:16628
54. Ishihara K, Kaneeda M, Yamamoto H (1994) J Am Chem Soc 116:11179
55. Ishihara K, Nakashima D, Hiraiwa Y, Yamamoto H (2003) J Am Chem Soc 125:24
56. Cheon CH, Yamamoto H (2008) J Am Chem Soc 130:9246

57. Lou S, Moquist PN, Schaus SE (2006) J Am Chem Soc 128:12660
58. Barnett DS, Moquist PN, Schaus SE (2009) Angew Chem Int Ed 48:8679
59. Lou S, Moquist PN, Schaus SE (2007) J Am Chem Soc 129:15398
60. Lou S, Schaus SE (2008) J Am Chem Soc 130:6922
61. Bishop JA, Lou S, Schaus SE (2009) Angew Chem Int Ed 48:4337
62. Akiyama T, Itoh J, Yokota K, Fuchibe K (2004) Angew Chem Int Ed 43:1566
63. Uraguchi D, Terada M (2004) J Am Chem Soc 126:5356
64. Akiyama T (2007) Chem Rev 107:5744
65. Terada M (2008) Chem Commun 4097
66. Terada M (2010) Bull Chem Soc Jpn 83:101
67. Nakashima D, Yamamoto H (2006) J Am Chem Soc 128:9626
68. Jiao P, Nakashima D, Yamamoto H (2008) Angew Chem Int Ed 47:2411
69. Cheon CH, Yamamoto H (2010) Org Lett 12:2476
70. Yang L, Zhu Q, Guo S, Qian B, Xia C, Huang H (2010) Chem Eur J 16:1638
71. Lv J, Li X, Zhong L, Luo S, Cheng J-P (2010) Org Lett 12:1096
72. Taylor MS, Jacobsen EN (2006) Angew Chem Int Ed 45:1520
73. Doyle AG, Jacobsen EN (2007) Chem Rev 107:5713
74. Shi Y-L, Shi M (2007) Adv Synth Catal 349:2129
75. Weil T, Kotke M, Kleiner CM, Schreiner PR (2008) Org Lett 8:1513
76. Klausen RS, Jacobsen EN (2009) Org Lett 11:887
77. Xu H, Zuend SJ, Woll MG, Tao Y, Jacobsen EN (2010) Science 327:986
78. Knowles RR, Lin S, Jacobsen EN (2010) J Am Chem Soc 132:5030
79. Taylor MS, Jacobsen EN (2004) J Am Chem Soc 126:10558
80. Taylor MS, Tokunaga N, Jacobsen EN (2005) Angew Chem Int Ed 41:6700
81. Raheem IT, Thiara PS, Peterson EA, Jacobsen EN (2007) J Am Chem Soc 129:13404
82. Raheem IT, Thiara PS, Jacobsen EN (2008) Org Lett 10:1577
83. Reisman SE, Doyle AG, Jacobsen EN (2008) J Am Chem Soc 130:7198
84. De CK, Klauber EG, Seidel D (2009) J Am Chem Soc 131:17060
85. Mita T, Jacobsen EN (2009) Synlett 1680
86. Peterson EA, Jacobsen EN (2009) Angew Chem Int Ed 48:6328
87. Zuend SJ, Coughlin MP, Lalonde MP, Jacobsen EN (2009) Nature 461:968
88. Brown AR, Kuo W-H, Jacobsen EN (2010) J Am Chem Soc 132:9286
89. Zuend SJ, Jacobsen EN (2007) J Am Chem Soc 129:15872
90. Zuend SJ, Jacobsen EN (2009) J Am Chem Soc 131:15358
91. Lu S-F, Du D-M, Zhang S-W, Xu J (2004) Tetrahedron Asymmetr 15:3433
92. Du D-M, Lu S-F, Fang T, Xu J (2005) J Org Chem 70:3712
93. Lu S-F, Du D-M, Xu J, Zhang S-W (2006) J Am Chem Soc 128:7418
94. Lu S-F, Du D-M, Xu J (2006) Org Lett 8:2115
95. Liu H, Xu J, Du D-M (2007) Org Lett 9:4725
96. Liu H, Lu S-F, Xu J, Du D-M (2008) Chem Asian J 3:1111
97. Liu H, Li W, Du D-M (2009) Sci China B 52:1321
98. Liu H, Du D-M (2010) Eur J Org Chem 2121
99. Mihara H, Xu Y, Shepherd NE, Matsunaga S, Shibasaki M (2009) J Am Chem Soc 131:8384
100. Shibasaki M, Kanai M, Matsunaga S, Kumagai N (2011) Multimetallic multifunctional catalysts for asymmetric reactions. Top Organomet Chem. doi:10.1007/3418_2011_1
101. Davies HML, Beckwith REJ (2003) Chem Rev 103:2861
102. Nakamura E, Yoshikai N, Yamanaka M (2002) J Am Chem Soc 124:7181
103. Covell DJ, White MC (2008) Angew Chem Int Ed 47:6448
104. Qi X, Rice GT, Lall MS, Plummer MS, White MC (2010) Tetrahedron 66:4816
105. Nakao Y, Yada A, Ebata S, Hiyama T (2007) J Am Chem Soc 129:2428
106. Watson MP, Jacobsen EN (2008) J Am Chem Soc 130:12594

Top Organomet Chem (2011) 37: 185–205
DOI: 10.1007/3418_2011_5

Published online: 12 March 2011

Bifunctional Phebox Complexes for Asymmetric Catalysis

Jun-ichi Ito and Hisao Nishiyama

Abstract Chiral bifunctional rhodium complexes bearing chiral bis(oxazolinyl) phenyl ligand (phebox) catalyzed asymmetric reactions were described. Meridional C_2-symmetric environment around the metal center in the chiral phebox ligand is a crucial structure factor for determining the catalytic performance in terms of reactivity and selectivity. The phebox–Rh chloro complex serves as a mild Lewis acid catalyst for asymmetric allylation, hetero-Diels–Alder, and Michael reaction. Although the catalytic precursor, Rh(III) complex, with octahedral geometry is a quite stable molecule, it can be readily activated by hydrosilane to generate the putative Rh(I) species, which is a key intermediate in asymmetric hydrosilylation of alkenes and conjugate reduction of α,β-unsaturated carbonyl compounds. The Rh enolate species generated in situ can be used for the C–C bond formation reaction, asymmetric reductive aldol reactions, and direct aldol reactions.

Keywords Asymmetric catalysis · Chiral tridentate ligand · Oxazoline ligand · Rhodium catalyst

Contents

J.-i. Ito and H. Nishiyama (✉)
Department of Applied Chemistry, Graduate School of Engineering, Nagoya University, Chikusa, Nagoya 464-8603, Japan
e-mail: hnishi@apchem.nagoya-u.ac.jp

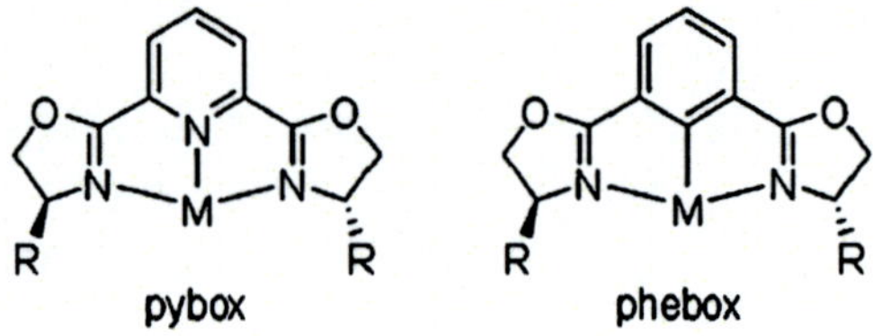

Fig. 1 Structure of pybox and phebox

1 Introduction

Modern organic synthesis requires high efficiencies and environmentally benign conditions for carbon–carbon bond formations and functional transformations. Multifunctional ability (concerto function) is considered to be one of the important features for molecular catalysts in asymmetric catalysis. Thus, design of ligands for metal catalysts is an important subject, because environment around the metal is significantly dependent on the structure of the ligand. In 1989, the pyridine-2,6-bis (oxazoline) (abbreviated as pybox) ligand as a C_2-symmetric and *N,N,N* meridional ligand was reported [1] (Fig. 1). Since then, the pybox ligands have been widely used in a lot of asymmetric catalysis [2]. The broad application of the pybox metal catalysts led to the development of the 2,6-bis(oxazolinyl)phenyl (abbreviated as phebox) ligand, which has a phenyl moiety in place of the pyridine fragment in the pybox ligand, reported by Nishiyama [3], Denmark [4], and Richards [5] (Fig. 1). The phebox metal complexes contain the metal–carbon covalent bond, which may stabilize the metal complex and change the electronic property of the metal center. Modification of substituents on the oxazoline and phenyl rings as well as other ligands could supply a variable character to the phebox metal complexes. To date, several transition metals, Ni, Pd, Pt, Rh, Ir, and Ru, have been used in the synthesis of the phebox metal complexes [6, 7]. In particular, phebox–Rh complexes achieved high asymmetric induction in asymmetric reactions.

The purpose of this chapter is to describe the concept of the chiral phebox complexes as *Bifunctional Catalysts* in terms of molecular design and mechanistic consideration. The applications of Rh complexes to asymmetric catalytic reactions are mainly discussed.

2 Chiral Phebox–Rh Complexes

The phebox–Rh(III) complexes **1** have been synthesized by the transmetalation of the phebox-$SnMe_3$ (**2**) with $[RhCl(coe)_2]_2$ in CCl_4 [8, 9] or by the C–H bond activation of phebox-H (**3**) with Rh(III) chloride in MeOH [6, 7] (Scheme 1). The chloro complex **1** contains the meridional coordination of the phebox ligand with a distorted octahedral geometry, in which two chlorides are located in *trans* positions and the H_2O ligand is coordinated in the phebox plane. In order to control the chiral

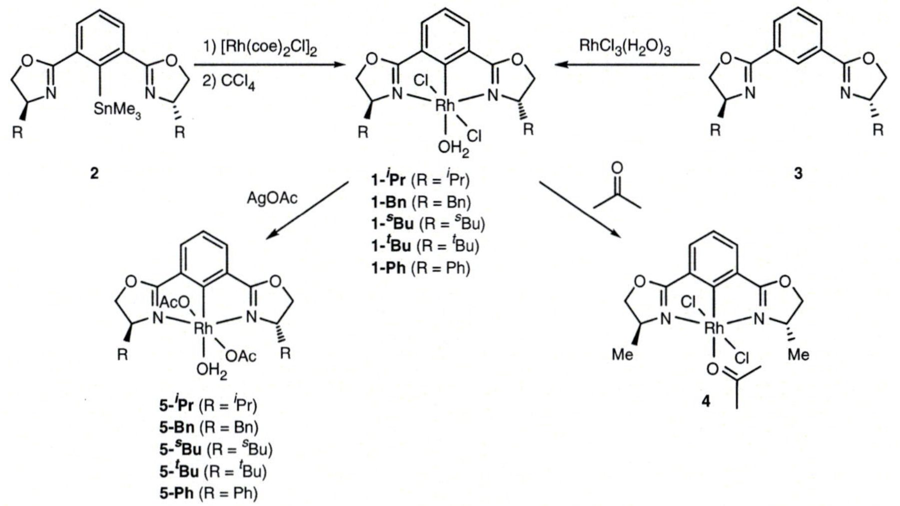

Scheme 1 Synthesis and reactions of phebox–Rh complexes

space, several substituents on the oxazoline ring, such as isopropyl, *sec*-butyl, *tert*-butyl, phenyl and benzyl, were introduced using β-aminoalcohols in the ligand synthesis. In the complexes **1**, H_2O ligand was replaced by 2-electron donors, such as acetone, to produce the acetone complex **4**. This reactivity implies that dissociation of H_2O can provide a vacant site on the Rh center that may work as a Lewis acid to promote activation of a carbonyl compound. The chloride complexes **1** were also converted to the acetate complexes **5** by treatment of silver acetate in CH_2Cl_2 at room temperature. The acetate complexes **5** exhibited the high reactivity toward hydrosilanes and diboron, which led to the development of catalytic C–C bond formation and functionalization reactions.

3 Asymmetric Allylation of Aldehydes with Allylstannane

Asymmetric allylation of aldehydes with allylic agents catalyzed by Lewis acid is a practical method for synthesizing optically active homoallylic alcohols [10]. The chloro complex **1** serves as an efficient catalyst for asymmetric allylation of aldehydes with allylstannane [8, 9, 11]. In the presence of 5 mol% of the benzyl-phebox–Rh complex **1-Bn**, the coupling reaction of benzaldehyde with allyltributyl-stannane in CH_2Cl_2 at room temperature proceeded smoothly to provide the corresponding homo-allyl alcohol **7a** in 88% yield with 61% ee (Scheme 2). When methallylstannane was subjected to the reaction, enantioselectivity of the allylated product **8** significantly increased to be over 90% ee.

The proposed mechanism is shown in Scheme 3. Dissociation of the H_2O ligand form **1** generates an unsaturated (phebox)$RhCl_2$ complex **9**, which interacts with the

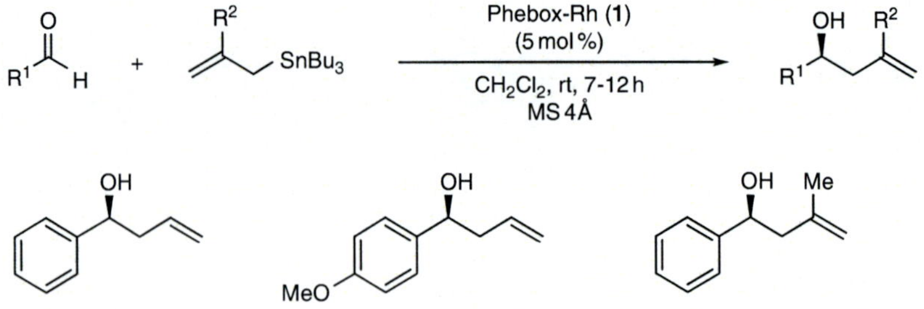

7a 88 %, 61 % ee (**1-Bn** cat.)

7b 99 %, 80 % ee (**1-Bn** cat.)

8 95 %, 93 % ee (**1-^{i}Pr** cat.)

Scheme 2 Asymmetric allylation of aldehydes with allylstannane

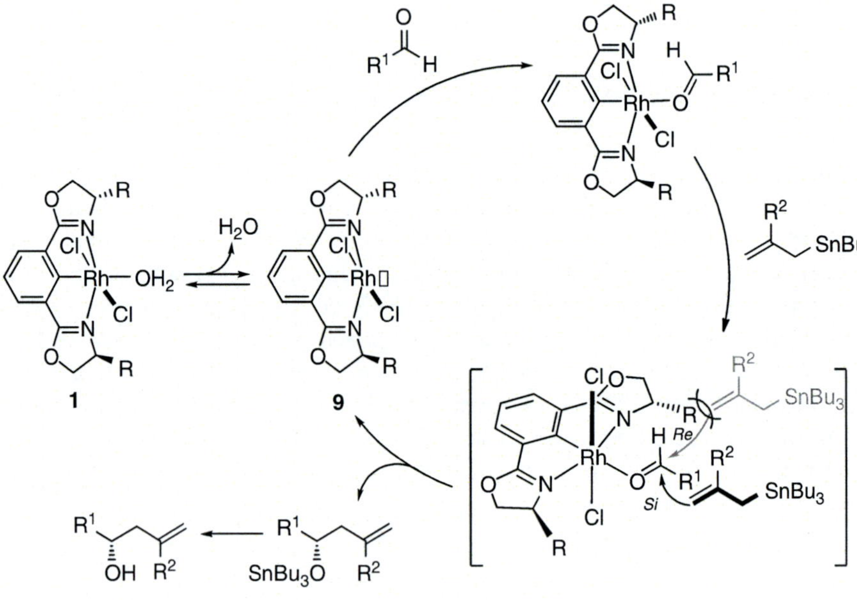

Scheme 3 Proposed mechanism for allylation

carbonyl group of an aldehyde on its Lewis acid vacant site (Scheme 3). The *Si* face attack of allylstannane to the aldehyde provides the C–C bond-forming product, desired allylic alcohol, and regenerates the catalyst intermediate. In the transition state, one of the substituents on the oxazoline rings plays an important role to shield the *Re* face of the carbonyl carbon. On the basis of the structural analysis of the acetone complex **4**, the C=O plane of the aldehyde might be shifted from the Cl–Rh–Cl axis to the phebox plane to avoid steric repulsion between the formyl proton and the chloride ligand. In the case of methallylstannane, the *Re* face attack to the aldehyde is disfavored because of the large steric repulsion between the methyl group (R^2 = Me) of methallylstannane and the oxazoline substituent. In the phebox–Rh chloro complex **1**, bifunctional metal–phebox ligand cooperation leading to the suitable chiral environment is crucial to discriminate the prochiral face of the substrate.

4 Asymmetric Hetero-Diels–Alder Reaction

Asymmetric hetero-Diels–Alder reaction of Danishefsky's diene with glyoxylate is efficiently catalyzed by the phebox–Rh(III) complexes **1** [12]. With 2 mol% of the benzyl phebox–Rh complex **1-Bn**, the reaction took place at −78°C for 1 h to give the corresponding dihydropyran derivatives **10** in high yields with good ee's (Scheme 4).

The low-temperature NMR experiment implied that the catalytic reaction proceeded via the concerned [4+2] cycloaddition pathway rather than the stepwise (Mukaiyama-aldol) mechanism. Furthermore, the phebox–Rh-catalyzed reaction proceeds via the *endo*-transition state on the basis of the *cis*-selectivity of **11** in the reaction of 2,4-dimethyl diene with *n*-butyl glyoxylate (Scheme 5). The absolute configurations of the dihydropyrans proved to be 2*R*, indicating that the *Re* face attack of the diene to the C=O group is a favorable pathway.

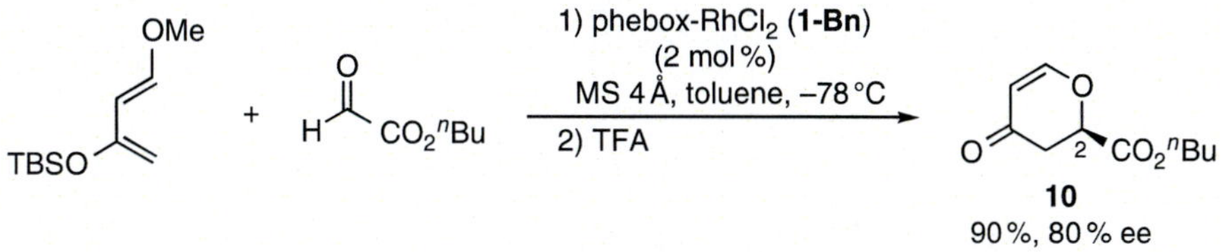

Scheme 4 Asymmetric Hetero-Diels–Alder reaction of Danishefsky's diene and glyoxylate

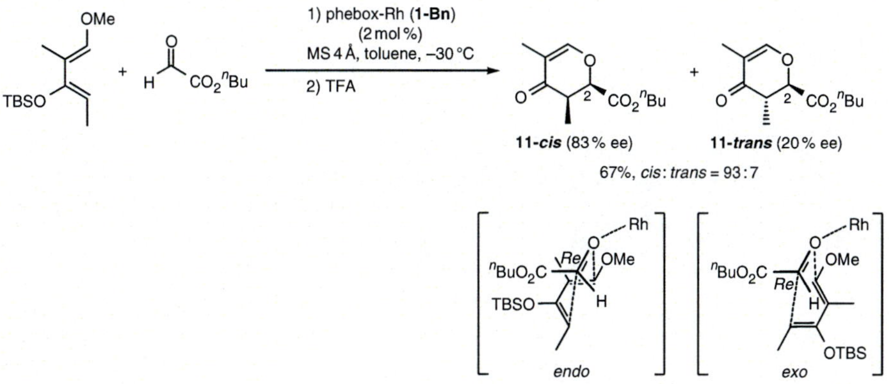

Scheme 5 Hetero-Diels–Alder reaction of 2,4-dimethyl diene and glyoxylate

5 Asymmetric Michael Addition

The reaction of Phebox-$SnMe_3$ **2** with $[RhCl(coe)_2]_2$ gave new phebox–Rh(III) complexes **12**, (phebox)$RhCl(SnMe_3)$ via oxidative addition of Rh(I) to the phebox-$SnMe_3$ bond (Scheme 6) [13].

The *tert*-butyl-phebox–Rh complex **12-^{t}Bu** generated in situ from the reaction of [(^{t}Bu-phebox)$SnMe_3$] (2 mol%) and $[RhCl(coe)_2]_2$ (1 mol%) effects efficiently the asymmetric Michael addition of α-cyanopropionates to acrolein under mild and neutral conditions (Scheme 7). The Michael adducts **13** were obtained in high yield with up to 86% ee. Enantioselectivity was improved by using α-cyanopropionates with bulky ester groups.

In the catalytic reaction, the high-valent Rh(III) complex **12** bearing the $SnMe_3$ and Cl ligands is postulated as a Lewis acid catalyst (Scheme 8). Coordination of the cyano group to the Rh center followed by enolization generates an N-bound enol. Then, the *Re* face of the enol attacks the acrolein to give the corresponding Michael adducts *R*-**13**. In this prochiral face determining step, the bulky $SnMe_3$ group and one of the oxazoline substituents cooperatively blocked the *Si* face of the enol. It is interesting to note that the chloro complex **1** did not catalyze the Michael addition in the absence of a base. Therefore, the bifunctionality originated from cooperative action between the $SiMe_3$ group and the phebox–Rh fragment might improve the catalytic activity as well as the enantioselectivity.

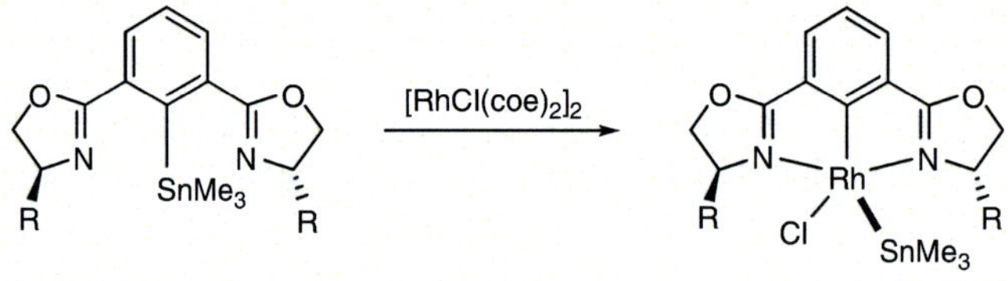

12

Scheme 6 Reaction of phebox-$SnMe_3$ and $[RhCl(coe)_2]_2$

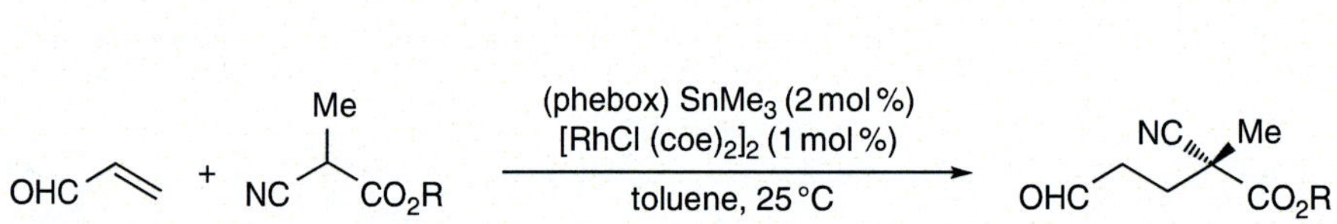

13

R = Et 96 %, 69 % ee
R = ^{i}Pr 97 %, 73 % ee
R = ^{t}Bu 82 %, 86 % ee

Scheme 7 Asymmetric Michael addition of α-cyanopropionates to acrolein

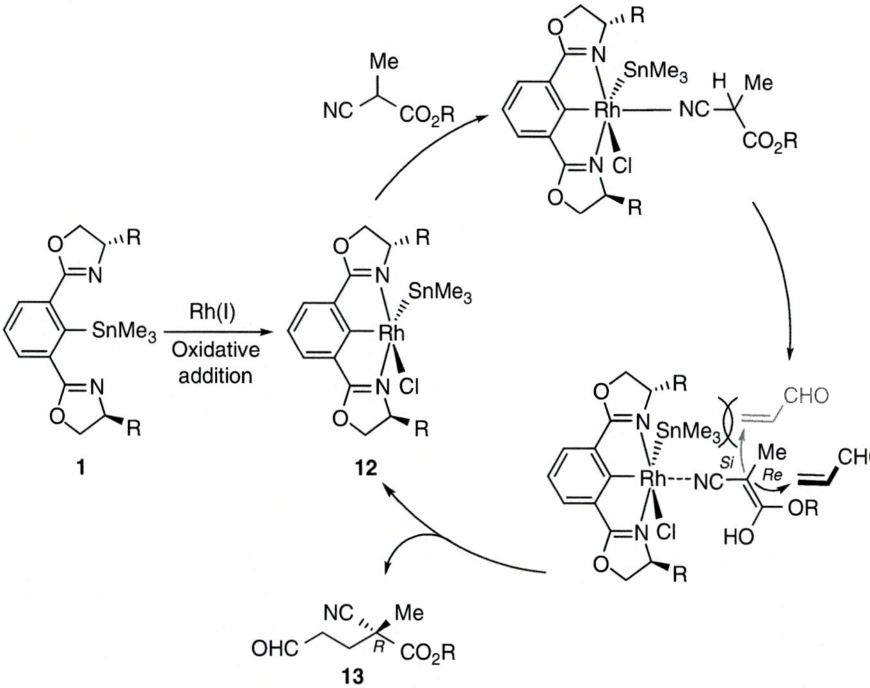

Scheme 8 Proposed mechanism for Michael addition

6 Asymmetric Hydrosilylation

Asymmetric hydrosilylation of alkenes is a versatile procedure to synthesize optically active organosilanes [14]. The phebox–Rh complex **1-^{i}Pr** also serves as an efficient catalyst for asymmetric hydrosilylation of styrene derivatives with $(EtO)_2MeSiH$. In fact, the reaction in toluene containing the complex **1-^{i}Pr** at 50°C for 3 days gave the corresponding α-silyl-adduct **14** and β-silyl-adduct **15** in high yield (Scheme 9) [15]. The successive Tamao oxidation with hydrogen peroxide afforded a mixture of 1-phenylethnaol and 2-phenylethanol in the ratio of ca. 50:50. Although the regioselectivity of the products was modest, the ee value of 1-phenylethanols was high enough. Use of Ag salts (2 mol%), such as $AgPF_6$, accelerated the catalytic reaction, in which the formation of a cationic complex was proposed.

A possible mechanism for the hydrosilylation reaction based on the Chalk–Harrod process is shown in Scheme 10 [16]. Initially, reduction of the phebox–Rh(III) with hydrosilane proceeds to give the Rh(I) intermediate **16**. Subsequent oxidative addition of hydrosilane gives a phebox–Rh(H)(SiR_3) species **17**. Coordination of an alkene to **17** produces two possible isomers **18a** and **18b**, which undergo insertion into the Rh–H bond followed by reductive elimination to give the α-silyl-adduct **14** and the β-silyl-adduct **15**. The absolute configuration of the α-silyl-adduct can be

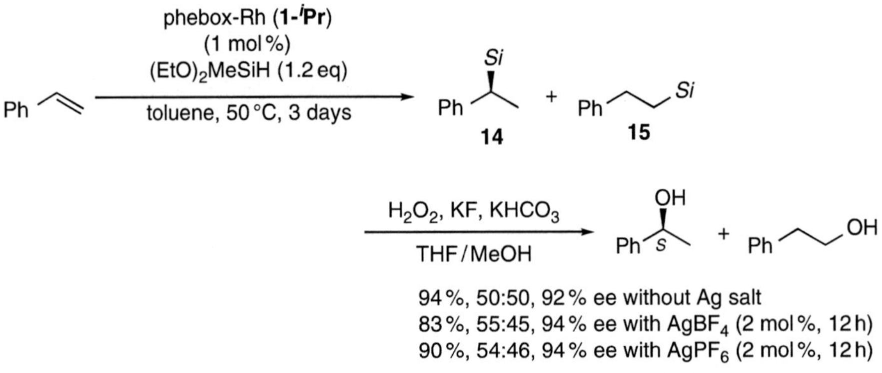

Scheme 9 Asymmetric hydrosilylation of styrene

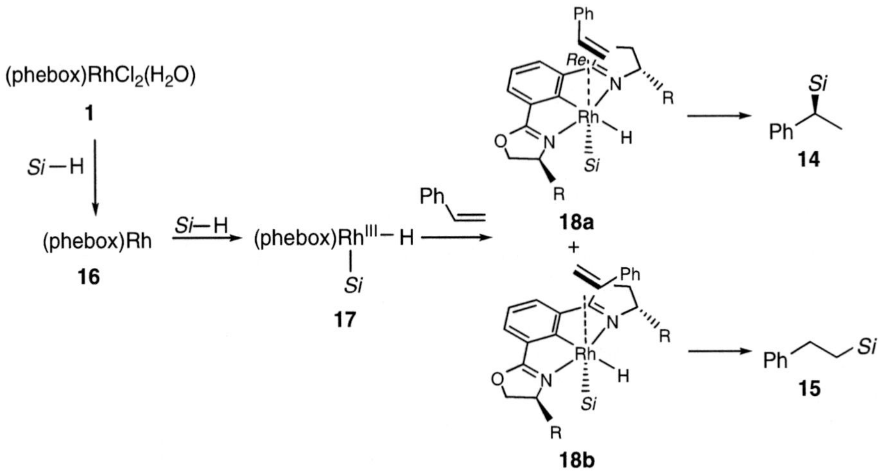

Scheme 10 Proposed mechanism for hydrosilylation

explained by the *Re* face coordination of alkene to the Rh center. In the catalytic reaction, the formation of vinylsilanes was not detected. This observation supports that the reaction with the phebox–Rh complex proceeds via the Chalk–Harrod process rather than the modified Chalk–Harrod process.

7 Asymmetric Conjugate Reduction

Asymmetric conjugate reduction of β,β-disubstituted α,β-unsaturated carbonyl compounds leads to a variety of optical active carbonyl compounds having a chiral center at the β-position. The phebox–Rh complexes are also applicable in the asymmetric conjugate reduction of α,β-unsaturated esters, ketones, and aldehydes.

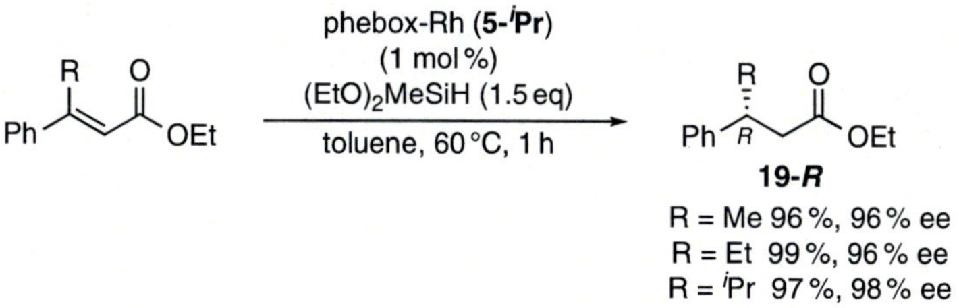

Scheme 11 Asymmetric conjugate reduction of α,β-unsaturated esters

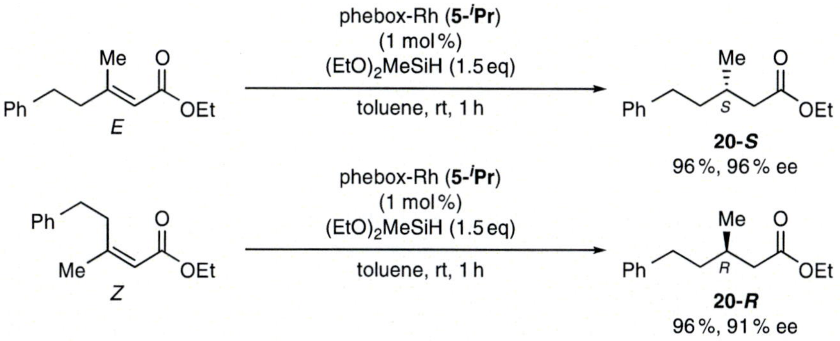

Scheme 12 Asymmetric conjugate reduction of (*E*)- and (*Z*)-stereoisomers

In facet, asymmetric reduction of (*E*)-3-phenylbut-2-enoate with hydrosilane at 60°C for 1 h in toluene containing 1 mol% of the isopropyl–phebox acetate complex **5-^{i}Pr** gave the corresponding *R*-products **19-*R*** in high yields with excellent ee's (Scheme 11) [17, 18]. As a hydrogen donor, alkoxysilanes, such as $(EtO)_3SiH$ and $(EtO)_2MeSiH$, are also effective, whereas the use of alkylsilanes retarded the catalysis giving the products with lower ee values.

The absolute configuration of the product depended on the geometry of the C=C bond. In the case of (*Z*)-3-phenylbut-2-enoate, the *S*-product **19-*S*** with a reverse absolute configuration was predominantly formed with moderate enantioselectivity (43% ee). Lower enantioselectivity was caused by *E*/*Z* isomerization of the α,β-unsaturated ester during the catalytic reaction. On the other hand, (*E*)-3-methyl-5-phenylpent-2-enoate gave the (*S*)-product **20-*S*** with 96% ee, while the (*Z*)-unsaturated esters gave the (*R*)-product **20-*R*** with 91% ee (Scheme 12).

The 1,4-reduction of α,β-unsaturated ketones with hydrosilane often competes with the 1,2-reduction. In the case of Wilkinson catalyst $Rh(PPh_3)_3Cl$, alkylsilane, Et_3SiH, leads to the 1,4-reduction, whereas arylsilane, Ph_2SiH_2 leads to the 1,2-reduction [19]. The phebox–Rh catalysts proved to be superior for the 1,4-reduction (Scheme 13). The use of alkoxysilanes, such as $(EtO)_3SiH$ and $(EtO)_2$-MeSiH, exclusively gave the conjugate reduction products **21** (1,4/1,2 = 100/0) with an excellent ee, up to 95% ee [18]. In contrast, the use of Et_3SiH and Me_2PhSiH caused a decrease in the selectivity to the 1,4-reduction. With the increase in

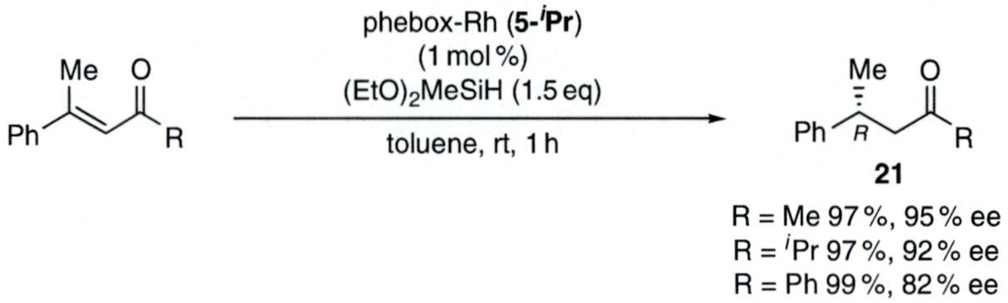

Scheme 13 Asymmetric conjugate reduction of α,β-unsaturated ketones

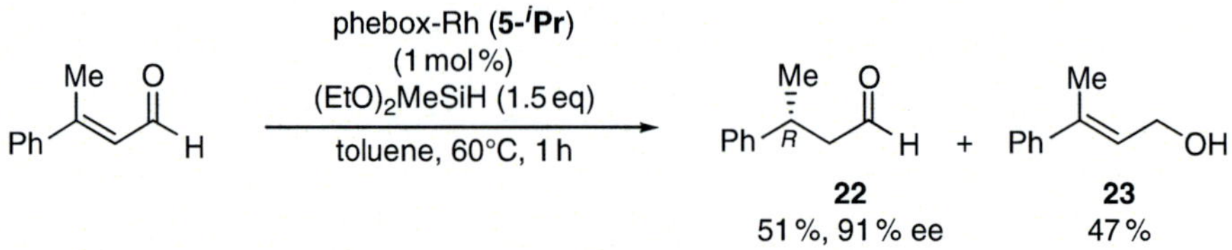

Scheme 14 Asymmetric conjugate reduction of α,β-unsaturated aldehydes

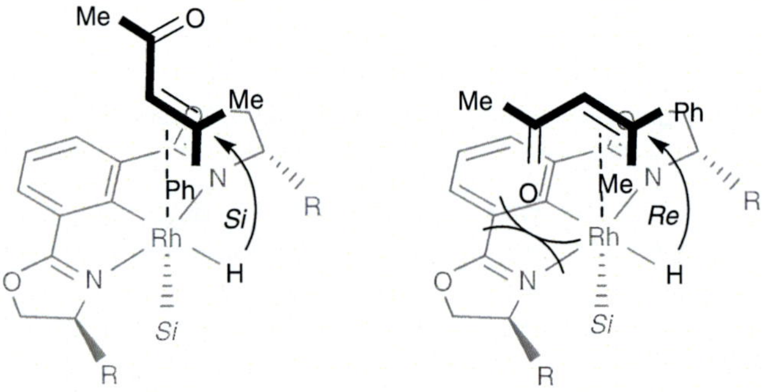

Fig. 2 Proposed mechanism for conjugate reduction

bulkiness of the ketone substituents from methyl to isopropyl and phenyl, the enantiomeric purity was declined.

Asymmetric conjugate reduction of α,β-unsaturated aldehyde with a chiral phebox–Rh complex resulted in the formation of 1,4-reduction product **22** and 1,2-reduction product **23** in 51% and 47% yields, respectively [20] (Scheme 14). The enantiomeric excess of **22** showed a high value of 91%.

A proposed mechanism and stereochemical course of the 1,4-reduction are described in Fig. 2. As in the case of hydrosilylation of alkenes described in Scheme 10, the phebox–Rh(I) intermediate **16** reacts with hydrosilane to give a hydro silyl Rh(III) species **17**. The hydride attacks to the *Si* face on the β-carbon atom of the α,β-unsaturated carbonyl compound to give the *R*-product. On the other hand, the stereochemistry with the *Re* face attack involves the steric repulsion between the oxazoline substituent and the carbonyl group. In the catalytic reaction,

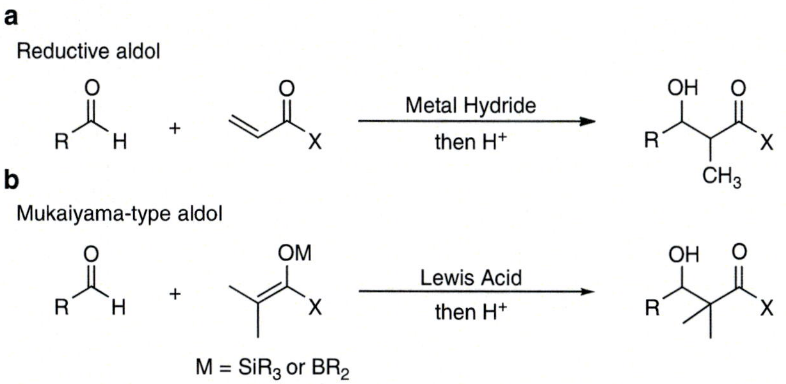

Scheme 15 Reductive aldol reaction and Mukaiyama-type aldol reaction

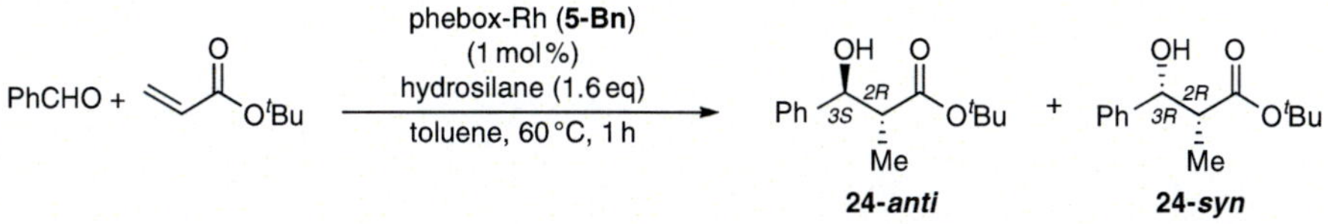

Si-H	Yield [%]	anti : syn	% ee anti	% ee syn
$(EtO)_2MeSiH$	98	98 : 2	94	57
Me_2PhSiH	93	94 : 6	96	54

Scheme 16 Asymmetric reductive aldol reaction of benzaldehyde and acrylate

combination between the Rh acetate fragments with alkoxyhydrosilane may be a key factor to generate the desired catalytically active Rh hydride species in asymmetric conjugate reduction.

8 Asymmetric Reductive Aldol Reaction

Reductive aldol reaction of α,β-unsaturated esters and enones with aldehyde mediated by a transition metal hydride complex and a hydride source, such as hydrosilane, is a versatile process to produce β-hydroxy carbonyl compounds (Scheme 15a) [21]. This reaction is thought to be an alternative transformation of Lewis acid-catalyzed Mukaiyama-type aldol reaction with silyl enol ethers or silyl ketene acetals (Scheme 15b).

The phebox–Rh acetate complex **5** is an efficient catalyst for the asymmetric reductive aldol reaction (Scheme 16) [22]. For example, the reaction of benzaldehyde with *t*-butyl acrylate in the presence of 1 mol% of the benzyl–phebox complex **5-Bn** proceeded in *anti*-selectivity and was completed within 0.5–1 h to form the

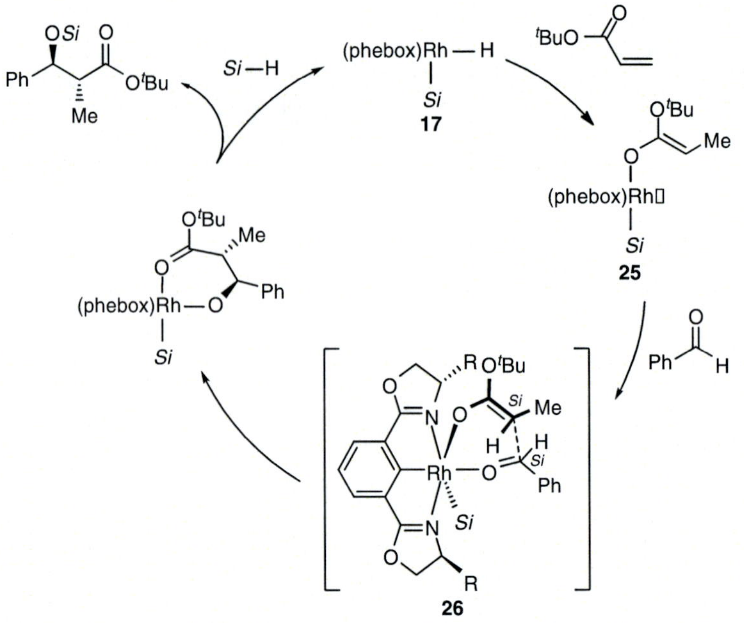

Scheme 17 Proposed mechanism for reductive aldol reaction

anti-coupling product **24** in high yields with high ee. In contrast to the conjugate reduction, both alkoxyhydrosilanes and alkylhydrosilanes were used as suitable hydride sources. Several aromatic and aliphatic aldehydes afforded the corresponding *anti*-β-hydroxyesters with high ee's.

The proposed mechanism was shown in Scheme 17. The 1,4-addition of acrylate to the Rh hydride intermediate **17** produces a Rh (*E*)-enolate intermediate **25**, in which the vacant site around the Rh center may act as a Lewis acid site for activating the carbonyl compound. The C–C bond formation proceeds though the chair-like transition state **26** following the Zimmerman–Traxler type. The stereochemical outcome of the *anti*-(2*R*,3*S*)-configuration can be explained by the *Si* face attack of the Rh enolate to the *Si* face of the aldehyde. Reductive elimination of the aldol product followed by oxidative addition of hydrosilane generates the Rh hydride intermediate. It is worthy to note that the stepwise reaction, which involves the formation of the silylketene acetal by the conjugate reduction of methyl acrylate with Me_2PhSiH and subsequent addition of benzaldehyde, did not give the desired aldol product. Although the use of $(EtO)_2MeSiH$ for the stepwise reaction provided the aldol product, the high *syn*-selectivity (5:95) was observed. These results clearly indicate that intermediate Rh enolate directly attacks the aldehyde without reductive elimination of the silylketene acetal. Therefore, the catalytic cycle is consisted of cooperative action of two key intermediates, namely the Rh hydride for conjugate reduction of acrylate and the Rh enolate for aldol type C–C bond formation. The Rh atom can capture aldehyde at an equatorial position as a Lewis acid.

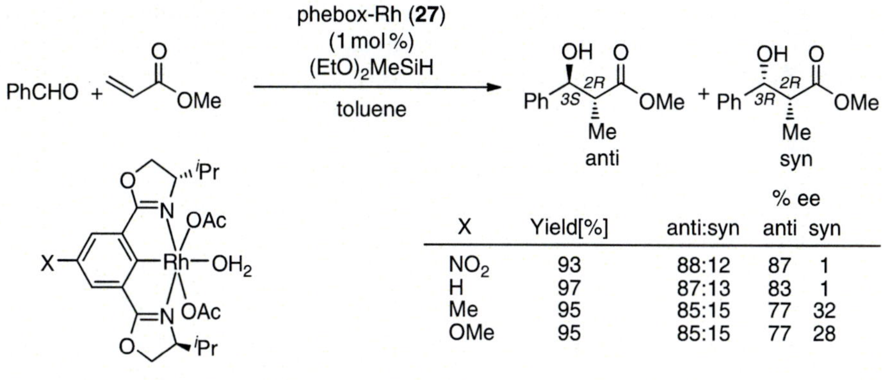

27-X

X	Yield[%]	anti:syn	% ee anti	% ee syn
NO_2	93	88:12	87	1
H	97	87:13	83	1
Me	95	85:15	77	32
OMe	95	85:15	77	28

Scheme 18 Asymmetric reductive aldol reaction with 4-substituted phebox–Rh catalysts

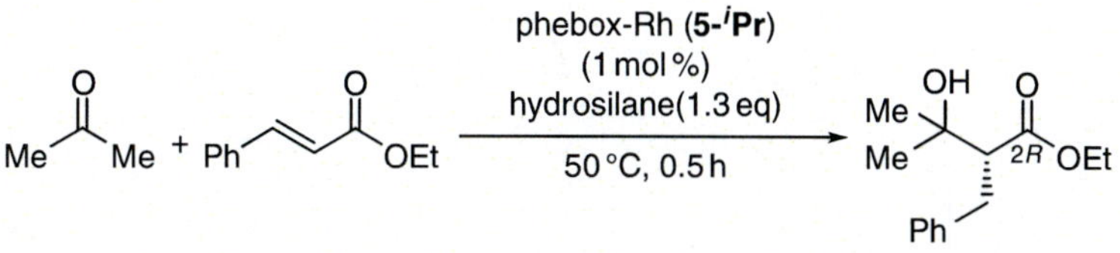

28

$MePh_2SiH$ 83 %, 97 % ee
Me_2PhSiH 78 %, 95 % ee
$(EtO)_2MeSiH$ 19 %, 91 % ee

Scheme 19 Asymmetric reductive aldol reaction of ketone and cinnamate

In order to elucidate the remote electronic effect, the phebox–Rh complexes containing several substituents at the 4-position on the phenyl ring were synthesized. It was found that these modified phebox–Rh complexes affected the enantioselectivity of the aldol reaction (Scheme 18) [23]. In the case of the reaction of methyl acrylate, the use of 4-NO_2-phebox–Rh complex **27-NO$_2$** improved the enantioselectivity compared with that of the 4-MeO-phebox–Rh complex **27-MeO**. In the molecular structure analysis of **27-NO$_2$**, the structural change with the elongated Rh-C bond and the wider N-Rh-N bite angle was observed.

The catalytic reductive aldol reaction of ketones as acceptor is a useful procedure to access ester derivatives bearing tertiary alcohol unit at the β-position. The phebox–Rh complex **5-^{i}Pr** effectively catalyzed the reductive aldol reaction of acetone with ethyl cinnamate in the presence of hydrosilane (Scheme 19) [24]. In fact, the use of $MePh_2SiH$ and Me_2PhSiH as hydrogen sources gave a β-tertiary alcohol **28** in the high yields with excellent ee's, while the reaction with $(EtO)_2$-$MeSiH$ resulted in the lower yield. Other ketones, cyclohexanone and acetophenone, were used in the catalytic reaction to provide the corresponding aldol products with high ee's.

The enantioselective intermolecular reductive aldol reaction with enone instead of ester is also an important C–C bond formation reaction. The coupling reaction of cyclopent-2-enone and benzaldehyde with hydrosilane in the presence of 1 mol% of the phebox–Rh complex **5-Ph** smoothly proceeded to give the corresponding *anti*-β-hydroxyketone **29** with the high ee value (Scheme 20) [25]. On the basis of the absolute configuration of **29**, the transition state of the reaction might involve the *Re* face attack of the Rh enolate to the *Re* face of the aldehyde captured on the rhodium atom leading to form 2*S*,1′*R*-diastereomeric product.

The reductive aldol reactions are also applicable to access optical active dihydrocinnamates having substituents at the α-position, which are important materials or intermediates for the synthesis of bioactive compounds. Combination of the phebox–Rh-catalyzed asymmetric reductive aldol reaction and subsequent dehydroxylation provided the corresponding α-substituted dihydrocinnamates **30** with a high ee (Scheme 21) [26].

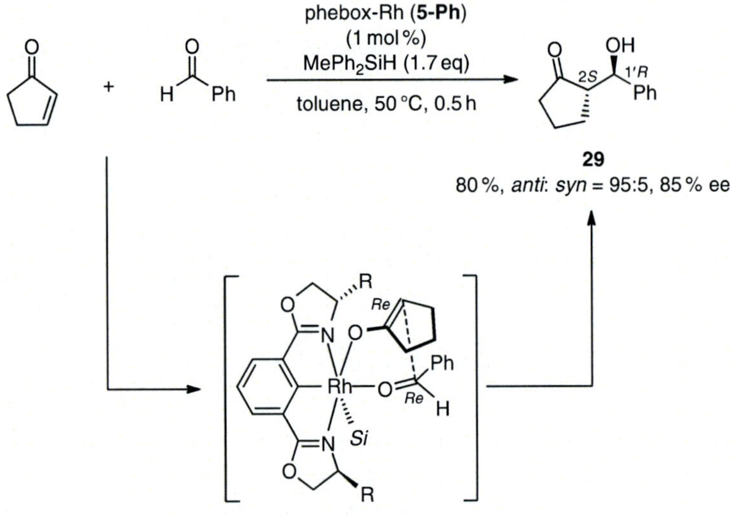

Scheme 20 Asymmetric reductive aldol reaction of enones

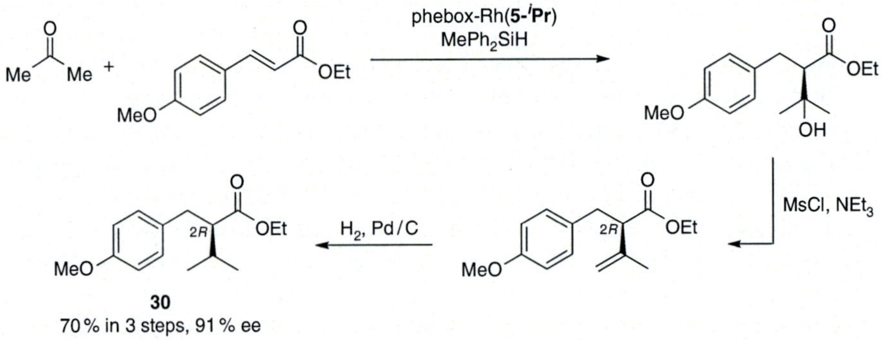

Scheme 21 Preparation of α-substituted dehydrocinnamate

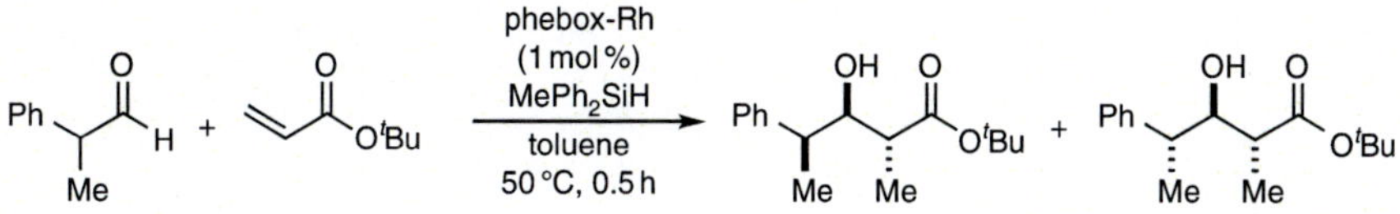

rac, S, R

31a
Felkin-Anh

31b
***anti*-Felkin-Anh**

+ other diastereomers

Cat.	Aldehyde	Yield (%)	dr	ee (%)
achiral (**32**)	*rac*	84	69^a: 21^b: 7:3	*rac*
achiral (**32**)	*S* (92 % ee)	84	68^a: 20^b:10:2	93 (2*R*, 3*R*, 4*S*)^a
S,S (**5-^{i}Pr**)	*rac*	70	55^a: 35^b: 8:2	76 (2*R*, 3*R*, 4*S*)^a
S,S (**5-^{i}Pr**)	*S* (92 % ee)	83	88^a: 4^b: 7:1	99 (2*R*, 3*R*, 4*S*)^a
S,S (**5-^{i}Pr**)	*R* (95 % ee)	75	20^a: 65^b:14:1	71 (2*R*, 3*R*, 4*R*)^b

^{a}Felkin-Anh product **31a**, ^{b}*anti*-Felkin-Anh product **31b**

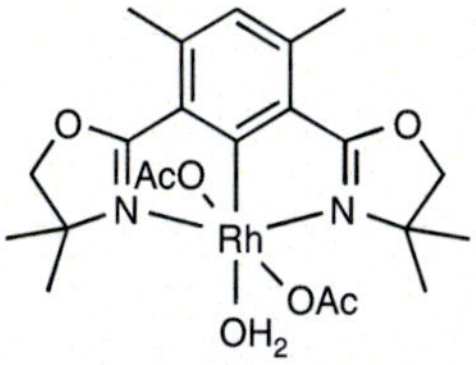

32

Scheme 22 Asymmetric reductive aldol reaction of 2-phenylpropanal with acrylate

Furthermore, the reductive aldol reaction can be used for the construction of α,β,γ-stereotriads. When the racemic phebox–Rh acetate complex **32** was subjected to the coupling reactions of (*S*)-2-phenylpropanal with acrylate, the Felkin–Anh product **31a** with (2*R*,3*R*,4*S*)-configuration was predominantly formed (Scheme 22) [27]. The *anti*-Felkin–Anh product **31b′** (enantiomer) was a minor diastereomer. The use of the chiral (*S*,*S*)-phebox–Rh complex **5-^{i}Pr** for the coupling reaction with (*S*)-2-phenylpropanal resulted in the formation of the Felkin–Anh product with high ee and de. On the other hand, the use of (*R*)-2-phenylpropanal afforded the *anti*-Felkin–Anh product **31b** as a major diastereomer with moderate enantioselectivity. Thus, a combination of (*S*)-2-phenylpropanal with the (*S*,*S*)-phebox–Rh complex **5-^{i}Pr** is a matched pair.

A possible transition state based on the Felkin–Anh model was shown in Scheme 23. Judging from the (2*R*,3*R*,4*S*)-configuration of the product **31a**, the major product is likely formed via the Felkin TS **33** showing the *Si* face attack of the Rh-(*E*)-enolate. This step could be the catalyst-controlled reaction with the chiral catalyst. According to the prochiral face discrimination in the phebox–Rh-catalyzed reductive aldol reaction with the linear substrate, the *Re* face attack of the Rh (*E*)-enolate in TS **34** is unfavorable. In the case of the (*R*)-aldehyde, the *anti*-Felkin–Anh's TS **35**, which gives the (2*R*,3*R*,4*R*)-product **31b**, takes the unfavorable conformation with the bulky phenyl group at the apical position.

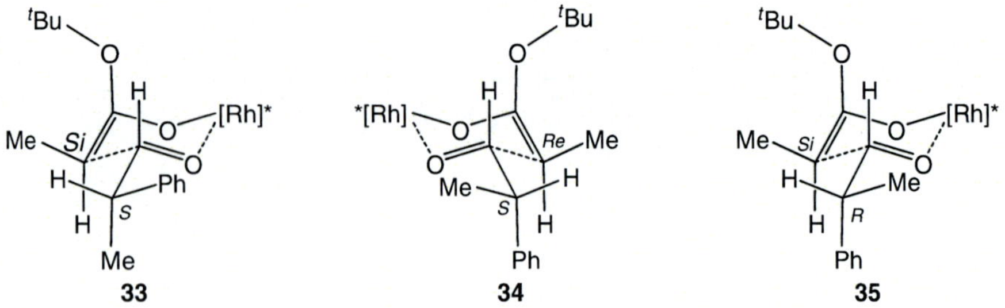

Scheme 23 Proposed mechanism via Felkin transition state

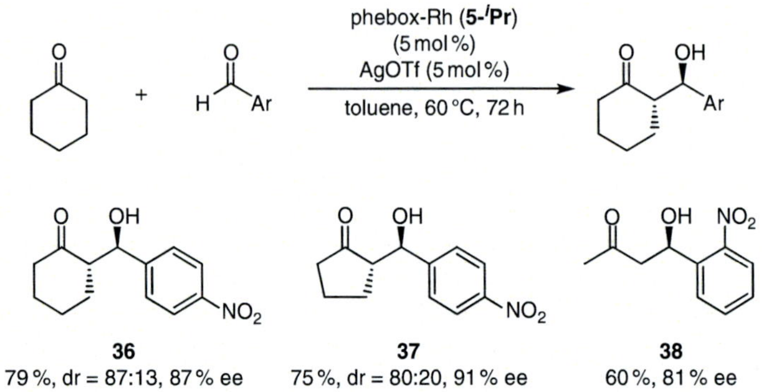

Scheme 24 Asymmetric direct aldol reaction of aldehydes with cyclic ketones

9 Asymmetric Direct Aldol Reaction

Asymmetric direct aldol reaction is a practical process to synthesize β-hydroxy-carbonyl compounds [28]. The phebox–Rh acetate complex **5-^{i}Pr** serves also as an excellent catalyst for asymmetric direct aldol reaction of cyclic ketones with benzaldehyde derivatives (Scheme 24) [29]. For example, Scheme 24 shows that the reaction of cyclohexanone with 4-nitrobenzaldehyde using 5 mol% of **5-^{i}Pr** and AgOTf gave the corresponding *anti*-β-hydroxyketone **36** preferentially in 79% yield with 87% ee for the *anti*-isomer. The coupling reaction of cyclopentanone and acetone afforded β-hydroxyketones **37** and **38**, respectively, with up to 91% ee.

The phebox–Rh(III) complex can work as a mild Lewis acid catalyst. In particular, its acetate complex has both Lewis acidic site on the metal center and the Brønsted basic site on the acetate ligand as shown in Scheme 25. This bifunctionality based on the acid–base synergy effect is crucial for determining the catalytic performance in the direct aldol reaction.

In a proposed mechanism illustrated in Scheme 26, the Rh enolate **39** is initially generated by abstraction of the α-proton of the ketone with the help of the basic carboxylate ligand. Next, the aldehyde is coordinated to the Lewis acidic Rh center

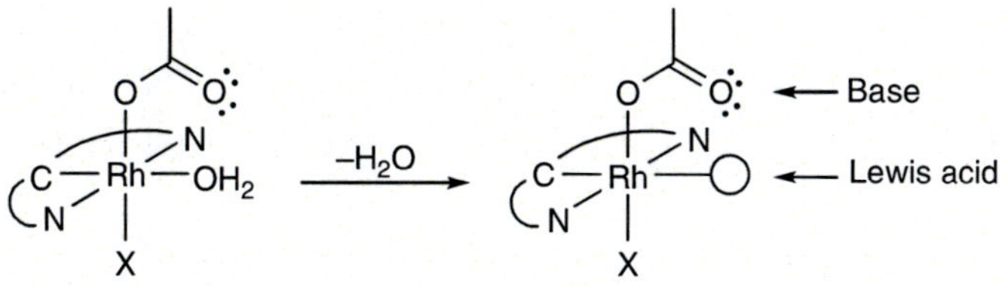

Scheme 25 Phebox–Rh complex as Brønsted base and Lewis acid

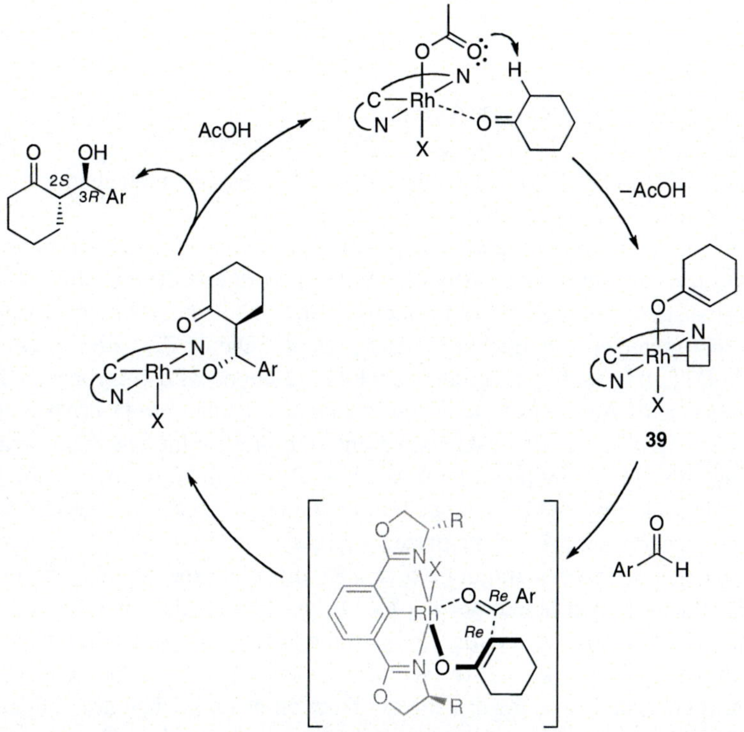

Scheme 26 Proposed mechanism for direct aldol reaction

and attacked by the Rh enolate to give the Rh aldolate. The C–C bond formation between *Re* face of the Rh enolate and the *Re* face of aldehyde through the cyclic transition state **40** constructs 2*S*,3*R* stereochemistry. This prochiral face discrimination in **40** is identical to that of the reductive aldol reaction of a cyclic enone with an aldehyde as described in Scheme 20. Finally, the release of the aldol product with AcOH regenerates the catalytically active acetate intermediate. The presence of OTf anion is essential to improve the enantioselectivity and diastereoselectivity. This result suggests that the anion ligand exchange reaction between OAc and OTf

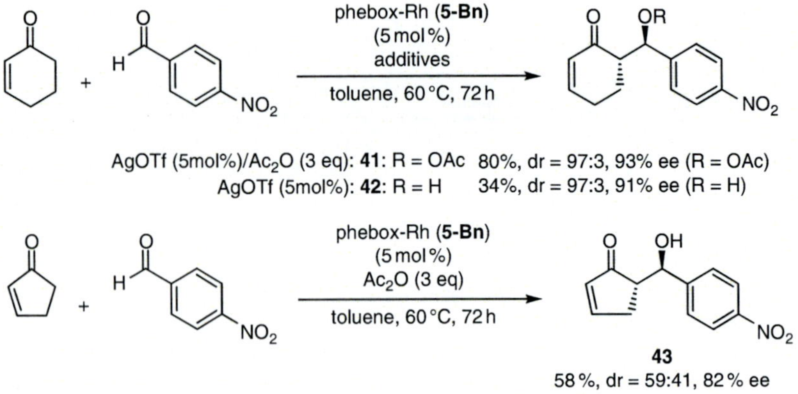

Scheme 27 Asymmetric direct aldol reaction of enones

generates a Rh-OTf fragment, which may influence the stereochemical course of the approaching aldehyde.

The vinylogous unsaturated aldol products derived from enones have high potential to serve as useful components for organic synthesis. The phebox–Rh acetate/AgOTf system also serves as an effective catalyst for the asymmetric direct aldol reaction of enones with aldehydes. As shown in Scheme 27, with 5 mol% of **5-Bn** and AgOTf, the coupling reaction of cyclohex-2-enone with 4-nitrobenzaldehyde in the presence of Ac_2O gave the acetylated aldol product *anti*-**41** in 80% yield with 93% ee [30]. Addition of Ac_2O was essential to prevent the retro-aldol reaction. In contrast, the aldol reaction without Ac_2O resulted in the formation of the β-hydroxyketone **42** in low yield. In the case of cyclopent-2-enone, the product **43** did not undergo acetylation even in the presence of Ac_2O.

The catalytic reaction might involve a Rh-dienolate species **44**, which is formed by the selective deprotonation of C(α′)-H rather than that of C-H, as shown in Scheme 28. The absolute configuration of the product suggested that the *Re* face of the Rh-dienolate attacks the *Re* face of the aldehyde. This hypothetical transition model is consistent with the direct aldol reaction of simple ketones with aldehydes. The Rh aldolate is trapped with Ac_2O or H^+ to produce **41** or **42**.

10 Asymmetric β-Boration

In 2008, Yun et al. first reported the asymmetric β-boration of α,β-unsaturated nitriles and esters using Cu catalysts with chiral phosphine ligands [31, 32]. The phebox–Rh acetate complex **5** works as an effective catalyst for asymmetric boration of α,β-unsaturated esters (Scheme 29) [33]. The catalytic reaction of ethyl (*E*)-cinnamate with bis(pinacolato)diboron (B_2pin_2, 1.1eq) in the presence of 1 mol % of the *sec*-butyl phebox complex **5-^{s}Bu** and 5 mol% of NaO^{t}Bu at 80°C afforded the borylated compound **46** in 86% yield. The product ee value of 97% was

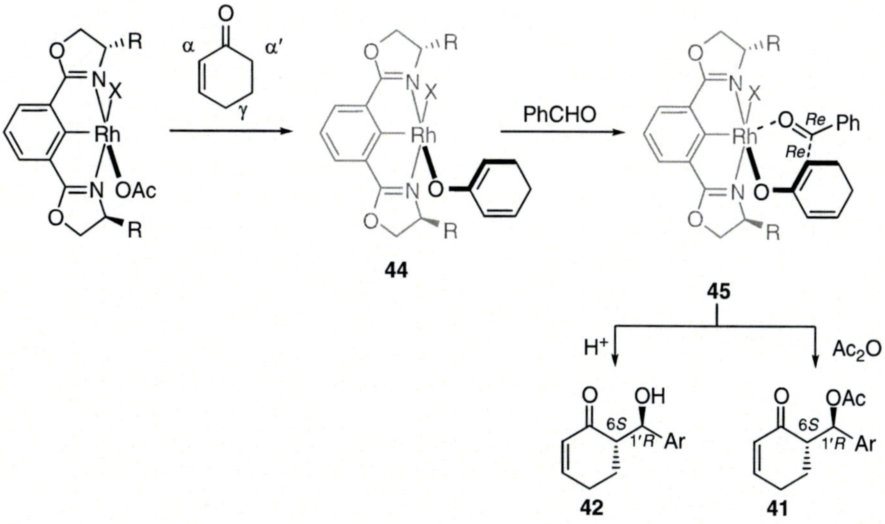

Scheme 28 Proposed mechanism for direct aldol reaction of enones

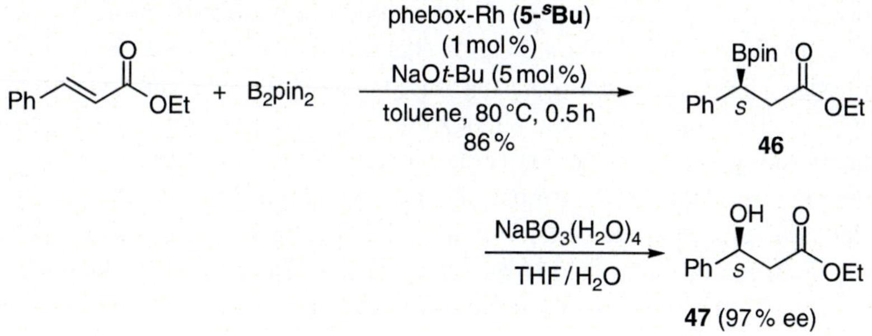

Scheme 29 Asymmetric β-boration of cinnamate with diboron

determined by the corresponding β-hydroxylester **47** derived from oxidation with sodium peroxoborate.

A proposed mechanism is shown in Scheme 30. As an initial step, the catalytically active species, a Rh(III) bory intermediate **48**, is generated by σ-bond metathesis of the catalyst precursor, Rh(III) acetate or butoxy complex with the B–B bond, but not *via* oxidative addition of the B–B bond. Then, insertion of α,β-unsaturated carbonyl into the boryl–Rh bond gives two possible Rh(III) enolate complexes which have the bory ligand, at the apical or equatorial position. The absolute configuration of the product **47** suggests that the *Si* face attack by the boryl ligand to the β-position likely proceeds to give the major (*S*)-enantiomer. Finally, the exchange reaction with diboron forms the product and the boryl Rh intermediate **48**.

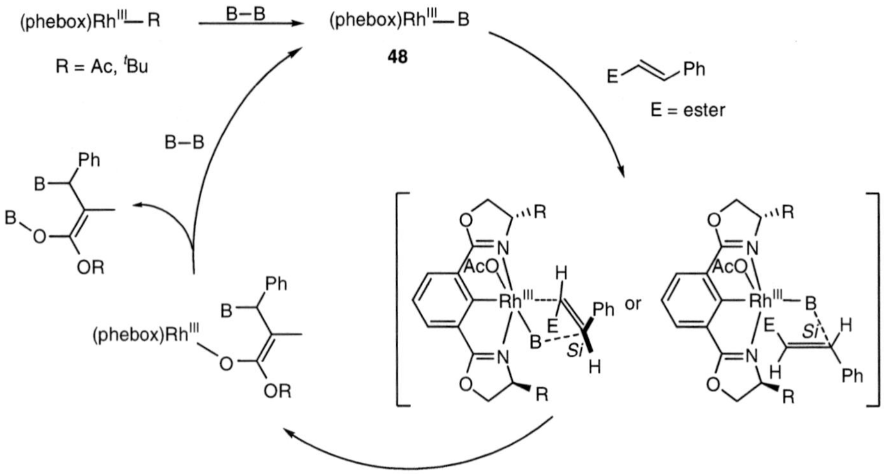

Scheme 30 Proposed mechanism for β-boration

11 Conclusion

In this chapter, we have successfully developed bifunctional chiral rhodium complexes bearing chiral phebox ligands that can be used in catalytic asymmetric reactions. The *N,C,N* meridional geometry with the rhodium–carbon covalent bond is the key character in the phebox complexes. The metal–phebox cooperative bifunctionality significantly contributes reactivity and selectivity in the catalytic asymmetric reactions. Furthermore, the prototype of the bifunctional catalyst can be explained to a wide range of asymmetric catalytic reactions promoted by the Lewis acids, hydrides, enolates, and bory active species. Their diversity further broadens the range of opportunities for asymmetric catalysis.

References

1. Nishiyama H, Sakaguchi H, Nakamura T, Horihata M, Kondo M, Itoh K (1989) Organometallics 8:846
2. Desimoni G, Faita G, Quadrelli P (2003) Chem Rev 103:3119
3. Motoyama Y, Makihara N, Mikami Y, Aoki K, Nishiyama H (1997) Chem Lett:951
4. Denmark SE, Stavenger RA, Faucher A-M, Edwards JP (1997) J Org Chem 62:3375
5. Stark MA, Richards CJ (1997) Tetrahedron Lett 38:5881
6. Nishiyama H, Ito J (2010) Chem Commun 46:203
7. Nishiyama H (2007) Chem Soc Rev 36:1133
8. Motoyama Y, Okano M, Narusawa H, Makihara N, Aoki K, Nishiyama H (2001) Organometallics 20:1580
9. Motoyama Y, Narusawa H, Nishiyama H (1999) Chem Commun:131
10. Denmark SE, Fu J (2003) Chem Rev 103:2763
11. Motoyama Y, Nishiyama H (2003) Synlett:1883

12. Motoyama Y, Koga Y, Nishiyama H (2001) Tetrahedron 57:853
13. Motoyama Y, Koga Y, Kobayashi K, Aoki K, Nishiyama H (2002) Chem Eur J 8:2968
14. Nishiyama H, Itoh K (2000) In: Ojima I (ed) Catalytic asymmetric synthesis, 2nd edn. Wiley-VCH, New York, p 111
15. Tsuchiya Y, Uchimura H, Kobayashi K, Nishiyama H (2004) Synlett:2099
16. Chalk AJ, Harrod JF (1965) J Am Chem Soc 87:16
17. Tsuchiya Y, Kanazawa Y, Shiomi T, Kobayashi K, Nishiyama H (2004) Synlett:2493
18. Kanazawa Y, Tsuchiya Y, Kobayashi K, Shiomi T, Ito J, Kikuchi M, Yamamoto Y, Nishiyama H (2006) Chem Eur J 12:63
19. Ojima I, Kogure T (1982) Organometallics 1:1390
20. Kanazawa Y, Nishiyama H (2006) Synlett:3343
21. Nishiyama H, Shiomi T (2007) Top Curr Chem 279:105
22. Nishiyama H, Shiomi T, Tsuchiya Y, Matsuda I (2005) J Am Chem Soc 127:6972
23. Shiomi T, Ito J, Yamamoto Y, Nishiyama H (2006) Eur J Org Chem:5594
24. Shiomi T, Nishiyama H (2007) Org Lett 9:1651
25. Shiomi T, Adachi T, Ito J, Nishiyama H (2009) Org Lett 11:1011
26. Hashimoto T, Shiomi T, Ito J, Nishiyama H (2007) Tetrahedron 63:12883
27. Hashimoto T, Ito J, Nishiyama H (2008) Tetrahedron 64:9408
28. Trost BM, Brindle CS (2010) Chem Soc Rev 39:1600
29. Inoue H, Kikuchi M, Ito J, Nishiyama H (2008) Tetrahedron 64:493
30. Mizuno M, Inoue H, Naito T, Zhou L, Nishiyama H (2009) Chem Eur J 15:8985
31. Lee JE, Yun J (2008) Angew Chem Int Ed 47:145
32. Sim HS, Feng X, Yun J (2009) Chem Eur J 15:1939
33. Shiomi T, Adachi T, Toribatake K, Zhou Li, Nishiyama H (2009) Chem Commun:5987

Index